T0212965

Lecture Notes in Computer Science   9939

Commenced Publication in 1973
Founding and Former Series Editors:
Gerhard Goos, Juris Hartmanis, and Jan van Leeuwen

More information about this series at http://www.springer.com/series/7409

Laurent Amsaleg · Michael E. Houle
Erich Schubert (Eds.)

Similarity Search and Applications

9th International Conference, SISAP 2016
Tokyo, Japan, October 24–26, 2016
Proceedings

 Springer

Editors
Laurent Amsaleg
CNRS–IRISA
Rennes
France

Erich Schubert
Ludwig-Maximilians-Universität München
Munich
Germany

Michael E. Houle
National Institute of Informatics
Tokyo
Japan

ISSN 0302-9743 ISSN 1611-3349 (electronic)
Lecture Notes in Computer Science
ISBN 978-3-319-46758-0 ISBN 978-3-319-46759-7 (eBook)
DOI 10.1007/978-3-319-46759-7

Library of Congress Control Number: 2016954121

LNCS Sublibrary: SL3 – Information Systems and Applications, incl. Internet/Web, and HCI

Printed on acid-free paper

This Springer imprint is published by Springer Nature
The registered company is Springer International Publishing AG
The registered company address is: Gewerbestrasse 11, 6330 Cham, Switzerland

Preface

This volume contains the papers presented at the 9th International Conference on Similarity Search and Applications (SISAP 2016) held in Tokyo, Japan, during October 24–26, 2016. SISAP is an annual forum for researchers and application developers in the area of similarity data management. It aims at the technological problems shared by numerous application domains, such as data mining, information retrieval, multimedia, computer vision, pattern recognition, computational biology, geography, biometrics, machine learning, and many others that make use of similarity search as a necessary supporting service.

From its roots as a regional workshop in metric indexing, SISAP has expanded to become the only international conference entirely devoted to the issues surrounding the theory, design, analysis, practice, and application of content-based and feature-based similarity search. The SISAP initiative has also created a repository (http://www.sisap.org/) serving the similarity search community, for the exchange of examples of real-world applications, source code for similarity indexes, and experimental test beds and benchmark data sets.

The call for papers welcomed full papers, short papers, as well as demonstration papers, with all manuscripts presenting previously unpublished research contributions. At SISAP 2016, all contributions were presented both orally and in a poster session, which facilitated fruitful exchanges between the participants.

We received 47 submissions, 32 full papers and 15 short papers, from authors based in 21 different countries. The Program Committee (PC) was composed of 62 members from 26 countries. Reviews were thoroughly discussed by the chairs and PC members: each submission received at least three to five reviews, with additional reviews sometimes being sought in order to achieve a consensus. The PC was assisted by 23 external reviewers.

The final selection of papers was made by the PC chairs based on the reviews received for each submission as well as the subsequent discussions among PC members. The final conference program consisted of 18 full papers and seven short papers, resulting in an acceptance rate of 38 % for full papers and 53 % cumulative for full and short papers.

The proceedings of SISAP are published by Springer as a volume in the Lecture Notes in Computer Science (LNCS) series. For SISAP 2016, as in previous years, extended versions of five selected excellent papers were invited for publication in a special issue of the journal Information Systems. The conference also conferred a Best Paper Award, as judged by the PC Co-chairs and Steering Committee.

The conference program and the proceedings are organized in several parts. As a first part, the program includes three keynote presentations from exceptionally skilled scientists: Alexandr Andoni, from Columbia University, USA, on the topic of "Data-Dependent Hashing for Similarity Search"; Takashi Washio, from the University of Osaka, Japan, on "Defying the Gravity of Learning Curves: Are More Samples

Better for Nearest Neighbor Anomaly Detectors?"; and Zhi-Hua Zhou, from Nanjing University, China, on "Partial Similarity Match with Multi-instance Multi-label Learning".

The program then carries on with the presentations of the papers, grouped in eight categories: graphs and networks; metric and permutation-based indexing; multimedia; text and document similarity; comparisons and benchmarks; hashing techniques; time-evolving data; and scalable similarity search.

We would like to thank all the authors who submitted papers to SISAP 2016. We would also like to thank all members of the PC and the external reviewers for their effort and contribution to the conference. We want to express our gratitude to the members of the Organizing Committee for the enormous amount of work they have done.

We also thank our sponsors and supporters for their generosity. All the submission, reviewing, and proceedings generation processes were carried out through the Easy-Chair platform.

August 2016   Laurent Amsaleg
                                                                   Michael E. Houle
                                                                   Erich Schubert

Organization

Program Committee Chairs

Laurent Amsaleg	CNRS-IRISA, France
Michael E. Houle	National Institute of Informatics, Japan

Program Committee Members

Giuseppe Amato	ISTI-CNR, Italy
Laurent Amsaleg	CNRS-IRISA, France
Hiroki Arimura	Hokkaido University, Japan
Ira Assent	Aarhus University, Denmark
James Bailey	University of Melbourne, Australia
Christian Beecks	RWTH Aachen University, Germany
Panagiotis Bouros	Aarhus University, Denmark
Leonid Boytsov	Carnegie Mellon University, USA
Benjamin Bustos	University of Chile, Chile
K. Selçuk Candan	Arizona State University, USA
Guang-Ho Cha	Seoul National University of Science and Technology, Korea
Edgar Chávez	CICESE, Mexico
Paolo Ciaccia	University of Bologna, Italy
Richard Connor	University of Strathclyde, UK
Michel Crucianu	CNAM, France
Bin Cui	Peking University, China
Vlad Estivill-Castro	Griffith University, Australia
Andrea Esuli	ISTI-CNR, Italy
Fabrizio Falchi	ISTI-CNR, Italy
Claudio Gennaro	ISTI-CNR, Italy
Magnus Lie Hetland	NTNU, Norway
Michael E. Houle	National Institute of Informatics, Japan
Yoshiharu Ishikawa	Nagoya University, Japan
Björn Þór Jónsson	Reykjavik University, Iceland
Ata Kabán	University of Birmingham, UK
Ken-ichi Kawarabayashi	National Institute of Informatics, Japan
Daniel Keim	University of Konstanz, Germany
Yiannis Kompatsiaris	CERTH – ITI, Greece
Peer Kröger	Ludwig-Maximilians-Universität München, Germany
Guoliang Li	Tsinghua University, China
Jakub Lokoč	Charles University in Prague, Czech Republic

Rui Mao	Shenzhen University, China
Stéphane Marchand-Maillet	Viper Group - University of Geneva, Switzerland
Henning Müller	HES-SO, Switzerland
Gonzalo Navarro	University of Chile, Chile
Chong-Wah Ngo	City University of Hong Kong, SAR China
Beng Chin Ooi	National University of Singapore, Singapore
Vincent Oria	New Jersey Institute of Technology, USA
M. Tamer Özsu	University of Waterloo, Canada
Deepak P	IBM Research, India
Apostolos N. Papadopoulos	Aristotle University of Thessaloniki, Greece
Marco Patella	DEIS – University of Bologna, Italy
Oscar Pedreira	Universidade da Coruña, Spain
Miloš Radovanović	University of Novi Sad, Serbia
Kunihiko Sadakane	The University of Tokyo, Japan
Shin'ichi Satoh	National Institute of Informatics, Japan
Erich Schubert	Ludwig-Maximilians-Universität München, Germany
Tetsuo Shibuya	Human Genome Center, Institute of Medical Science, The University of Tokyo, Japan
Yasin Silva	Arizona State University, USA
Matthew Skala	IT University of Copenhagen, Denmark
John Smith	IBM T.J. Watson Research Center, USA
Nenad Tomašev	Google, UK
Agma Traina	University of São Paulo at São Carlos, Brazil
Takeaki Uno	National Institute of Informatics, Japan
Michel Verleysen	Université Catholique de Louvain, Belgium
Takashi Washio	ISIR, Osaka University, Japan
Marcel Worring	University of Amsterdam, The Netherlands
Pavel Zezula	Masaryk University, Czech Republic
De-Chuan Zhan	Nanjing University, China
Zhi-Hua Zhou	Nanjing University, China
Arthur Zimek	Ludwig-Maximilians-Universität München, Germany
Andreas Züfle	George Mason University, USA

Additional Reviewers

Tetsuya Araki
Konstantinos Avgerinakis
Nicolas Basset
Michal Batko
Jessica Beltran
Hei Chan
Elisavet Chatzilari
Anh Dinh
Alceu Ferraz Costa

Karina Figueroa
David Novak
Ninh Pham
Nora Reyes
José Fernando
 Rodrigues Jr
Ubaldo Ruiz
Manos Schinas
Pascal Schweitzer

Diego Seco
Francesco Silvestri
Eleftherios
 Spyromitros-Xioufis
Eric S. Tellez
Xiaofei Zhang
Yue Zhu

Keynotes

Data-Dependent Hashing for Similarity Search

Alexandr Andoni

Columbia University, New York, USA

The quest for efficient similarity search algorithms has lead to a number of ideas that proved successful in both theory and practice. Yet, the last decade or so has seen a growing gap between the theoretical and practical approaches. On the one hand, most successful theoretical methods rely on data-indepependent hashing, such as the classic Locality Sensitive Hashing scheme. These methods have provable guarantees on correctness and performance. On the other hand, in practice, methods that adapt to the given datasets, such as the PCA-tree, often outperform the former, but provide no guarantees on performance or correctness.

This talk will survey the recent efforts to bridge this gap between theoretical and practical methods for similarity search. We will see that data-dependent methods are provably better than data-independent methods, giving, for instance, the first improvements over the Locality Sensitive Hashing schemes for the Hamming and Euclidean spaces.

Defying the Gravity of Learning Curves: Are More Samples Better for Nearest Neighbor Anomaly Detectors?

Takashi Washio

Osaka University, Suita, Japan

Machine learning algorithms are conventionally considered to provide higher accuracy when more data are used for their training. We call this behavior of their learning curves "the gravity", and it is believed that no learning algorithms are "gravity-defiant". A few scholars recently suggested that some unsupervised anomaly detector ensembles follow the gravity defiant learning curves. One explained this behavior in terms of the sensitivity of the expected k-nearest neighbor distances to the data density. Another discussed the former's incorrect reasoning, and demonstrated the possibilities of both gravity-compliance and gravity-defiant behaviors by applying the statistical bias-variance analysis. However, the bias-variance analysis for density estimation error is not an appropriate tool for anomaly detection error. In this talk, we argue that the analysis must be based on the anomaly detection error, and clarify the mechanism of the gravity-defiant learning curves of the nearest neighbor anomaly detectors by applying analysis based on computational geometry to the anomaly detection error. This talk is based on collaborative work with Kai Ming Ting, Jonathan R. Wells, and Sunil Aryal from Federation University, Australia.

Partial Similarity Match with Multi-Instance Multi-Label Learning

Zhi-Hua Zhou

Nanjing University, Nanjing, China

In traditional supervised learning settings, a data object is usually represented by a single feature vector, called an instance. Such a formulation has achieved great success; however, its utility is limited when handling data objects with complex semantics where one object simultaneously belongs to multiple semantic categories. For example, an image showing a lion besides an elephant can be recognized simultaneously as an image of a lion, an elephant, "wild" or even "Africa"; the text document "Around the World in Eighty Days" can be classified simultaneously into multiple categories such as scientific novel, Jules Verne's writings or even books on traveling, etc. In many real tasks it is crucial to tackle such data objects, particularly when the labels are relevant to partial similarity match of input patterns. In this talk we will introduce the MIML (Multi-Instance Multi-Label learning) framework which has been shown to be useful for these scenarios.

Contents

Time-Evolving Data

Scalable Similarity Search

Graphs and Networks

BFST_ED: A Novel Upper Bound Computation Framework for the Graph Edit Distance

Karam Gouda[1,2]([✉]), Mona Arafa[1], and Toon Calders[2]

[1] Faculty of Computers and Informatics, Benha University, Benha, Egypt
{karam.gouda,mona.arafa}@fci.bu.edu.eg
[2] Computer and Decision Engineering Department,
Universit Libre de Bruxelles, Brussels, Belgium
{karam.gouda,toon.calders}@ulb.ac.be

Abstract. Graph similarity is an important operation with many applications. In this paper we are interested in graph edit similarity computation. Due to the hardness of the problem, it is too hard to exactly compare large graphs, and fast approximation approaches with high quality become very interesting. In this paper we introduce a novel upper bound computation framework for the graph edit distance. The basic idea of this approach is to picture the comparing graphs into hierarchical structures. This view facilitates easy comparison and graph mapping construction. Specifically, a hierarchical view based on a breadth first search tree with its backward edges is used. A novel tree traversing and matching method is developed to build a graph mapping. The idea of spare trees is introduced to minimize the number of insertions and/or deletions incurred by the method and a lookahead strategy is used to enhance the vertex matching process. An interesting feature of the method is that it combines vertex map construction with edit counting in an easy and straightforward manner. This framework also allows to compare graphs from different hierarchical views to improve the upper bound. Experiments show that tighter upper bounds are always delivered by this new framework at a very good response time.

Keywords: Graph similarity · Graph edit distance · Upper bounds

1 Introduction

Due to its ability to capture attributes of entities as well as their relationships, graph data model is currently used to represent data in many application areas. These areas include but are not limited to Pattern Recognition, Social Network, Software Engineering, Bio-informatics, Semantic Web, and Chem-informatics. Yet, the expressive power and flexibility of graph data representation model come at the cost of high computational complexity of many basic graph data tasks. One of such tasks which has recently drawn lots of interest in the research community is computing the graph edit distance. Given two graphs, their graph edit distance computes the minimum cost graph editing to be performed on one of them to

© Springer International Publishing AG 2016
L. Amsaleg et al. (Eds.): SISAP 2016, LNCS 9939, pp. 3–19, 2016.
DOI: 10.1007/978-3-319-46759-7_1

get the other. A graph edit operation is a kind of vertex insertion/deletion, edge insertion/deletion or a change of vertex/edge's label (relabeling) in the graph.

A close relationship exists between graph editing and graph mapping. Given a graph editing one can define a graph mapping and vice versa. The problem of graph edit distance computation is then reduced to the problem of finding a graph mapping which induces a minimum edit cost. Graph edit distance computation methods such as those based on A* [6,12,13] exploit this relationship and compute graph edit distance by exploring the vertex mapping space in a best first fashion in order to find the optimal graph mapping. Unfortunately, since computing graph edit distance is NP-hard problem [16] those methods can not scale to large graphs. In practice, to be able to compare large graphs, fast algorithms seeking suboptimal solutions have been proposed. Some of them deliver unbounded solutions [1,14,15,17], while others compute either upper and/or lower bound solutions [2,4,9,16].

Recent interesting upper bounds and the one introduced in this paper are obtained based on graph mapping. The intuition is that the better the mapping between graphs, the better the upper bound on their edit distance. In [10] a graph mapping method is developed, which first constructs a cost matrix between the vertices of the two graphs, and then uses a cubic-time bipartite assignment algorithm, called Hungarian algorithm [8], to optimally match the vertices. The cost matrix holds the matching costs between the neighbourhoods of corresponding vertices. The idea behind this heuristic being that a mapping between vertices with similar neighborhoods should induce a graph mapping with low edit cost. A similar idea is used in [16]. The main problem with these heuristics is that the pairwise vertex cost considers the graph structure only locally. Thus, in cases where neighborhoods do not differentiate the vertices, e.g., as with unlabeled graphs, these methods work poorly. To enhance the graph mapping obtained by these methods and tighten the upper bound, additional search strategies were deployed, however, at the cost of extra computation time. For example, an exhaustive vertex swapping procedure is used in [16]. A greedy vertex swapping is used in [11]. Even though much time is needed by these improvements, the resulted graph mapping is prone to local optima, which is susceptible to initialization.

This paper presents a novel linear-time upper bound computation framework for the graph edit distance. The idea behind this approach is to picture the comparing graphs into hierarchical structures. This view facilitates easy comparison and graph mapping construction. To implement the framework, the breadth first search tree (BFST) representation is adopted as a hierarchical view of the graph, where each comparing graph is represented by a breadth first search tree with its backward edges. A pre-order BFST traversing and matching method is then developed in order to build a graph mapping. A slight drift from the pure pre-order traversal is that for each visited source vertex in the traversal, all its children and those of its matching vertex are matched before visiting any of these children. This facilitates for a vertex to find a suitable correspondence to match among various options. In addition, the idea of spare trees is introduced to decrease the number of insertions and/or deletions incurred by the method, and a lookahead strategy is used to enhance the vertex matching process. An interesting feature of the matching method is that it combines map construction with edit counting in easy and straightforward manner. This novel framework allows to explore a quadratic space of graph mappings to tighten the bound, where for each two corresponding vertices it is possible to run the tree traversing and matching method on the distinct hierarchical view imposed by these two vertices. Moreover, this quadratic space can be explored in parallel to speed up the process, a feature which is not offered by any of the previous methods. Experiments show that tighter upper bounds are always delivered by this framework at a very good response time.

2 Preliminaries

2.1 Graphs

In this section, we first give the basic notations. Let Σ be a set of discrete-valued labels. A labeled graph G can be represented as a triple (V, E, l), where V is a set of vertices, $E \subseteq V \times V$ is a set of edges, and $l\colon V \to \Sigma$ is a labeling function. $|V|$ is the numbers of vertices in G, and is called the *order* of G. The degree of a vertex v, denoted $deg(v)$, is the number of vertices that are directly connected to v. A labeled graph G is said to be *connected*, if each pair of vertices $v_i, v_j \in V$, $i \neq j$, are directly or indirectly connected. In this paper, we focus on simple and connected graphs with labeled vertices. A simple graph is undirected graph with neither self-loops nor multiple edges. Hereafter, a labeled graph is simply called a graph unless stated otherwise.

A graph $G = (V, E, l)$ is a *subgraph* of another graph $G' = (V', E', l')$, denoted $G \subseteq G'$, if there exists a *subgraph isomorphism* from G to G'.

Definition 1 (Sub-)graph isomorphism. *A subgraph isomorphism is an injective function $f\colon V \to V'$, such that (1) $\forall\, u \in V$, $l(u) = l'(f(u))$. (2) $\forall\, (u, v) \in E$, $(f(u), f(v)) \in E'$, and $l((u,v)) = l'((f(u), f(v)))$. If $G \subseteq G'$ and $G' \subseteq G$, G and G' are graph isomorphic to each other, denoted as $G \cong G'$.*

Definition 2 (Maximum) common sub-graph. *Given two graphs G_1 and G_2. A graph $G = (V, E)$ is said to be a common sub-graph of G_1 and G_2 if $\exists\, H_1 \subseteq G_1$ and $H_2 \subseteq G_2$ such that $G \cong H_1 \cong H_2$. A common sub-graph G is a maximum common edge (resp. vertex) sub-graph if there exists no other common sub-graph $G' = (V', E')$ such that $|E'| > |E|$ (resp. $|V'| > |V|$).*

2.2 Graph Editing and Graph Edit Distance

Given a graph G, a graph edit operation p is a kind of vertex or edge deletion, a vertex or edge insertion, or a vertex relabeling. Notice that vertex deletion occurs only for isolated vertices. Each edit operation p is associated with a cost $c(p)$ to do it depending on the application at hand. It is clear that a graph edit

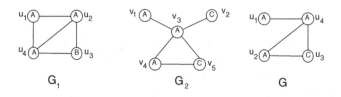

Fig. 1. Two comparing graphs G_1 and G_2.

operation transforms a graph into another one. A sequence of edit operations $\langle p_i \rangle_{i=1}^k$ performed on a graph G to get another graph G' is called *graph editing*, denoted $G^{edit} = \langle p_i \rangle_{i=1}^k$. The cost of graph editing is, thus, the sum of its edit operation's costs, i.e. $\mathcal{C}(G^{edit}) = \sum_{i=1}^k c(p_i)$.

Given two graphs G_1 and G_2 there could be multiple graph editings of G_1 to get G_2. The optimal graph editing is defined as the one associated with the minimal cost among all other graph editings transforming G_1 into G_2. The cost of an optimal graph editing defines the *edit distance* between G_1 and G_2, denoted $GED(G_1, G_2)$. That is, $GED(G_1, G_2) = \min_{G^{edit}} \mathcal{C}(G^{edit})$. In this paper we assume the unit cost model, i.e. $c(p) = 1$, $\forall p$. Thus, the optimal graph editing is the one with the minimum number of edit operations.

Example 1. *Figure 1 shows two graphs G_1 and G_2. An optimal graph editing of G_1 to get G_2 can be obtained as follows: A deletion operation of the edge (u_1, u_2), a relabeling operation of the vertex u_3 from label B into label C, an insertion of a new vertex u_5 with label C, and an insertion of a new edge (u_5, u_4). Thus, $GED(G_1, G_2) = 4$.*

2.3   Graph Mapping

Given two graphs G_1 and G_2, a *graph mapping* aims at finding correspondence between the vertices and edges of the two graphs. Every vertex map f: $V_1 \cup \{u^n\} \to V_2 \cup \{v^n\}$, where u^n and v^n are dummy vertices with special label ϵ, defines a graph mapping, where the vertex $u \in V_1$ or $v \in V_2$ has no correspondence at the other graph if $f(u) = v^n$ or $f(u^n) = v$, resp. The edge $(u, v) \in E_1$ has no correspondence if $(f(u), f(v)) \notin E_2$. Also, the edge $(v, v') \in E_2$ has no correspondence if $(u, u') \notin E_1$ such that $v = f(u)$ and $v' = f(u')$.

There exists a relationship between graph editing and graph mapping. More generally any graph mapping induces a graph editing which relabels all mapped vertices, and inserts or delete the non-mapped vertices/edges of the two graphs [5]. Conversely, given a graph editing, the maximum common subgraph isomorphism (MCSI) between G and G_2 defines a graph mapping between G_1 and G_2, where G is the graph obtained from G_1 after applying the deletion and relabeling operations in the graph editing.

Example 2. *Given the graph editing of Example 1. The graph G obtained from G_1 after applying the edge deletion and vertex relabeling operations of this graph*

editing is shown in Fig. 1. The MCSI $f = \{(u_1, v_1), (u_2, v_4), (u_3, v_5), (u_4, v_3)\}$
between G *and* G_2 *defines a graph mapping. On the other hand, consider the vertex map* $f = \{(u_1, v_2), (u_2, v_4), (u_3, v_1), (u_4, v_3), (u^n, v_5)\}$. *A graph editing can be defined from* f *as: two relabeling operations on* u_1 *and* u_3. *Two deletion operations of the edges* (u_1, u_2) *and* (u_2, u_3). *An insertion operation of a vertex corresponding to the unmatched vertex* v_5. *Two insertion operations of the edges* (u_4, u_5) *and* (u_2, u_5).

In view of the relationship between graph editing and graph mapping, the problem of graph edit distance computation is reduced to the problem of finding an optimal graph mapping – a mapping which induces a minimum edit cost. Due to the hardness of obtaining such a graph mapping (computing graph edit distance is known to be NP-hard problem [16]), approximate graph mapping methods become very popular, especially when large graphs are under investigation [3,7,11,16]. Any of those mapping methods overestimates the graph edit distance. The intuition behind those methods is that the better the mapping between the comparing graphs, the better the upper bound on their edit distance. In this paper we present an efficient upper bound computation framework for the graph edit distance which is also based on graph mapping. We first sketch the framework and then present the details of the implementing algorithm. Hereafter, the comparing graphs G_1 and G_2 are called the source and target graphs, resp; their edges (resp. vertices) are called the source and target edges (resp. vertices).

3 A Novel Upper Bound Computation Framework

The main idea of our approach is to picture the graphs to be compared into hierarchical structures. This view allows easy comparison and fast graph mapping construction. It also facilitates counting of the induced edit operations. Breadth first search (BFS) is a graph traversing method allowing a hierarchical view of the graph through the breadth first search tree it constructs. This view is defined as follows.

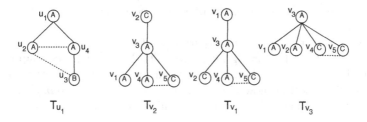

Fig. 2. One BFST view for G_1, $G_1^{u_1} = \langle T_{u_1}, E_{u_1} \rangle$, and three different for G_2, namely, $G_2^{v_2} = \langle T_{v_2}, E_{v_2} \rangle$, $G_2^{v_1} = \langle T_{v_1}, E_{v_1} \rangle$ and $G_2^{v_3} = \langle T_{v_3}, E_{v_3} \rangle$. Black edges constitute BFSTs and backward edges are shown by dashed lines.

Definition 3 (BFST representation of a graph). *Given a graph G and a vertex u ∈ G. Let T_u be the breadth first search tree (BFST) rooted at u. The BFST representation of G given u, denoted as G^u, is defined by the BFST-Edges pair $G^u = \langle T_u, E_u \rangle$, where E_u is the set of graph edges which are not part of T_u, called backward edges.*

Example 3. *Consider the graphs G_1 and G_2 of Fig. 1. Figure 2 shows some of their hierarchical representations using breadth first search trees.*

Algorithm 1: BFST_ED(G_1, G_2)

1:   Let T_u and T_v be the breadth first trees rooted at $u \in G_1$ and $v \in G_2$, resp;
2: $f = \{0, \ldots, 0\}$; $f_{cost} = 0$;   /*f is a vertex map*/
3:   BFST_Mapping_AND_Cost(T_u, T_v, f, f_{cost});
4:   **for** each source or target backward edge **do**
5:   **if** the matching vertices of its end points have no backward edge **then** f_{cost}++;
6:   output f and f_{cost};

Fig. 3. BFST_ED: An upper bound computation framework of $GED(G_1, G_2)$.

Given the source and target graphs G_1 and G_2. Let T_u and T_v be the breadth first trees rooted at $u \in G_1$ and $v \in G_2$, resp. Based on the BFST view of the graph, an upper bound computation framework of the graph edit distance can be developed. First a tree mapping between T_u and T_v is constructed. This tree mapping determines a vertex map between the vertex sets of the two graphs. Using this vertex map, the edit cost on backward edges is calculated and then added to the tree mapping edit cost to produce an upper bound of the graph edit distance. Note that it is possible as a result of the tree matching method an edge is inserted at the position of a source backward edge. If it is the case the final edit cost should be decremented because an edge is already there and this insertion should not be occurred. This framework, named BFST_ED (which stands for the bold letters in: Breadth First Search Tree based Edit Distance), is outlined in Fig. 3. The vector f holds the map on graph vertices. The value $f_i \neq 0$ indicates that the ith vertex of V_1 has been mapped. f_{cost} is the graph mapping cost.

The most important step in this framework is the tree mapping and edit counting method BFST_Mapping_AND_Cost. The better the tree mapping produced by this routine, the better the overall graph edit cost returned by the framework. The question now is *how to build a good tree mapping between two breadth first search trees?* In the following subsections we answer this question.

3.1   Random and Degree-Based BFSTs Matching

The simplest and most direct answer to the previous question is to randomly match vertices at corresponding tree levels. That is, a source vertex at a given tree level l can match any target vertex at the corresponding level. This matching, however, may incur a huge edit cost between the two trees as a vertex having no correspondence has to be deleted as well as its subtree if it is a source one, or to be inserted with its subtree if it is a target one.[1] Moreover, any of these subtree insertions or deletions entails the insertion or deletion of an edge connecting the subtree with its parent. Unfortunately, the number of vertices that have no correspondence will increase as we go down the tree using this matching method. Suppose that at a given tree level the number of source vertices is equal to the number of target ones, and at one of its preceding levels, there exist vertices with no correspondence. Deletions or insertions of subtrees made at the preceding tree level will change the equality at the given level and entail extra deletions and/or insertions.

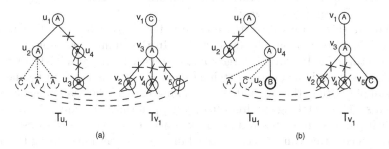

Fig. 4. A picture of the edit operations performed on two comparing BFSTs (a) using random assignment (b) using OUT degree assignment. Vertex/edge insertions are shown by dashed vertices/edges. Vertex relabeling is done on blacked source vertices.

Example 4. *Given the source and target trees T_{u_1} and T_{v_1} of Fig. 2. The edit cost returned by BFST_ED is 13. The random matching in* BFST_Mapping_AND_Cost *induces 10 edit operations, and 3 edit operations are required for backward edge modifications. The vertex map returned by* BFST_Mapping_AND_Cost *is as: $f = \{(u_1, v_1), (u_2, v_3), (u_3, v^n), (u_4, v^n), (u^n, v_2), (u^n, v_4), (u^n, v_5)\}$. This map includes 2 vertex deletions, 2 edge deletions, 3 vertex insertions, and 3 edge insertions. Figure 4(a) gives a picture on how the* mapping_AND_cost *method based on random assignment matches the source T_{u_1} with the target T_{v_1} and computes their graph edit cost.*

An idea to decrease the number of insertions and/or deletions caused by random assignment, and thus decrease the overestimation of GED, is based on the **OUT** degree of a BFST vertex defined as follows.

---

[1] Since all edit modifications usually occur at the source tree to get the target one, any deletion at the target tree is equivalent to an insertion at the source tree in our model.

Definition 4 (OUT degree of a BFST vertex). *Given a graph G. Let T_u be the BFST rooted at $u \in G$. For each tree vertex $w \in T_u$, the OUT degree of w, denoted $OUT(w)$, is defined as the number of its children in the tree.*

The idea is to match the vertices at corresponding tree levels which have near OUT degrees. According to this matching, vertices which have no correspondence will decrease and consequently the edit cost returned by the method as well. Based on this idea, the edit cost in Example 4 is decreased from 13 to 10 edit operations as the vertex map returned by BFST_Mapping_AND_Cost has four less insertion and deletion operations, two on vertices and two on edges, at the cost of one extra vertex relabeling operation for matching the source vertex at the bottom level. The associated vertex map is given as follows: $f = \{(u_1, v_1), (u_4, v_3), (u_2, v^n), (u_3, v_2), (u^n, v_4), (u^n, v_5)\}$. This map incurs 7 edit operations on the BFSTs and 3 on backward edges. Figure 4(b) pictures the tree editing based on OUT degree assignment.

Although this matching method is very fast,[2] still the overall edit cost returned is far from the graph edit distance. In the running example, the best edit cost returned is 10 which is large compared with 4 – the graph edit distance. Another important issue of this matching method which is not seen by the running example is that the method is not taking care of the matching occurred for parents while matching children. It may happen that for many matched children, their parents are matched differently which requires extra edit operations. Though this counting can be accomplished in a subsequent phase using the associated vertex map, the tree mapping cost will be very high. Next, we present a tree mapping and matching method addressing all previous issues.

3.2   An Efficient BFSTs Matching Method

The bad overestimation of the graph edit distance returned by the previous method is due to two reasons. One lies at the simple tree traversing method which does not take previous vertex matching into account and blindly processes the trees level by level. The second reason lies at the vertex matching process itself: Vertices are randomly matched or in the best case are matched based on their OUT degrees which offer a very narrow lookahead view for the comparing vertices. Not to mention the very large number of insertions and/or deletions produced by this matching method. Below we introduce a new tree traversal and vertex matching method which addresses all previous issues.

Traversing the comparing BFSTs in pre-order can offer a solution to the first issue as vertices can be matched in the traversal order. This matching order guarantees that vertices can be matched only if their parents are matched. Though the pre-order traversal removes the overhead of any subsequent counting phase as in the previous method, it limits the different options for matching a given

---

[2] No computations are soever required for random assignment; only climbing the source tree and at each tree level the corresponding vertices are randomly matched. For OUT degree assignment, extra computations are required to match vertices with the closest OUT degrees.

vertex, where only one option is allowed which is based on the visited vertex. To overcome this, one can compare and match all corresponding children of both an already visited source vertex and its matching target before visiting any of these children. This in turn facilitates for a child to find a suitable correspondence to match among various options.

What is the suitable correspondence for a vertex to match? It could be based on the OUT degree as in the previous method. However, the OUT degree gives a very narrow view as we have already noticed. Fortunately, the BFST structure offers a wider lookahead view which is adopted by our method. This view is represented by a tuple, called feature vector, consisting of three values attached with each vertex. These values are calculated during the building process of the BFSTs.

Definition 5 (A feature vector of a BFST vertex). *Given a graph G and let T_u be the BFST rooted at $u \in G$. For each tree vertex $w \in T_u$, the feature vector of w, denoted $f(w)$, is a tuple $f(w) = \langle SUB(w), BW(w), l(w) \rangle$, in which:*

- *$SUB(w)$ is the number of vertices and edges of the subtree rooted at w.*
- *$BW(w)$ is the number of backward edges incident on w.*
- *$l(w)$ is the vertex label.*

Obviously, all tree leaves have SUB count zero. $BW(w)$ is defined for each tree vertex w as: $BW(w) = deg(w) - (OUT(w) + 1)$. Based on Definition 5, a source vertex favors a target vertex to match which has near vertex distance, defined as follows.

Definition 6 (Vertex distance). *Given two source and target tree vertices w and w' with their feature vectors $f(w)$ and $f(w')$. The distance between w and w', denoted $d(w, w')$, is defined based on feature vectors as:*

$$d(w, w') = |SUB(w) - SUB(w')| + |BW(w) - BW(w')| + c(w, w'), \quad (1)$$

where the cost function c returns 0 if the two matching items, i.e. vertices w and w', have identical labels, and 1 otherwise.

By considering the difference $|BW(w) - BW(w')|$ in calculating the vertex distance, the method partially takes care of the backward edges while matching vertices. In fact, $BW(w)$ is introduced to minimize the number of edit operations required for matching backward edges. Formally, let $C_u = \{u_1, \ldots, u_k\}$ and $C_v = \{v_1, \ldots, v_l\}$ be the children of two matched source and target vertices u and v, in the given order. A child u_i of u favors a child v_k of v to match based on the following equation.

$$k = argmin_{v_j \in C_v}(d(u_i, v_j)). \quad (2)$$

That is, the distance between a vertex u_i and its matching vertex v_k should be minimal among other vertices. In cases where there are more than one candidate for a vertex to match, then the method selects the one with the smallest vertex id.

So far the preorder traversal with Eq. 2 addresses some of the previous issues: No subsequent counting phase is required by the method and the method also offers a wider lookahead view to better match the corresponding vertices. Unfortunately, this traversal may worsen the other issues. In fact it may increase the number of insertions and/or deletions because it could happen that for a visited vertex the number of its children differs from the number of children of its matching vertex, though the total number of vertices might be equal at the children level. To overcome this issue the idea of *spare trees* is brought to the method.

Definition 7 (spare subtrees). *Given two comparing BFSTs T_u and T_v rooted at $u \in G_1$ and $v \in G_2$, resp. Any subtree of T_u or T_v rooted at a vertex w is called spare subtree if the vertex w has no correspondence while pre-order traversing T_u and T_v.*

The idea of spare subtrees has been introduced in order to answer the following question: *Why do we get rid of each unmatched vertex with its subtree and pay a high edit cost for doing so, though it could be beneficial later on instead of being costly right now.* The pre-order traversing and matching method is developed by building a *spare-parts store ST_u* at each comparing BFST T_u in order to preserve these unmatched vertices and their subtrees. During tree traversal, when an encountered source or target vertex has no correspondence, the method asks the spare-parts store for a suitable counterpart. If such a spare-part does exist it is matched and removed from the store, otherwise the new vertex itself with its subtree goes to the relevant spare-parts store. This idea guarantees that each vertex will get a counterpart as long as the other tree has this counterpart, i.e., if the number of vertices of the other tree has at least the number of vertices of the tree where the vertex belongs to. At the end of the tree traversal the spare-parts store associated with the tree of small order will be empty and the other store will contain a number of spare subtrees equal to the vertex difference $||V_1| - |V_2||$. Finally, the number of vertices and edges in each remaining spare subtree will be added to the tree mapping cost. Fortunately, the size of each remaining spare subtree will be very small.

Algorithm 2 in Fig. 5 is a recursive encoding of the method. In fact we do not put the whole spare subtrees in the store, references to their roots are the only information that is maintained (refer to line 12). Also, if a vertex and its subtree is characterized as a spare part, the connecting edge with its parent vertex (the vertex where it hangs on) is deleted and the tree mapping cost is updated (see line 3: All edges connecting children which have no correspondence are deleted if they are source vertices and inserted otherwise). Moreover, if this vertex is a source one, it is temporary blocked, i.e., it is temporary removed from the pre-order traversal (line 13). Alternatively, if a spare source subtree is matched and removed from the store, it goes directly into the pre-order traversal again (line 28). It means that the root of this subtree will be hung on and become a child of the currently processing parent vertex. For hanging this spare vertex no edge insertion is required since the matching vertex, whether it comes from the other spare store or as a corresponding child, has already charged by an equivalent deletion operation at line 3 of the edge connecting it with its parent.

Algorithm 2: mapping_AND_cost($T_{r_1}, T_{r_2}, f, f_{cost}$)

1: let $C_{r_1} = \{u_{11}, \ldots, u_{1n_1}\}$ be the children of r_1 in the given order;
2: let $C_{r_2} = \{u_{21}, \ldots, u_{2n_2}\}$ be the children of r_2 in the given order;
3: $f_{cost}+ = ||C_{r_1}| - |C_{r_2}||$; /* edge deletion if $|C_{r_1}| > |C_{r_2}|$, insertions otherwise */
4: Let C_{r_l} be the smallest set of children, $l = 1$ or 2 (in the following steps $m = 1$ or 2, $m \neq l$)
5:     **for** each child $u_{li} \in C_{r_l}$ **do** /*find a suitable correspondence from C_{r_m}.*/
6:        $k = \text{argmin}_{u_{mj} \in C_{r_m}} d(u_{li}, u_{mj})$;
7:        **if** $l = 2$ **then** $f[u_{mk}] = u_{li}$;
8:        **else** $f[u_{li}] = u_{mk}$; u_{mk} is matched;
9:        **if** $l(u_{li}) \neq l(u_{mk})$ **then** f_{cost}++;
10:    **for** each remaining $u_{mi} \in C_{r_m}$ **do** /* each u_{mi} searches for correspondence at ST_{r_l}.*/
11:       **if** $ST_{r_l} = \emptyset$ **then** /*u_{mi} is spared if the other store is empty */
12:         $ST_{r_m} = ST_{r_m} \cup \{u_{mi}\}$;
13:         $C_{r_m} = C_{r_m} \setminus \{u_{mi}\}$;
14:       **else** /*u_{mi} tries to find a suitable correspondence.*/
15:         $k = \text{argmin}_{u_j \in ST_{r_l}} d(u_{mi}, u_j)$;
16:         **if** $l = 2$ **then** $f[u_{mi}] = u_k$;
17:         **else**
18:           $f[u_k] = u_{mi}$; u_{mi} is matched;
19:           $C_{r_l} = C_{r_l} \cup \{u_k\}$; /*if u_k is a source one, it goes into the preorder traversal again.*/
20:         **if** $l(u_k) \neq l(u_{mi})$ **then** f_{cost}++;
21:         $ST_{r_l} = ST_{r_l} \setminus \{u_k\}$;
22: **for** each $u_{1i} \in C_{r_1}$ **do**
23:    mapping_AND_cost($T_{u_{1i}}, T_{f[u_{1i}]}, f, f_{cost}$);

Fig. 5. Pre-order traversing and matching method.

Example 5. *Figure 6 explains how the traversing method (Algorithm 2) matches T_{u_1} with T_{v_1} of Fig. 2 and computes the tree edit cost. The graph edit cost produced by BFST_ED is 5: 6 edit operations are required to transform T_{u_1} into T_{v_1}; one of the tree edge insertions is removed because it is occurred at the position of the backward edge (u_2, u_4), and finally zero edit operations are required on the remaining backward edges.*

Theorem 1 *(Time Complexity). The procedure BFST_mapping_AND_cost (Algorithm 2) returns the vertex map f and its induced edit cost f_{cost} in $O(d^2|V_1|)$, where d is the maximum vertex degree in both graphs.*

Theorem 2 *(Correctness). The value f_{cost} returned by BFST_ED(G_1, G_2) with Algorithm 2 at Fig. 5 is the edit cost induced by the returned vertex map f.*

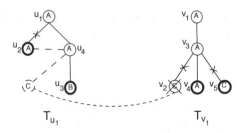

Fig. 6. A possible tree editing transforming T_{u_1} into T_{v_1}, which is produced by the preorder method (Algorithm 2). Vertex and edge insertions are shown by dashed lines and vertex relabeling is shown by heavy-blacked lines. This tree editing has the following 6 edit operations given in order according to the algorithm: deletion of the edge (u_1, u_2), deletion of two target edges which is equivalent to two edge insertions at the source tree, relabeling of u_3, deletion of v_2 which is equivalent to vertex insertion at the source tree, and relabeling of u_2. The vertex map returned by this algorithm in the order of its construction is as follows: $f = \{(u_1, v_1), (u_4, v_3), (u_3, v_4), (u_2, v_5), (u^n, v_2)\}$.

3.3   Improving the Overestimation: BFST_ED_ALL

Previously, based on the chosen graph vertex, a hierarchical representation of the graph could be given. Thus, for each graph G, it is possible to construct $|V|$ distinct hierarchical views, each of which starts from a different vertex. The multi-hierarchical views of a graph gives us the opportunity to compare two graphs from different hierarchical perspectives and choose the best obtained graph mapping, instead of restricting ourselves to a single view comparison. This multi-view comparison is implemented and called BFST_ED_ALL. In fact BFST_ED_ALL explores $|V_1| \times |V_2|$ possible graph mappings and returns the one with the least overestimation.

4   Experimental Evaluation

In this section, we aim at empirically studying the proposed method. We conducted several experiments, and all experiments were performed on a 2.27 GHz Core i3 PC with 4 GB memory running Linux. Our method is implemented in standard C++ using the STL library and compiled with GNU GCC.

Benchmark Datasets: We chose several real graph datasets for testing the method.

(1) **AIDS** (http://dtp.nci.nih.gov/docs/aids/aidsdata.html) is a DTP AIDS Antiviral Screen chemical compound dataset. It consists of 42, 687 chemical compounds, with an average of 46 vertices and 48 edges. Compounds are labelled with 63 distinct vertex labels but the majority of these labels are H, C, O and N.

(2) **Linux**   (http://www.comp.nus.edu.sg/~xiaoli10/data/segos/linux_segos. zip) is a Program Dependence Graph (PDG) dataset generated from the

Linux kernel procedure. PDG is a static representation of the data flow and control dependency within a procedure. In the PDG graph, an vertex is assigned to one statement and each edge represents the dependency between two statements. PDG is widely used in software engineering for clone detection, optimization, debugging, etc. The Linux dataset has in total 47,239 graphs, with an average of 45 vertices each. The graphs are labelled with 36 distinct vertex labels, representing the roles of statements in the procedure, such as declaration, expression, control-point, etc.

(3) **Chemical** is a chemical compound dataset. It is a subset of PubChem (https://pubchem.ncbi.nlm.nih.gov) and consists of one million graphs. It has 24 vertices and 26 edges on average. The graphs are labelled with 81 distinct vertex labels.

4.1   Comparison with Exact Methods

We first evaluate the performance of our methods, BFST_ED and BFST_ED_All, against exact GED computation methods. We want to see how much speed up can be achieved by our methods at the cost of how much loss in accuracy of GED. In this experiment, we use the recent exact GED computation method named CSI_GED [5], and randomly choose two source and target vertices to run BFST_ED. As the exact computation of GED is expensive on large graphs, to make this experiment possible, graphs with acceptable order were randomly selected from the data sets. From these graphs, four groups of ten graphs each were constructed. The graphs in each group have the same number of vertices, and the number of vertices residing in each graph among different groups varies from 5 to 20. In this experiment, each group is compared with the one having the largest graph order. Thus, we have 100 graph matching operations in each group comparison. For estimating the errors, the mean relative overestimation of the exact graph edit distance, denoted ϕ_o, is calculated.[3] Figure 7 plots the value ϕ_o of each method on each group for the different data sets, where the horizontal axis shows the order of the comparing group. It is clear that $\phi_o = 0$ for CSI_GED. Figure 7 also plots the mean run time ϕ_t taken by each method on each group for each data set.

First, we observe that on the different data sets the accuracy loss of BFST_ED_All is very small on small order groups and increases with increasing graph order. It is between 10–20% on large groups. Accuracy loss of BFST_ED, on the other hand, is even worse and exhibits the same trend. It is about 3–4 times larger than that of BFST_ED_All. Looking at the run time of the three methods. We observe that on large groups comparisons, BFST_ED_All outperforms CSI_GED by 2–5 orders of magnitude and it is outperformed by BFST_ED from 1–2 orders of magnitude. One thing that should be noticed is that on the very small order group, the one with order 5, CSI_GED is faster than BFST_ED_All on all real data sets.

---

[3] ϕ_o is defined for a pair of graphs matching as: $\phi_o = \frac{|\lambda - GED|}{GED}$, where λ and GED are the approximate and exact graph edit distances, resp.

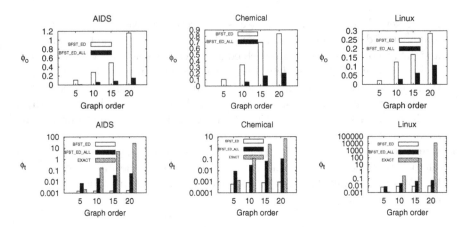

Fig. 7. Comparative accuracy and time with exact method.

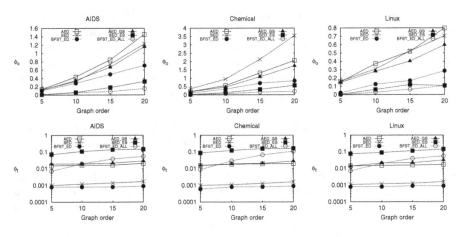

Fig. 8. Comparative accuracy and time with different methods: small order graphs.

4.2   Comparison with Approximation Methods

In this set of experiments, we compare our methods against the state-of-the-art upper bound computation methods such as Assignment Edit Distance (AED) method [10], the Star-based Edit Distance (SED) method [16], and their extensions. These methods are extended by applying a postprocessing vertex swapping phase to enhance the obtained graph mapping. In [5], a greedy vertex swapping procedure is applied on the map obtained from AED, and is abbreviated as "AED_GS", and in [16] an exhaustive vertex swapping is applied on the map obtained from SED and is abbreviated as "SED_ES". The executables for competitor methods were obtained from their authors.

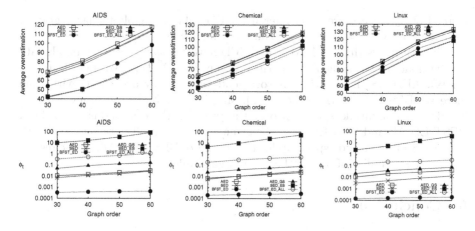

Fig. 9. Comparative accuracy and time with different methods: large order graphs.

Comparison with Respect to GED. First we compare the different methods on graphs where the exact graph edit distance is known. Therefore, we use the groups of graphs from the previous experiment. To look at bound tightness, ϕ_o is calculated for each of these methods. Obviously, the smaller the mean relative overestimation, the better is the approximation method. We also aim at investigating ϕ_t for each method.

Figure 8 plots ϕ_o and ϕ_t for each method on the different data sets. It shows that BFST_ED_All always produces smaller ϕ_o values than the ones produced by other methods on all data sets. The gap between ϕ_o values is remarkable on the AIDS and Chemical data sets, where ϕ_o values of BFST_ED_All are almost half of those produced by SED_ES, the best competitor. On Linux data set, those produced by SED_ES are comparable with ours on the largest group comparison. In addition to the good results on bound tightness, the average run time of BFST_ED_ALL is better than that of other methods. It is about 2 times faster than the best competitor. Looking at each method individually, there is a clear trade-off between bound tightness and speed. The first map is always come at high speed but at the cost of accuracy loss. In conclusion, we can see that the upper bound obtained by BFST_GED_ALL provides near approximate solutions at a very good response time compared with current methods.

Comparison on Large Graphs. In this set of experiments we evaluate the different methods on large graphs. In each data set, four groups of ten graphs each are selected randomly, where each group has a fixed graph order chosen as: 30, 40, 50, and 60. Each of these groups is compared using the different methods with a database of 1000 graphs chosen randomly from the same data set. Figure 9 shows the average edit overestimation returned by each method per graph matching on each group. The average edit overestimation is adopted instead of ϕ_o since there is no reference GED value available for large graphs. The figure also shows the average running time for all data sets.

Figure 9 shows that both AED and SED have the same accuracy on all data sets with almost the same running time (except that AED is two times faster on Linux). AED_GS shows little improvements of accuracy over AED with time increase. BFS_ED, on the other hand, shows much better accuracy with 2–3 orders of magnitude speed up over the previous three methods. Also, both BFST_ED_All and SED_ES show the same accuracy on all data sets; but with two orders of magnitude speed up for the benefit of BFST_ED_All. These results shows the scalability of our methods on large graphs.

5   Conclusion

In this paper, the computational methods approximating the graph edit distance are studied; in particular, those overestimating it. A novel overestimation approach is introduced. It uses breadth first hierarchical views of the comparing graphs to build different graph maps. This approach offers new features not present in the previous approaches, such as the easy combination of vertex map construction and edit counting, and the possibility of constructing graph maps in parallel. Experiments show that near overestimation is always delivered by this new approach at a very good response time.

References

1. Conte, D., Foggia, P., Sansone, C., Vento, M.: Thirty years of graph matching in pattern recognition. Int. J. Pattern Recogn. Artif. Intell. **18**, 265–298 (2004)
2. Fischer, A., Suen, C., Frinken, V., Riesen, K., Bunke, H.: Approximation of graph edit distance based on hausdorff matching. Pattern Recogn. **48**(2), 331–343 (2015)
3. Gaüzère, B., Bougleux, S., Riesen, K., Brun, L.: Approximate graph edit distance guided by bipartite matching of bags of walks. In: Fränti, P., Brown, G., Loog, M., Escolano, F., Pelillo, M. (eds.) S+SSPR 2014. LNCS, vol. 8621, pp. 73–82. Springer, Heidelberg (2014)
4. Gouda, K., Arafa, M.: An improved global lower bound for graph edit similarity search. Pattern Recogn. Lett. **58**, 8–14 (2015)
5. Gouda, K., Hassaan, M.: CSI_GED: an efficient approach for graph edit similarity computation. In: ICDE, pp. 265–276 (2016)
6. Hart, P., Nilsson, N., Raphael, B.: A formal basis for the heuristic determination of minimum cost paths. IEEE Trans. SSC **4**(2), 100–107 (1968)
7. Justice, D., Hero, A.: A binary linear programming formulation of the graph edit distance. IEEE Trans. PAMI **28**(8), 1200–1214 (2006)
8. Munkres, J.: A network view of disease and compound screening. J. Soc. Ind. Appl. Math. **5**, 32–38 (1957)
9. Neuhaus, M., Bunke, H.: Edit distance-based kernel functions for structural pattern classification. Pattern Recogn. **39**, 1852–1863 (2006)
10. Riesen, K., Bunke, H.: Approximate graph edit distance computation by means of bipartite graph matching. Image Vis. Comput. **27**(7), 950–959 (2009)

11. Riesen, K., Fischer, A., Bunke, H.: Computing upper and lower bounds of graph edit distance in cubic time. In: El Gayar, N., Schwenker, F., Suen, C. (eds.) ANNPR 2014. LNCS, vol. 8774, pp. 129–140. Springer, Heidelberg (2014)

12. Riesen, K., Emmenegger, S., Bunke, H.: A novel software toolkit for graph edit distance computation. In: Kropatsch, W.G., Artner, N.M., Haxhimusa, Y., Jiang, X. (eds.) GbRPR 2013. LNCS, vol. 7877, pp. 142–151. Springer, Heidelberg (2013)

13. Riesen, K., Fankhauser, S., Bunke, H.: Speeding up graph edit distance computation with a bipartite heuristic. In: MLG, pp. 21–24 (2007)

14. Riesen, K., Neuhaus, M., Bunke, H.: Bipartite graph matching for computing the edit distance of graphs. In: Escolano, F., Vento, M. (eds.) GbRPR. LNCS, vol. 4538, pp. 1–12. Springer, Heidelberg (2007)

15. Serratosa, F.: Fast computation of bipartite graph matching. Pattern Recogn. Lett. **45**, 244–250 (2014)

16. Zeng, Z., Tung, A., Wang, J., Feng, J., Zhou, L.: Comparing stars: on approximating graph edit distance. PVLDB **2**(1), 25–36 (2009)

17. Zhao, X., Xiao, C., Lin, X., Wang, W., Ishikawa, Y.: Efficient processing of graph similarity queries with edit distance constraints. VLDB J. **22**, 727–752 (2013)

Pruned Bi-directed K-nearest Neighbor Graph
for Proximity Search

Masajiro Iwasaki[✉]

Yahoo Japan Corporation, Tokyo, Japan
miwasaki@yahoo-corp.jp

Abstract. In this paper, we address the problems with fast proximity searches for high-dimensional data by using a graph as an index. Graph-based methods that use the k-nearest neighbor graph (KNNG) as an index perform better than tree-based and hash-based methods in terms of search precision and query time. To further improve the performance of the KNNG, the number of edges should be increased. However, increasing the number takes up more memory, while the rate of performance improvement gradually falls off. Here, we propose a pruned bi-directed KNNG (PBKNNG) in order to improve performance without increasing the number of edges. Different directed edges for existing edges between a pair of nodes are added to the KNNG, and excess edges are selectively pruned from each node. We show that the PBKNNG outperforms the KNNG for SIFT and GIST image descriptors. However, the drawback of the KNNG is that its construction cost is fatally expensive. As an alternative, we show that a graph can be derived from an approximate neighborhood graph, which costs much less to construct than a KNNG, in the same way as the PBKNNG and that it also outperforms a KNNG.

1 Introduction

How to conduct fast proximity searches of large-scale high dimensional data is an inevitable problem not only for similarity-based image retrieval and image recognition but also for multimedia data processing and large-scale data mining. Image descriptors, especially local descriptors, are used for various image recognition purposes. Since a large number of local descriptors are extracted from just one image, shortening the query time is crucial when handling a huge number of images. Thus, indices are indispensable in this regard for large-scale data, and as a result, various indexing methods have been proposed. In recent years, an approximate proximity search method that does not guarantee exact results has been the prevailing method used in the field because the query time rather than search accuracy is prioritized.

Hash-based and quantization-based methods are approximate searches without original objects. LSH [1], which is one of the hash-based methods, searches for proximate objects by using multiple hash functions, which compute the same

---

The original version of this chapter was revised: The presentation of Fig. 5(b) was incorrect. The erratum to this chapter is available at 10.1007/978-3-319-46759-7_26

© Springer International Publishing AG 2016
L. Amsaleg et al. (Eds.): SISAP 2016, LNCS 9939, pp. 20–33, 2016.
DOI: 10.1007/978-3-319-46759-7_2

hash value for objects that are close to each other. Datar et al. [2] applied LSH to L_p spaces so that it could be used in various applications. Spectral hashing [3] was proposed as a method that optimizes the hash function by using a statistical approach for datasets. Quantization-based methods [4,5] quantize objects and search for quantized objects. For example, the product quantization method (PQ) [5] splits object vectors into sub vectors and quantizes the sub vectors to improve the search accuracy. While recent hash-based and quantization-based methods drastically reduce memory usage, the search accuracies are significantly lower than those of proximity searches using original objects.

Proximity searches using original objects are broadly classified into tree-based and graph-based. In the tree-based method, a whole space is hierarchically and recursively divided into sub spaces. As a result, the sub spaces form a tree structure. Various kinds of methods have been proposed, including kd-tree [6], SS-tree [7], vp-tree [8], and M-tree [9]. While these methods provide exact search results, tree-based approximate search methods have also been studied. ANN [10] is a method that applies an approximate search to a kd-tree. SASH [11] is a tree that is constructed without dividing a space. FLANN [12] is an open source library for approximate proximity searches. It provides randomized kd-trees wherein multiple kd-trees are searched in parallel [12,13] and k-means trees that are constructed by hierarchical k-means partitioning [12,14].

Graph-based methods use a neighborhood graph as a search index. Arya et al. [15] proposed a method that uses randomized neighbor graphs as a search index. Sebastian et al. [16] used a k-nearest neighbor graph (KNNG) as a search index. Each node in the KNNG has directed edges to the k-nearest neighboring nodes. Although a KNNG is a simple graph, it can reduce the search cost and provides a high search accuracy. Wang et al. [17] improved the search performance by using seed nodes, which are starting nodes for exploring a graph, obtained with a tree-based index depending on the query from an object set. Hajebi et al. [18] showed that searches using KNNGs outperform LSH and kd-trees for image descriptors. Therefore, in this paper, we focused on a graph-based approximate search for image descriptors to acquire higher performance.

Let $G = G(V, E)$ be a graph, where V is a set of nodes that are objects in a d-dimensional vector space $\mathbb{R}^d$. E is the set of edges connecting the nodes. In graph-based proximity searches, each of the nodes in a graph corresponds to an object to search for. The graph that these methods use is a neighborhood graph where neighboring nodes are associated with edges. Thus, neighboring nodes around any node can be directly obtained from the edges. The following is a simple nearest neighbor search for a query object that is not a node of a graph using a neighborhood graph in a best-first manner.

An arbitrary node is selected from all of the nodes in the graph to be the target. The closest neighboring node to the query is selected from the neighboring nodes of the target. If the distance between the query and the closest neighboring node is shorter than the distance between the query and the target node, the target node is replaced by the closest node. Otherwise, the target node is the nearest node (the search result), and the search procedure is terminated.

The search performance of a KNNG improves as the number of edges for each node increases. However, the rate of improvement gradually tapers off while the edges occupy more and more memory. To avoid this problem, we propose a pruned bi-directed k-nearest neighbor graph (PBKNNG). First, it adds reversely directed edges to all of the directed edges in a KNNG. While it can improve the search performance, the additional edges tend to concentrate on some of the nodes. Such excess edges obviously reduce the search performance because the number of accesses to unnecessary nodes to search increases. Therefore, second, the long edges of each node holding excess edges are simply pruned. Third, edges that have alternative paths for exploring the graph are selectively pruned. Thus, we show that the PBKNNG outperforms not only the KNNG but also the tree- and quantization-based methods.

As the number of objects grows, the brute force construction cost of a KNNG exponentially increases because the distances between all pairs of objects in the graph need to be computed. Thus, Dong et al. [19] reduced the construction cost by constructing an approximate KNNG. Here, the ANNG [20] is not an approximate KNNG but an approximate neighborhood graph that is incrementally constructed using approximate k-nearest neighbors that are searched for by using the partially constructed ANNG. Such approximate neighborhood graphs can drastically reduce construction costs. In this paper, we also show that the search performance of a graph (PANNG) derived from an ANNG instead of a KNNG in the same way as a PBKNNG can be close to that of a PBKNNG.

The contributions of this paper are as follows.

- We propose a PBKNNG derived from a KNNG and show that it outperforms not only the KNNG but also the tree- and quantization-based methods.
- We show the effectiveness of a PANNG derived from an approximate neighborhood graph instead of a KNNG derived in the same way as a PBKNNG.

2   KNNG-Based Proximity Search

2.1   Proximity Search Algorithm

Most applications including image search and recognition require more than one object to be the result for a specific query. Therefore, we decided to focus on k-nearest neighbor (KNN) searches in this study. The search procedure with a graph-based index generally consists of two steps: obtaining seed nodes and exploring the graph with the seed nodes. Seed nodes can be obtained by random sampling [18,20], clustering [16], or finding nodes that neighbor a query by using a tree-based index [17,21]. Although the methods using a tree-based index perform the best, we used the simplest method, random sampling, in order to evaluate the graph structure without the effect of the tree-structure or clustering. As far as the second step goes, there are two methods of exploring a graph. In the first, the neighbors of the query are traced from seed objects in the best-first manner in Sect. 1, and this is done repeatedly using different seeds to improve the search accuracy [16,18]. In the second, nodes within the search

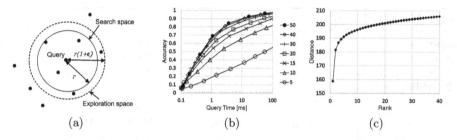

Fig. 1. (a) Relationship between the search space, exploration space, and query. (b) Search accuracy vs. query time of KNNG for different numbers of edges k for 10 million SIFT image descriptors. (c) Average distance of objects for each rank of nearest neighbors vs. rank of nearest neighbors.

space, which is narrowed down as the search progresses, are explored [17,20]. The former method has a drawback in that the same nodes are accessed multiple times because it performs the best-first procedure repeatedly. As a result, search performance deteriorates. Therefore, we use the latter to evaluate graphs in this paper.

During KNN search, the distance of the farthest object in the search result from the query object is set as the search radius r. The actual explored space is wider than the search space defined by r. The radius of the exploration space r_e is defined as $r_e = r(1 + \epsilon)$, where ϵ expands the exploration space to improve the search accuracy. As ϵ increases, the accuracy improves; however, the search cost increases because more objects within the expanded space must be accessed. Figure 1(a) shows how the search space, exploration space, and query are related. Algorithm 1 is the pseudo code of the search. Here, KnnSearch returns a set of resultant objects R. Let q be a query object, k_s be the number of resultant objects, C be the set of already evaluated objects, $d(x, y)$ be the distance between objects x and y, and $N(G, x)$ be the set of neighboring nodes associated with the edges of node x in graph G. The function Seed(G) returns seed objects sampled randomly from graph G. In a practical implementation, sets S and R are priority queues. While making set C a simple array would reduce the access cost, the initializing cost is expensive for large-scale data. For this reason, a hash set is used instead.

2.2   Problem Definition

For simplicity, we will analyze the nearest neighbor search instead of a k-nearest neighbor search. If Condition 1 is satisfied, the nearest neighbor is obtained in a best-first manner from an arbitrary node on the neighborhood graph [22].

Condition 1. $\forall a \in G, \forall q \in \mathbb{R}^d$, if $\forall b \in N(G, a), d(q, a) \leq d(q, b)$, then $\forall b \in G, d(q, a) \leq d(q, b)$.

Delaunay triangulation, which satisfies Condition 1, has absolutely fewer edges than a complete graph that also satisfies Condition 1. The number of edges,

---

Algorithm 1. KnnSearch

Input: G, q, k_s, ϵ
Output: R
1: $S \leftarrow \text{Seed}(G), r \leftarrow \infty, R \leftarrow \emptyset$
2: **while** $S \neq \emptyset$ **do**
3:     $s \leftarrow \underset{x \in S}{\text{argmin}}\, d(x, q), S \leftarrow S - \{s\}$
4:     **if** $d(s, q) > r(1 + \epsilon)$ **then**
5:       **return** R
6:     **end if**
7:     **for all** $o \in N(G, s)$ **do**
8:       **if** $o \notin C$ **then**
9:         $C \leftarrow C \cup \{o\}$
10:       **if** $d(o, q) \leq r(1 + \epsilon)$ **then**
11:         $S \leftarrow S \cup \{o\}$
12:       **end if**

13:       **if** $d(o, q) \leq r$ **then**
14:         $R \leftarrow R \cup \{o\}$
15:         **if** $|R| > k_s$ **then**
16:           $R \leftarrow R - \{\underset{x \in R}{\text{argmax}}\, d(x, q)\}$
17:         **end if**
18:         **if** $|R| = k_s$ **then**
19:           $r \leftarrow \max_{x \in R} d(x, q)$
20:         **end if**
21:       **end if**
22:     **end if**
23:     **end for**
24: **end while**
25: **return** R

---

however, increases drastically as the dimension of the objects increases. Therefore, a Delaunay triangulation is impractical in terms of the index size due to a huge number of the edges. As a result, most of the graph-based methods instead use a KNNG, where the number of edges can be arbitrarily specified. The search results of KNNG, however, are approximate because this graph does not satisfy Condition 1.

Figure 1(b) shows the accuracy versus query time for different numbers of edges k in a KNNG. The dataset consisted of 10 million SIFT image descriptors (128-dimensional data). The search was conducted with Algorithm 1. The curves of the figure are depicted by varying ϵ. Being closer to the top-left corner of the figure means better performance in terms of query time and accuracy. In this paper, accuracy is measured in terms of precision. In fact, precision and recall are identical in the KNN search. From Fig. 1(b), one can see that the search performance improves as the number of edges k in the KNNG increases. However, the rate of improvement gradually decreases. The memory needed for storing over 50 edges is large, whereas the improvement brought by storing so many edges is not so great.

We examined the distribution of neighboring objects around a query object. 1,000 objects were randomly selected as queries from 10 million objects, and the 40 nearest neighbors for each query object were sorted by distance. Figure 1(c) shows the average distance of the objects for each rank of the nearest neighbors. The distance of the highest ranking object that is the nearest to the query object is significantly shorter than the distances of lower ranked objects. Thus, the neighboring region around an arbitrary object is extremely sparse, while outside the neighboring region is extremely dense.

Therefore, the case in Fig. 2(a) frequently occurs in high-dimensional spaces. The figure depicts the space of distances from node o_1. The number of edges in KNNG is three. The rank of o_2 in ascending order of the distance from o_1 is much

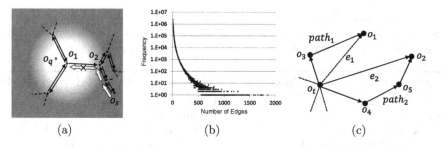

Fig. 2. (a) Relationship between nodes and edges in the case of problem conditions. (b) Frequency of nodes vs. number of edges for each node in a BKNNG. (c) Selective edge removal. The target node is o_t, which has excess edges. If $p = 3$, e_1 is removed, and e_2 is not.

higher than the rank of o_1 in ascending order of the distance from o_2. Thus, while the directed edge from o_1 to o_2 is generated, an edge from o_2 to o_1 is not generated. Therefore, during a search, when the query o_q is close to node o_1 and the seed object o_s is near object o_2, node o_1 cannot be reached through o_2 from node o_s because there is no path from o_2 to o_1. As a result, search accuracy is reduced for high-dimensional data where such conditions frequently occur. Increasing the number of edges helps to avoid such disconnections between neighboring nodes. Figure 1(b) shows that increasing the number of edges improves performance until around 30 edges, after which the improvement rate tapers off. While more edges can reduce such disconnections, more than enough edges increase the number of accessed nodes that are ineffective for searching. As a result, the query time increases.

3   Our Approach

To resolve the problem that increasing the number of edges to improve accuracy causes the query time to increase, we propose two types of graph structures: the pruned bi-directed k-nearest neighbor graph and pruned ANNG.

3.1   Pruned Bi-directed K-nearest Neighbor Graph

For a first step of our proposal, a reversely directed edge can be added for each directed edge instead of increasing the number of edges of each node. Furthermore, if a corresponding reversely directed edge already exists, it is not added. This solution can connect disconnected pairs of nodes and suppress any increase in ineffective long edges. We refer to the resultant graph as a bi-directed k-nearest neighbor graph (BKNNG). It theoretically has up to twice as many edges as a KNNG. However, since a KNNG likely has some node pairs with directed edges pointing to each other, the number of edges in a BKNNG is typically less than twice that of a KNNG. In the case of 10 million SIFT objects, the number of edges in a BKNNG generated from a KNNG wherein each node has 10 edges is about 186 million. Therefore, the number of cases in Fig. 2(a), where one pair of nodes has one directed edge between two nodes, is about 86 million. 14 million pairs of nodes have two different directed edges between each other.

Algorithm 2. ConstructPBKNNG

Input: G, k_p, k_r, p
Output: G
1: for all $o \in V$ do
2:    for all $n \in N(G, o)$ do
3:       if $N(G, n) \cap \{o\} = \emptyset$ then
4:          $N(G, n) \leftarrow N(G, n) \cup \{o\}$
5:       if $|N(G, n)| > k_p$ then
6:          $N(G, n) \leftarrow N(G, n) - \{\underset{x \in N(G,n)}{\mathrm{argmax}}\, d(x, n)\}$
7:       end if
8:    end if
9:    end for
10: end for
11: RemoveEdgesSelectively(G, k_r, p)
12: return G

Algorithm 3. RemoveEdgesSelectively

Input: G, k_r, p
Output: G
1: for all $o \in V$ do
2:    for all $n \in N(G, o)$ do
3:       if $\mathrm{Rank}(N(G, o), n) > k_r$ then
4:          if PathExists$(G, o, n, p) = true$ then
5:             $N(G, o) \leftarrow N(G, o) - n$
6:          end if
7:       end if
8:    end for
9: end for
10: return G

Figure 2(b) shows the frequency of nodes versus the number of edges in a BKNNG that was generated from a KNNG in which each node had 10 edges. The number of edges is widely distributed from 10 up to 1,851. The number of edges having the highest frequency is 10. The average number of edges is about 18.6. Since excess edges for some of the nodes reduce the search performance as a result of the computations for all the excess edges, the excess edges should be pruned. Too long edges of nodes holding excess edges are obviously not effective for exploring a graph because they do not connect to neighboring nodes. For a second step, to prune such edges, the edges are sorted in ascending order of length while reversely directed edges are being added. Here, let k_p be the maximum number of edges for each node after pruning. Edges whose rank is larger than k_p are forcedly removed (forced edge removal). Even though the processing cost is small enough, excess edges can be effectively reduced. Nevertheless, long and excess edges still remain. Since some of the long edges are effective for exploring the graph because they connect clusters and some are not, these edges should be selectively pruned to maintain the connections. If an edge from a source node to a destination node has an alternative path from the source node to the destination node, even if the edge is removed, the destination can be descended from the source through the path instead of the removed edge. Note that as the number of edges on the alternative path increases, the distance computation cost also increases during a search. Therefore, the shortest path should be found, and fewer edges on the path is better. For a third step, if the edges are ranked lower than k_r, where $k_r < k_p$, and have alternative paths that consist of less than p edges that should be all ranked higher than k_r for each node, they are removed (selective edge removal). Figure 2(c) shows the selective edge removal.

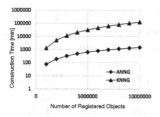

Fig. 3. Construction time vs. number of nodes in ANNG and KNNG

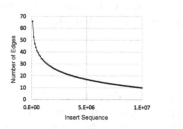

Fig. 4. Average number of edges for every 100,000 consecutive nodes of ANNG

Algorithm 4. ConstructPANNG

Input: $O, k_c, k_p, k_r, \epsilon_c, p$
Output: G
1: **for all** $o \in O$ **do**
2:     $N(G, o) \leftarrow$ KnnSearch(G, o, k_c, ϵ_c)
3:     **for all** $n \in N(G, o)$ **do**
4:         $N(G, n) \leftarrow N(G, n) \cup \{o\}$
5:         **if** $|N(G, n)| > k_p$ **then**
6:             $N(G, n) \leftarrow N(G, n) - \{\underset{x \in N(G,n)}{\arg\max} \, d(x, n)\}$
7:         **end if**
8:     **end for**
9: **end for**
10: RemoveEdgesSelectively(G, k_r, p)
11: **return** G

$Path_1$ is the shortest alternative path of the edge e_1. $Path_2$ is the shortest alternative path of e_2. If $p = 3$, then e_1 is removed because the number of edges on $path_1$ is two. However e_2 is not removed because that on $path_2$ is three. Although finding the shortest path is time consuming, the limitation p of the number of edges on the alternative paths contributes to reducing the processing time to find the shortest alternative path. We refer to the resultant graph as a pruned bi-directed k-nearest neighbor graph (PBKNNG). Algorithm 2 shows the pseudo code for constructing a PBKNNG. Here, a KNNG is the input graph $G = G(V, E)$. Algorithm 2 calls Algorithm 3, which is the selective edge removal. Rank$(N(G, o), n)$ returns the rank of a node n by the distance (edge length) to the neighboring nodes $N(G, o)$ of a node o. PathExists(G, o, n, p) exhaustively explores the graph G from o within p edges and returns whether the shortest alternative path from o to n exists.

3.2   Pruned ANNG

While the KNNG is extremely expensive to construct, an approximate neighborhood graph is much less costly. An ANNG [20], which is one of the approximate neighborhood graphs, has high search performance. To create an ANNG incrementally, approximate k-nearest neighbor objects for edges are searched for by using the partially created ANNG. Figure 3 shows construction times for KNNG and ANNG. KNNG construction times for more than two million objects were estimated from the construction time for one million objects. The figure shows that an ANNG has significantly lower construction times compared with a KNNG. However, the initially inserted nodes of the ANNG tend to have a huge number of edges compared with the subsequently inserted nodes. Figure 4 shows the average number of edges for every 100,000 nodes along the insertion sequence of 10 million SIFT image descriptors, wherein 10 nearest neighbors are added as edges for each node during insertion. The number of edges for the first sequence exceeds 60. The excess edges are pruned in the same way as in PBKNNG, and we refer to the resultant graph as a pruned ANNG (PANNG). Algorithm 4 is the pseudo code for creating a PANNG. Let O be the set of inserted objects and k_c be the initial number of edges for an inserted node. ϵ_c is for the expansion factor of the explored space of the KNN search. KnnSearch(G, o, k_c, ϵ_c) is a KNN search function that returns the k_c nearest neighbors to the query object o. In this study, Algorithm 1 is also used as the KnnSearch in Algorithm 4.

4   Experimental Results

The experiments used 128-dimensional SIFT image descriptors [23] and 960-dimensional GIST image descriptors [24]. SIFT is a local descriptor, and GIST is a global descriptor. The descriptors were extracted from about 1 million images downloaded from Flickr[1]. The SIFT descriptors were extracted from the image set by using OpenCV[2]. Since just one GIST descriptor is extracted from an image by using Lear's GIST C implementation[3], the GIST descriptors were extracted from 4 by 4 block images into which each image in the image set was divided in order to extract 10 million descriptors. Duplicates were removed from the descriptors. A 10-million-object dataset and 500-object query set were randomly selected from each of the descriptors. A Euclidean distance function was used. Each SIFT element was stored in memory as a 1-byte integer, and each GIST element was stored as a 4-byte floating point number. The resulting size of the KNN search was 20 nearest neighbors. We conducted the experiments on an Intel Xeon E5-2630L (2.0 GHz and 64 GB of memory). Although the CPU had six cores, the experimental software was not processed in parallel.

Parameter Determination: First, we evaluated the search performance to determine the number of seed nodes. The search performance was assessed in terms of the query time and the search accuracy while varying the number of the seed nodes from 1 to 100 using the SIFT dataset. The results indicated that the query times for all seed node numbers were almost the same when the accuracy was over 0.5. The query time for 10 seed nodes was slightly shorter when the accuracy was less than 0.5. Thus, 10 seed nodes were used in the experiment.

Figure 5(a) plots the search performance of a BKNNG using the SIFT dataset with a varying number of edges k_o, which represents not the actual number of edges but the number of edges for the original KNNG from which the BKNNG is derived. It can be seen that excess edges tended to reduce performance, and the plot for $k_o = 10$ indicates that 10 edges gave the best performance almost overall.

---

[1] https://www.flickr.com/.

[2] http://opencv.org/.

[3] http://people.csail.mit.edu/torralba/code/spatialenvelope/.

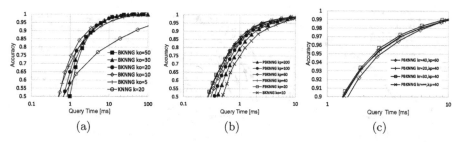

Fig. 5. Accuracy vs. query time for SIFT dataset. (a) BKNNG for different numbers of edges k_o and BKNNG with $k_o = 10$. (b) PBKNNGs with $k_r = \infty$ derived from BKNNG with $k_o = 10$ for different values of k_p. (c) PBKNNGs derived from BKNNG with $k_o = 10$ for different values of k_p and k_r.

Since search performance largely depends on the number of edges, to equitably compare the different graphs, the total numbers of edges in the graphs should be as close to equal as possible. The total number of edges in the BKNNG is up to twice that of the original KNNG. Therefore, to compare them, Fig. 5(a) also shows the performance for a KNNG with $k = 20$. The KNNG and the BKNNG with $k_o = 10$ had almost identical numbers of edges. The actual average number of edges in the BKNNG was about 18.6 because edges were not added to node pairs that already had two different directed edges between them. In spite of it having fewer edges than the KNNG, the BKNNG performed considerably better than the KNNG in a higher accuracy range. For example, the query time of the BKNNG was more than 10 times shorter than that of the KNNG at an accuracy of 0.9. These results indicate that adding bi-directed edges to the KNNG significantly improved performance in this range of accuracy. Figure 5(b) shows the performance of PBKNNGs with $k_r = \infty$ derived from a KNNG with $k_o = 10$ for different values of k_p. The parameter $k_r = \infty$ disables the selective edge removal. The BKNNG has many edges, and these edges increase the query time. Therefore, while pruning edges improves performance, pruning too many edges reduces it. From Fig. 5(b), it can be seen that the PBKNNGs where k_p is 20 and 40 show almost identical performance and are better than the others. Figure 5(c) shows the performance of PBKNNGs derived from KNNG with $k_o = 10$ and $p = 3$ for different values of k_r and k_p in a higher accuracy range. Since the performance obviously decreased where $p > 3$, we adopted $p = 3$ in all of the experiments. The PBKNNG with $k_p = 40$ and $k_r = 30$ was slightly better than the others. This shows that selective edge removal is more effective at improving performance.

Figure 6(a) shows the performance of PANNG[4] for different values of k_p and k_r. The PANNG was constructed by using Algorithm 4 with $k_c = 10$ and $\epsilon_c = 0.1$. The curves in the figure show a similar tendency to those for PBKNNG in Fig. 5(b) and (c). The nodes inserted in the initial stage tend to have a huge number of edges, as Fig. 2(b) shows. Therefore, pruning contributes to improving search performance. However, pruning too many edges reduces performance in

---

[4] http://research-lab.yahoo.co.jp/software/ngt/.

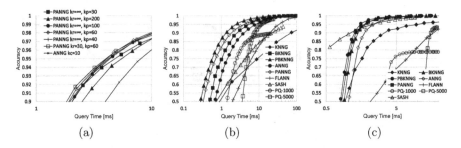

(a)   (b)   (c)

Fig. 6. Accuracy vs. query time. (a) ANNG with $k_c = 10$ and PANNGs derived from the ANNG for different k_p and k_r for SIFT. (b) Comparison of KNNG ($k = 20$), BKNNG ($k_o = 10$), PBKNNG ($k_r = 30$, $k_p = 40$), ANNG ($k_c = 10$), PANNG ($k_r = 3$, $k_p = 60$), FLANN, SASH, PQ ($R = 1000$), and PQ ($R = 5000$) for SIFT. (c) Comparison of KNNG ($k = 20$), BKNNG ($k_o = 10$), PBKNNG ($k_r = 30$, $k_p = 40$), ANNG ($k_c = 10$), PANNG ($k_r = 30$, $k_p = 60$), FLANN, SASH, PQ ($R = 1000$), and PQ ($R = 5000$) for GIST.

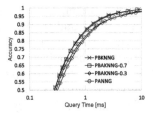

Fig. 7. PBKNNG, PBAKNNGs, and PANNG with various edge precisions indicated in Table 1 for SIFT

Table 1. Average edge precision and average rank of top 10 shortest edges for PBKNNG, PBAKNNGs, and PANNG

Graph	Average precision	Average rank
PBKNNG	1.00	5.50
PBAKNNG-0.7	0.706	7.79
PBAKNNG-0.3	0.303	17.9
PANNG	0.567	29.4

the same way as with PBKNNG. From the figure, the PANNG with $k_p = 60$ and $k_r = 30$ performed the best overall in a higher accuracy range.

Comparison of Graph-Based Indexes, FLANN, SASH, and PQ: Figures 6(b) and (c) compare the performances of KNNG, BKNNG, PBKNNG, ANNG, PANNG, FLANN, SASH[5], and PQ for SIFT and GIST using the determined parameters. KNNG performed the worst, and PBKNNG performed the best among the graph-based methods. Although PANNG was slightly worse than PBKNNG, it is practically advantageous because the construction cost of an ANNG is considerably lower than that of an exact KNNG.

FLANN automatically selects the best algorithm for the dataset. For the SIFT and GIST dataset, it selected hierarchical k-means partitioning. For constructions of SASH, we used the number of parent nodes $p = 4$. Even though approximate searches without original objects are not our target, we compared it with PQ. While PQ does not require the objects in the memory, the search accuracy is significantly lower. To compare fairly, we added a verification step

---

[5] http://research.nii.ac.jp/~meh/sash/.

after the PQ search, which computes distances for the results of the PQ using the objects in the memory and returns the k nearest neighbors. According to the experiment of PQ [5], the best parameters were explored and determined. We used the number of codewords for the product quantization $k^* = 256$, the number of subvectors $m = 8$, and the number of codewords $k' = 1024$. The curves of PQ were plotted by varying the number of the nearest neighbors of the coarse quantizer w for the number of the nearest neighbors of PQ $R = 1000$ and 5000. It can be seen that PBKNNG and PANNG outperformed FLANN and PQs overall and outperformed SASH excluding at lower accuracy for GIST in Fig. 6(b) and (c).

Edge Precision Effect Analysis: In spite that a KNNG does not satisfy Condition 1, the PBKNNG derived from the KNNG works well. This suggests that it might be unnecessary to use an exact KNNG to generate a PBKNNG. To clarify this, just as we derived the PBKNNG from the KNNG, we derived a pruned bi-directed approximate k-nearest neighbor graph (PBAKNNG) from an approximate KNNG, which is intentionally generated by pruning the edges of the KNNG according to a specific probability, called "edge precision." Fig. 7 compares the performances of PBKNNG, PBAKNNGs, and PANNG to clarify the effect of varying the edge precision. The PBKNNG was constructed with $k_o = 10$. The PBAKNNGs were derived from AKNNG with $k_o = 10$ for the edge precisions 0.7 and 0.3. The PANNG was constructed with $k_c = 10$ for $\epsilon_c = 0.1$. All of them were constructed with $k_p = 40$ and $k_r = 30$. Table 1 shows the average edge precision and the average rank of the top 10 shortest edges for each of 1,000 randomly sampled nodes of the indexes in Fig. 7. From the order of the average precisions in Table 1, the search performances of PANNG should be between PBAKNNG-0.7 and PBAKNNG-0.3. It is, however, almost the same as PBAKNNG-0.3 at higher accuracy. We suppose that performance is affected by both the precision and the average rank of edges. Since the performance decreases for low edge precision are all rather small, these results show that an exact KNNG is dispensable in order to make an approximate search.

Memory Usage: Since our search algorithm needs a large number of distance computations, all of the objects should be placed in memory to reduce the search cost. Here, we will discuss the memory usage of a logical index structure instead of an actual structure since our actual implementation uses a standard template library (STL) including a non-negligible amount of memory overhead. The logical index structure has an array of nodes consisting of objects, a pointer to the edge array for each node, and the size of the edge array. Its memory usage is as follows.

$$
\begin{aligned}
\text{memory usage} &= \text{node array usage} + \text{edge array usage} \\
\text{node array usage} &= (\text{object dimensionality} \cdot \text{size of object element variable} \\
&\quad +\text{size of pointer to edge array} \\
&\quad +\text{size of edge array size variable}) \cdot \text{total number of objects} \\
\text{edge array usage} &= \text{size of node ID variable} \cdot \text{total number of edges}
\end{aligned}
\tag{1}
$$

The length of each edge is used to prune excess edges, so the memory usage for the edge array for index construction is as follows.

$$\text{edge array usage} = (\text{size of node ID variable} + \text{size of distance variable}) \cdot \text{total number of edges} \tag{2}$$

The total numbers of edges for the PBKNNG for $k_o = 10$, $k_p = 40$, and $k_r = 30$ are 165,529,883 for the SIFT dataset and 163,959,473 for the GIST dataset. The logical memory usage derived from Formula 1 amounts to 1.90 GB for SIFT and 36.5 GB for GIST. Since the experimental implementation included additional information for the evaluations and memory overhead of the STL, the actual amount of the memory was not measured.

5   Conclusion

We derived a PBKNNG from a KNNG as an index of high-dimensional data such as image descriptors. The experiment showed that the PBKNNG outperforms not only the KNNG but also the FLANN, SASH, and PQ in most cases on SIFT and GIST datasets. The drawback of the KNNG is its high construction cost, and an approximate neighborhood graph is much less costly. The experiment also showed a PANNG derived from the approximate neighborhood graph instead of a KNNG in the same way as the PBKNNG outperforms the KNNG, FLANN, SASH, and PQ in most cases and performs only a little worse than the PBKNNG.

References

1. Gionis, A., Indyk, P., Motwani, R.: Similarity search in high dimensions via hashing. In: Proceedings of 25th International Conference on Very Large Data Bases, pp. 518–528 (1999)
2. Datar, M., Immorlica, N., Indyk, P., Mirrokni, V.: Locality-sensitive hashing scheme based on p-stable distributions. In: Proceedings of the 20th Annual Symposium on Computational Geometry, pp. 253–262. ACM (2004)
3. Weiss, Y., Torralba, A., Fergus, R.: Spectral hashing. In: Advances in Neural Information Processing Systems, pp. 1753–1760 (2009)
4. Gong, Y., Lazebnik, S.: Iterative quantization: a procrustean approach to learning binary codes. In: 2011 IEEE Conference on Computer Vision and Pattern Recognition (CVPR), pp. 817–824. IEEE (2011)
5. Jegou, H., Douze, M., Schmid, C.: Product quantization for nearest neighbor search. IEEE Trans. Pattern Anal. Mach. Intell. **33**(1), 117–128 (2011)
6. Bentley, J.: Multidimensional binary search trees used for associative searching. Commun. ACM **18**, 509–517 (1975)
7. White, D., Jain, R.: Similarity indexing with the SS-tree. In: Proceedings of 12th International Conference on Data Engineering, pp. 516–523 (1996)
8. Yianilos, P.: Data structures and algorithms for nearest neighbor search in general metric spaces. In: Proceedings of the 4th Annual ACM-SIAM Symposium on Discrete Algorithms, pp. 311–321 (1993)

9. Ciaccia, P., Patella, M., Zezula, P.: M-tree: An efficient access method for similarity search in metric spaces. In: Proceedings of International Conference on Very Large Data Bases, pp. 426–435 (1997)
10. Arya, S., Mount, D., Netanyahu, N., Silverman, R., Wu, A.: An optimal algorithm for approximate nearest neighbor searching fixed dimensions. J. ACM **45**(6), 891–923 (1998)
11. Houle, M.E., Sakuma, J.: Fast approximate similarity search in extremely high-dimensional data sets. In: 21st International Conference on Data Engineering (ICDE 2005), pp. 619–630. IEEE (2005)
12. Muja, M., Lowe, D.G.: Scalable nearest neighbor algorithms for high dimensional data. IEEE Trans. Pattern Anal. Mach. Intell. **36**(11), 2227–2240 (2014)
13. Silpa-Anan, C., Hartley, R.: Optimised KD-trees for fast image descriptor matching. In: IEEE Conference on Computer Vision and Pattern Recognition, CVPR 2008, pp. 1–8. IEEE (2008)
14. Nister, D., Stewenius, H.: Scalable recognition with a vocabulary tree. In: IEEE Computer Society Conference on Computer Vision and Pattern Recognition, vol. 2, pp. 2161–2168. IEEE (2006)
15. Arya, S., Mount, D.M.: Approximate nearest neighbor queries in fixed dimensions. In: Proceedings of the Fourth Annual ACM-SIAM Symposium on Discrete Algorithms. SODA 1993, Philadelphia, PA, USA, pp. 271–280. Society for Industrial and Applied Mathematics (1993)
16. Sebastian, T., Kimia, B.: Metric-based shape retrieval in large databases. In: Proceedings of 16th International Conference on Pattern Recognition, vol. 3. 291–296 (2002)
17. Wang, J., Li, S.: Query-driven iterated neighborhood graph search for large scale indexing. In: Proceedings of the 20th ACM International Conference on Multimedia, MM 2012, pp. 179–188. ACM, New York (2012)
18. Hajebi, K., Abbasi-Yadkori, Y., Shahbazi, H., Zhang, H.: Fast approximate nearest-neighbor search with k-nearest neighbor graph. In: Proceedings of the 22nd International Joint Conference on Artificial Intelligence, pp. 1312–1317 (2011)
19. Dong, W., Moses, C., Li, K.: Efficient k-nearest neighbor graph construction for generic similarity measures. In: Proceedings of the 20th International Conference on World Wide Web, WWW 2011, pp. 577–586. ACM, New York (2011)
20. Iwasaki, M.: Proximity search in metric spaces using approximate k nearest neighbor graph (in Japanese). IPSJ Trans. Database **3**(1), 18–28 (2010)
21. Iwasaki, M.: Proximity search using approximate k nearest neighbor graph with a tree structured index (in Japanese). IPSJ J. **52**(2), 817–828 (2011)
22. Navarro, G.: Searching in metric spaces by spatial approximation. VLDB J. **11**(1), 28–46 (2002)
23. Lowe, D.G.: Object recognition from local scale-invariant features. In: The Proceedings of the Seventh IEEE International Conference on Computer Vision, vol. 2, pp. 1150–1157. IEEE (1999)
24. Oliva, A., Torralba, A.: Modeling the shape of the scene: a holistic representation of the spatial envelope. Int. J. Comput. Vis. **42**(3), 145–175 (2001)

A Free Energy Foundation of Semantic Similarity in Automata and Languages

Cewei Cui$^{(\boxtimes)}$ and Zhe Dang$^{(\boxtimes)}$

School of Electrical Engineering and Computer Science,
Washington State University, Pullman, WA 99164, USA
{ccui,zdang}@eecs.wsu.edu

Abstract. This paper develops a free energy theory from physics including the variational principles for automata and languages and also provides algorithms to compute the energy as well as efficient algorithms for estimating the nondeterminism in a nondeterministic finite automaton. This theory is then used as a foundation to define a semantic similarity metric for automata and languages. Since automata are a fundamental model for all modern programs while languages are a fundamental model for the programs' behaviors, we believe that the theory and the metric developed in this paper can be further used for real-word programs as well.

1 Introduction

Semantic similarity between two software systems plays a central role in studying software evolution, software plagiarism and in a more general context of semantic mining of an executable object. Clearly, a syntactic metric of such similarity defined on the source codes of two programs (i.e., the source codes look similar) is far from being enough to catch the semantic similarity. The reasons are not hard to see: plagiarized software can have a completely different "look" even though its semantics is not much modified from the original true copy. We shall also notice that the semantic metric that we are looking for shall be an invariant on the dynamic behaviors (instead of the source codes) of the software systems.

Automata are a fundamental model for all modern software systems while languages (sets of words) are a model for the requirements (i.e., behaviors specifications as sequences or words of events) of the systems. Hence, it is meaningful to study the semantic similarity metric from the fundamental; e.g., finite automata and regular languages. Such studies may provide a hint of inspiration for studying more general software systems and more importantly, finite automata themselves are useful as well in software design (i.e., statecharts [12]).

We now take a new and physical view on a run of a finite-state program (i.e., a finite automaton M). When M runs, it receives input symbols while each input symbols "drives" M to transit from the current state to the next. We now imagine the automaton as a gas molecule while each state that the run passes resembles an observation of the molecule's physical position, speed, etc. Hence, a

© Springer International Publishing AG 2016
L. Amsaleg et al. (Eds.): SISAP 2016, LNCS 9939, pp. 34–47, 2016.
DOI: 10.1007/978-3-319-46759-7_3

run of M corresponds to an observation sequence, called a microstate in physics, of the molecule. Clearly, the semantic similarity metric that we look for must rely on an invariant on the runs of an automaton M. This invariant can be found in the thermodynamic formalism that provides a mathematical structure to rigorously deduce, in a many-particle system (e.g. gas), from a microscopic behavior to a macroscopic characteristic [11,18,19,23]. Even though a microstate is highly random, the formalism resolves the challenge of how the highly random microstates in the system demonstrate almost stable macroscopic characteristic such as free energy. Returning back to the automaton, the free energy computed can be treated as an invariant on the runs (i.e., dynamic behaviours) of the automaton M. Notice that the free energy is measured on an equilibrium; being interpreted in a software system, a program's long-term behavior is still highly dynamic and even random, however the dynamics itself is invariant ! This thermodynamic view of software systems provides the physical foundation of our semantic similarity metric. However, in order to build up the foundation, we need first develop a free energy theory for automata and languages, and the semantic similarity metric is then a direct and natural by-product of the theory. Notice that the purpose of developing the theory is not limited to semantic similarity; it shall be considered part of a new development in the traditional automata theory that is of more than 60 years of history. We shall expect a wider range of applications of the theory itself, in addition to the similarity metric.

Related work. Delvenne [9] and Koslicki and Thompson [14,15] are among the first works to directly use the thermodynamic formalism in Computer Science. In particular, Delvenne [9] computes free energy rank, using free energy on a random walk on a graph, for web search while a "missing link" from a page to another is present. This work inspires us to study the free energy of finite automata. Koslicki and Thompson [14,15] study topological pressure of a DNA sequence using a thermodynamic formalism, where an interested pattern on a DNA sequence is given a weight. This work inspires us to study free energy for formal languages (sets of words). Shannon's entropy has widely been considered to have a root in thermodynamics, e.g., Boltzmann equation. In fact, the notion of entropy rate or information rate, originally by Shannon [21] and later by Chomsky and Miller [2], that computes the information quantity in a string (word), can be treated as a special case of free energy or topological pressure in thermodynamic formalism, as pioneered in Ruelle [18], Walters [23], Gurevich [11], Sarig [19], etc. Recently, the classical formalism of Shannon's information rate has been used by the authors in software analysis, testing, and program security analysis [6,7,13,16]. In particular, our previous paper [3] on the information rate of a random walk of a graph has been recently used by Naval, Laxmi, Rajarajan, Gaur and Conti in malware detection [17]. We are confident that thermodynamic formalism can find its own applications in various areas of computer science, such as similarity in programs and graphs [1,4,8,10,22,24]. In particular, our previous work on program similarity [4] is based on information rate and the Jaccard index.

The rest of the paper is organized as follows. We first briefly introduce an existing result in thermodynamic formalism, the variational principle [18], in its simplest form in physics. Then, we develop a free energy theory for automata and languages, and, in various cases, provide algorithms to compute the quantity. Finally, we provide a semantic similarity metric based on the free energy theory.

2   A Free Energy Theory of Automata and Languages

In Computer Science, a program's semantics can be interpreted as the set of all of its behaviors. In various settings, such a behavior can be simply a run (such as a sequence of state transitions), or a sequence of I/O events. Bearing thermodynamic formalism in mind, we understand the sequence as a microstate of the software system and ask a question:

> Given the fact that there are so many (even infinite) microstates in a program, and these microstates are highly "probabilistic" or random, can we use thermodynamic formalism to help us understand some of the program's macroscopic properties that are stable and *free of randomness*?

The answer is yes but need some work. The set of behaviors is a language and we now consider a language $L \subseteq \Sigma^*$ on a finite and nonempty alphabet Σ and a function $\psi : \Sigma^\omega \to \mathbb{R}$. In this paper, we consider a simplest form of ψ: for each $x_0 x_1 x_2 \cdots \in \Sigma^\omega$, $\psi(x_0 x_1 x_2 \cdots) = U(x_0, x_1)$. Herein, $U : \Sigma \times \Sigma \to \mathbb{R}$ is a given *cost function* over words of length 2. The intended purpose of the cost function is to assign a cost to a pattern in a word while the cost can be interpreted as, in a practical setting, an amount of a resource, a priority level, etc.

The cost function is abstracted from "potential" in the thermodynamic formalism, which we briefly explain now. An infinite word $\underline{x} = x_0 x_1 x_2 \cdots \in \Sigma^\omega$ is a microstate in thermodynamics. The microstate evolves into $\sigma(\underline{x}) = x_1 x_2 \cdots$, $\sigma^2(\underline{x}) = x_2 \cdots$, $\cdots$, $\sigma^n(\underline{x}) = x_n x_{n+1} \cdots$, etc., as discrete time n evolves. Herein, σ is the shift-to-left operator defined in an obvious way. The evolution from $\underline{x}$ to $\sigma(\underline{x})$ is to break off the first symbol x_0 in $\underline{x}$ from the rest. The break-off needs energy which is given by a pre-defined potential $\psi(\underline{x})$. This explanation comes from Sarig's lecture notes [20] and we think it is a best way to illustrate the physics behind the mathematics. Hence, for an n-step evolution, the total energy needed or the total potential possessed is $(S^n \psi)(\underline{x}) =_{\text{def}} \sum_{0 \leq i \leq n-1} \psi(\sigma^i(\underline{x}))$, where $\sigma^0(\underline{x}) = \underline{x}$.

In automata theory, this can also be analogously understood as follows. When an infinite word $\underline{x} = x_0 x_1 x_2 \cdots$ is read (symbol by symbol, from left to right), the reader reads the first symbol x_0, then the second symbol x_1, etc. Each such symbol-read consumes energy since essentially it performs a shift-to-left operation in the view of thermodynamics.

In thermodynamics, a particle has high energy if it tends to evolve into one of many choices of different microstates. Similarly, a microstate is of high energy if it is the result of being chosen from many microstates. Using the Boltzmann equation, the asymptotic total energy needed per step for all microstates $\underline{x}$ in n-step evolution is the *(Gurevich) free energy*

$$\lim_{n \to \infty} \frac{1}{n} \ln \sum_{\underline{x}:\sigma^n(\underline{x})=\underline{x}} e^{(S^n \psi)(\underline{x})}, \tag{1}$$

where the $\underline{x}$ is called a periodic orbit (The limit exists with a weak side condition (see Proposition 3.2 in [20]). For the mathematics, see page 63 of [11,20]).

Coming back to Computer Science, we modify the formula in (1) so that the summation is over the (finite) words w of length n in a language L instead of ω-words $\underline{x}$. To do this, we first consider $w = x_0 \cdots x_{n-1}$ which is the prefix (with length n) of $\underline{x} = x_0 \cdots x_{n-1} \cdots$. For the term $(S^n \psi)(\underline{x})$ in (1), we modify it slightly into

$$(U)(w) =_{\text{def}} \sum_{0 \leq i < n-1} U(x_i, x_{i+1}) \tag{2}$$

while, as we have mentioned earlier, the ψ now takes the special form $\psi(x_0 x_1 \cdots) = U(x_0, x_1)$. Then the periodic orbits in (1) can be safely replaced with words of length n. We thereby obtain the definition of *(Gurevich) free energy of language L*:

$$G_U(L) = \limsup_{n \to \infty} \frac{1}{n} \ln \sum_{w \in L, |w|=n} e^{(U)(w)}. \tag{3}$$

(By convention, $0 \ln 0 = 0$. The limit superior can be replaced by limit in many cases, e.g., when L is prefix-closed and regular, as we show later.)

Intuitively, the free energy $G_U(L)$ characterizes the average "cost" per symbol of words in L with respect to the cost function U. A particularly interesting example is to interpret the cost as "uncertainty", as shown in a later example. A special case is when $U = 0$. In this case, the free energy is simply (modulo a constant) the information rate of L that was originally proposed by Shannon [21] and Chomsky and Miller [2], and more recently studied in [5]. We shall notice that, once U is given, the free energy of L is a constant and its definition does not involve any probabilistic arguments.

We explore how to compute the free energy defined in (3) for various classes of languages. We shall first point out that the free energy is not computable in general, even for simple classes of languages.

Theorem 1. *The free energy for language $L' = (\Sigma^* - L)\Sigma^*$, where L is a context-free language, is not computable.*

Later, we will show that the free energy is computable for regular languages. The proof needs the variational principle in thermodynamics, which is briefly introduced as follows. In a thermodynamic system, nature tends to make particles move in a way that maximizes the free energy [20] as

$$\sup_{\mu} \left\{ H_\mu + \int \psi d\mu \right\}, \tag{4}$$

which is called the *free energy* of the aforementioned σ (Markov shift in thermo-dynamic formalism [20]) with potential ψ. We shall point out that the Markov shift itself does not contain any probabilities. Consider the compact metric space whose topology is generated from the cylinder sets in the form of $[x_0 \cdots x_{n-1}] = \{\underline{x} : \underline{x} = x_0 \cdots x_{n-1} \cdots\}$, and uses metric $d(\underline{x}, \underline{y}) = 2^{-min\{i : x_i \neq y_i\}}$, where $\underline{x} = x_0 \cdots x_{n-1} \cdots$ and $\underline{y} = y_0 \cdots y_{n-1} \cdots$. In the definition, the μ is a prob-ability measure over the space that is invariant under the Markov shift σ. The measure μ, in plain English, is a Markov chain defined on the Markov shift σ and H_μ is the Kolmogorov-Sinai entropy of the Markov chain μ (intuitively, it quantifies the average randomness, called entropy, on one step of the Markov chain), and $\int \psi d\mu$ is the average potential on one step of the Markov chain. Due to space limitation, we omit the mathematics behind the definition which can be found in [20].

One of the most important achievements in thermodynamic formalism estab-lishes the variational principle [18]. When interpreted on periodic orbits, it says that the free energy, defined in (1), is indeed the free energy on the Markov shift (again, with a side condition–see Sarig's notes [20]– which we omit here for simplicity.):

$$\lim_{n \to \infty} \frac{1}{n} \ln \sum_{\underline{X} : \sigma^n(\underline{X}) = \underline{X}} e^{(S^n \psi)(\underline{X})} = \sup_\mu \{H_\mu + \int \psi d\mu\}. \tag{5}$$

The supremum on the RHS of (5) is achieved by an equilibrium probability measure μ^* (called Parry measure), which can be computed from a nonnegative matrix, called Gurevich Matrix [11], constructed from the definition of ψ when ψ is defined as U, mentioned earlier. In particular, the LHS of (5) can be computed from the Perron-Frobenius eigenvalue of the matrix [11].

We now generalize the free energy from Markov shift to a finite automaton. Let M be a nondeterministic finite automaton (NFA) with finitely many states specified by Q and with alphabet Σ. Transitions in M are specified by a set $T \subseteq Q \times \Sigma \times Q$, where each transition $t \in T$ in the form of (p, a, q), or simply written $p \xrightarrow{a} q$, means that M moves from state p to state q on reading input symbol a. A run is a sequence of transitions

$$\tau = (p_0, a_0, p_1)(p_1, a_1, p_2) \cdots (p_{n-1}, a_{n-1}, p_n), \tag{6}$$

for some n, satisfying $p_0 \xrightarrow{a_0} p_1 \xrightarrow{a_1} \cdots \xrightarrow{a_{n-1}} p_n$. In M, there is a designated initial state q_{init} and a number of designated accepting states $q_{\text{accept}} \in F \subseteq Q$. The run τ is initialized if p_0 in (6) is the initial state. It is an accepting run if it is initialized and the last state p_n in (6) is an accepting state. In this case, we say that the word $a_0 \cdots a_{n-1}$ is accepted by M. As usual, we use $L(M)$ to denote the language accepted by M. M is deterministic (i.e. a DFA) if, for each p and a, there is at most one q such that $p \xrightarrow{a} q$. It is well-known that an NFA can be converted into a DFA such that both automata accept the same language.

Throughout the paper, we assume that M is cleaned up. That is, all the states are dropped from M whenever it cannot be reached from the initial state or it cannot reach an accepting state.

We now associate a cost function $V : T \to \mathbb{R}$ which assigns a cost value to every transition in M. We write M_V for the M associated with cost function V, called an NFA with cost.

We first assume that M is strongly connected. That is, every state can reach every state in M. More precisely, for each p and q, there is a run in the form of (6) with $p_0 = p$ and $p_n = q$. We now define the free energy of M_V. We note that the results in [11] are defined on strongly connected graphs only. However, a finite automaton M is not, strictly speaking, a graph since there could be multiple transitions from one state to another (while in a graph there is at most one edge from a node to another). Therefore, we first need to carefully translate the automaton M into a Markov shift (a graph) as follows. Let Θ be the set of all infinite sequences in the form of

$$\underline{t} = p_0(p_0, a_0, p_1)p_1(p_1, a_1, p_2)p_2 \cdots p_{n-1}(p_{n-1}, a_{n-1}, p_n)p_n \cdots \qquad (7)$$

or

$$\underline{t} = (p_0, a_0, p_1)p_1(p_1, a_1, p_2)p_2 \cdots p_{n-1}(p_{n-1}, a_{n-1}, p_n)p_n \cdots \qquad (8)$$

satisfying $p_0 \xrightarrow{a_0} p_1 \xrightarrow{a_1} \cdots \xrightarrow{a_{n-1}} p_n \cdots$ (we shall note that $\underline{t}$ may not start from the initial state of M). In terms of the thermodynamics formalism, we define a potential function ψ such that for each $\underline{t}$ in Θ, its potential $\psi(\underline{t}) = V(p_0, a_0, p_1)$ if $\underline{t}$ is in the form of (7); $\psi(\underline{t}) = 0$ if $\underline{t}$ is in the form of (8). We can similarly define a compact metric space over Θ whose topology is generated by cylinders and the metric d defined earlier. Let μ be a σ-invariant probability measure. Now the Markov shift σ on Θ defines a graph $\hat{M}$ as follows. For all p, a, and q, $\hat{M}$ has node p, node q, node (p, a, q), and edges from node p to node (p, a, q) and from node (p, a, q) to node q, iff (p, a, b) is a transition in M. Clearly, $\hat{M}$ is a strongly connected graph.

The free energy $\mathcal{E}(\hat{M})$ is defined as the quantity in (4). It is known that [11] the free energy can be computed as follows. Suppose that the graph $\hat{M}$ has k nodes indexed with $1, \cdots, k$. For a node q (resp. node (p, a, q)) in $\hat{M}$, we use $[q]$ (resp. $[(p, a, q)]$) for its index. We now construct a $k \times k$ matrix $\mathbf{M}$, called the Gurevich matrix, as follows. For each i and j,

- $\mathbf{M}_{ij} = 0$ if there is no edge from node i to node j in $\hat{M}$;
- $\mathbf{M}_{ij} = e^{V(p,a,q)}$ if there is an edge from node i to node j in $\hat{M}$ and $i = [p]$ and $j = [(p, a, q)]$ for some p, a, q;
- $\mathbf{M}_{ij} = e^0$ if there is an edge from node i to node j in $\hat{M}$ and $i = [(p, a, q)]$ and $j = [q]$ for some p, a, q.

Clearly, $\mathbf{M}$ is a non-negative and irreducible (since M is strongly connected) matrix. Let λ denote the spectral radius of $\mathbf{M}$, which is obtained as the largest positive real eigenvalue of $\mathbf{M}$, which is called the Perron-Frobenius eigenvalue of $\mathbf{M}$. Finally, according to [11], the free energy $\mathcal{E}(\hat{M})$ can be efficiently computed as $\ln \lambda$.

Lemma 1 *(Gurevich theorem).* $\mathcal{E}(\hat{M}) = \ln \lambda$, *where λ is the Perron-Frobenius eigenvalue of the Gurevich matrix $\mathbf{M}$.*

Finally, *the free energy $\mathcal{E}(M_V)$ for finite automaton M* with cost function V is defined as $2 \cdot \mathcal{E}(\hat{M})$[1].

We shall note that the definition of $\mathcal{E}(M_V)$ does not mention the initial and accepting states of M. We now bring in those states and prove the variational principle for finite automata. To do this, we first use a finite sequence to represent a periodic orbit, as we did in (2). That is, for a run τ in (6) with length $|\tau| = n$, we define

$$(V)(\tau) =_{\text{def}} \sum_{0 \leq i < n} V(p_i, a_i, p_{i+1}). \tag{9}$$

In particular, we use R to denote the set of all runs of M and A to denote the set of all accepting runs of M; herein, both sets are languages on transitions in M.

Theorem 2 *(Variational Principle for Strongly Connected NFA). Let M be an NFA that is strongly connected and with cost function V on transitions. Then the following equations hold:*

$$G_V(R) = \lim_{n \to \infty} \frac{1}{n} \ln \sum_{\tau \in R, |\tau| = n} e^{(V)(\tau)} = \mathcal{E}(M_V), \tag{10}$$

and

$$G_V(A) = \limsup_{n \to \infty} \frac{1}{n} \ln \sum_{\tau \in A, |\tau| = n} e^{(V)(\tau)} = \mathcal{E}(M_V). \tag{11}$$

In particular when every state in M is an accepting state (hence $L(M)$ is prefix closed),

$$G_V(A) = \lim_{n \to \infty} \frac{1}{n} \ln \sum_{\tau \in A, |\tau| = n} e^{(V)(\tau)} = \mathcal{E}(M_V). \tag{12}$$

In general, M is not necessarily strongly connected. However, it is well-known (using the linear-time Tarjan algorithm) that M can be uniquely partitioned into a number of components $M^1, \cdots, M^k$, for some $k \geq 1$, such that

- each M^i is strongly connected, or it is singleton (i.e. it contains only one state that does not have a self-loop transition), and
- each M^i is maximal (i.e. the above condition is no longer true if it is enlarged).

---

[1] The factor 2, intuitively, comes from the fact that we "stretch", by a factor of 2, a run in finite automaton M to correspond it to a walk in graph $\hat{M}$. A somewhat more efficient way to compute $\mathcal{E}(M_V)$ is to construct an $m \times m$ Gurevich matrix $\mathbf{M}'$ where m is the number of states in M such that $\mathbf{M}'_{ij} = 0$ if there is no transition from p_i to p_j in M, else $\mathbf{M}'_{ij} = \sum_{a:(p_i,a,p_j) \in T} e^{V(p_i,a,p_j)}$. Herein, $p_1, \cdots, p_m$ are all states in M. One can show that $\mathcal{E}(M_V) = \ln \lambda'$ where λ' is the Perron-Frobenius eigenvalue of $\mathbf{M}'$. We omit the details.

From Lemma 1, the free energy $\mathcal{E}(M_V^i)$ for each component M^i can be computed from its graph representation $\hat{M}^i$ (when M^i is singleton, its free energy is defined to be 0). We now define the free energy of M_V to be $\mathcal{E}(M_V) = \max_i \mathcal{E}(M_V^i)$. Clearly, using the definition of $\mathcal{E}(M_V^i)$ and Lemma 1, we immediately have:

Theorem 3. *When NFA M is cleaned-up, $\mathcal{E}(M_V)$ is computable.*

This definition of $\mathcal{E}(M_V)$ is valid, since, from (11), we can easily show:

Theorem 4 *(Variational Principle for NFA). When NFA M is cleaned-up, we have $G_V(A) = \mathcal{E}(M_V)$, where A is the set of all accepting runs of M, and V is a cost function that assigns a cost in $\mathbb{R}$ to a transition in M.*

We now consider a regular language L associated with a cost function defined earlier as $U : \Sigma \times \Sigma \rightarrow \mathbb{R}$. Let M be an NFA with a cost function V. We say that (M, V) *implements* (L, U) if M accepts L, and, for each $w \in L$ with length at least 2 and each accepting run τ for w in M, we have $(U)(w) = (V)(\tau)$ (i.e. the total cost on w defined by U is kept exactly the same as the total cost of each accepting run τ defined by V).

Theorem 5. *For each regular language L associated with a cost function U: (1) For each NFA M and cost function U for transitions in M such that (M, V) implements (L, U), we have $G_U(L) \leq \mathcal{E}(M_V)$. (2) There is a DFA M and cost function U for transitions in M such that (M, V) implements (L, U). Furthermore, we have $G_U(L) = \mathcal{E}(M_V)$. (3) When L is prefix closed, the limit (instead of limsup) in the definition of $G_U(L)$ in formula (3) exists.*

It is difficult to show whether a nonregular language L associated with a cost function U has a computable free energy (see also Theorem 1). Below, we show a class of languages where the energy is computable. Let $P(\theta_1, \cdots, \theta_k)$ be a Presburger formula over nonnegative integer variables $\theta_1, \cdots, \theta_k$, for some k. For each $1 \leq i \leq k$, we associate P with a regular language r_i. We use $\mathbf{r}$ to denote $\langle r_1, \cdots, r_k \rangle$. We then define a language $L_{P,\mathbf{r}}$ as the set of all words in the form of $w_1 \cdots w_k$ such that each $w_i \in r_i$, and the lengths $|w_i|$ of w_i's satisfy $P(|w_1|, \cdots, |w_k|)$. A *linear-length* language L is specified by a regular language L' along with P and $\mathbf{r}$ such that $L = L' \cap L_{P,\mathbf{r}}$. Intuitively, L is a subset of a given regular language such that each word in the subset is the concatenation of k subwords, each of which is drawn from a regular language and the lengths of the subwords are constrained by a Presburger formula. For instance, $L = \{a^n b^{2n} a^{3n} : n \geq 0\}$ is a linear-length language, which is not context-free. We can show:

Theorem 6. *For linear-length language L and cost function U, the free energy $G_U(L)$ is computable (from L's specification and U). The computability remains even when L is a finite union of linear-length languages.*

We turn back to NFA M (which is cleaned-up) with a cost function V on transitions. We show an application of the free energy of (M, V) in estimating nondeterminism in an NFA. There are applications for such an estimation. In software engineering, NFAs can be used to specify a highly nondeterministic system such as a concurrent system where the input symbols are the observable events. Being nondeterministic, the same sequence of input symbols can have many different execution sequences. Hence, an ideal measure for the non-determinism would be the asymptotic growth rate of the ratio $f(n)/g(n)$ where $f(n)$ is the total number of executions of input sequences of length n while $g(n)$ is the total number of input sequences of length n. More precisely, we define (slightly different from the above) $g(n)$ to be the number of words α of length $\leq n$ in $L(M)$, and $f(n)$ to be the number of initialized runs of all α's. Then, the nondeterminism in M is defined by

$$\lambda_M = \lim_{n \to \infty} \frac{\log f(n) - \log g(n)}{n}.$$

Clearly, the limit in $\lambda_M \geq 0$ exists and is finite (since runs are prefix closed, and M has no ϵ-transitions). In particular, when M is deterministic, $\lambda_M = 0$.

In reality, such a metric is relevant. For instance, it can be used to estimate a form of coverage of extensive testing of a nondeterministic system (e.g. how many paths have already been exercised for an average input sequence). The estimation is important since it is well-known that nondeterministic software systems are notoriously difficult and costly to test. However, in computing λ_M, the difficult part is the asymptotic growth rate of $\log g(n)$ since currently available algorithms must convert NFA M into a DFA and use the Chomsky-Miller algorithm [2]. The conversion may cause an exponential blow-up in the number of states in the DFA, which is not tractable. We need a practically efficient algorithm to give an estimation of λ_M.

We propose an efficient estimation approach based on free energy. For the given NFA M, we define a cost function V on transitions of M as follows. Let $k(p, a)$ be the total number of p'' such that $p \xrightarrow{a} p''$ is a transition in M. For each transition $p \xrightarrow{a} p'$ in M, we define $V(p, a, p') = \ln k(p, a)$. Notice that if $k(p, a) = 1$ (in this case, M is deterministic at state p on input a), $V(p, a, p') = 0$. Otherwise (i.e., $k(p, a) > 1$. In this case, M is nondeterministic at state p on input a), $V(p, a, p') = \ln k(p, a) > 0$.

Example 1. Figure 1 shows an NFA with the V assigned on transitions.

We now use the free energy difference $\lambda_M^+ = \mathcal{E}(M_V) - \mathcal{E}(M_0)$ to estimate λ_M. Herein, M_0 is the M where each transition is assigned with cost 0. Notice that $\lambda_M^+ \geq 0$ (roughly, from the Perron-Frobenius theorem applied on the Gurevich matrices for M_V and for M_0) and, when M is deterministic, $\lambda_M^+ = \lambda_M = 0$.

We first intuitively explain the meaning behind λ_M^+. In M_V, each transition $p \xrightarrow{a} p'$ is assigned a cost which is the number of nats (information units in natural logarithm) needed to code the transition when p and a are given. Hence, the total cost of an average (initialized) run on a word α will be the total number of nats to code the run (which starts from the known initial state) when α is known. This total cost divided by the length n of the word α will result in nat rate δ of

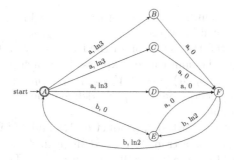

Fig. 1. The finite automaton for Example 1 with costs labeled.

the code. Notice that, in the definition of free energy of M, there are two parts: the average cost per step (which roughly corresponds to the nat rate δ) and the metric entropy (which roughly corresponds to the "average" natural logarithm of the branching factor at a state in M). Now, in M_0, the free energy is the "average" natural logarithm of the branching factor at a state in M. Hence, the $\lambda_M^+ = \mathcal{E}(M_V) - \mathcal{E}(M_0)$ is roughly equal to the nat rate δ (to encode the transition per step for a given input), which is the intended meaning in λ_M. When M is deterministic, the input α decides the run and hence the nat rate is of course zero (no extra nat is needed to encode the run).

We now compute the estimation λ_M^+ for the example NFA M in Fig. 1 and obtain $\lambda_M^+ = \mathcal{E}(M_V) - \mathcal{E}(M_0) = 1.0850 - 0.5857 = 0.4993$. Next, we convert the NFA into a DFA (using an online tool) M' and compute $\lambda_M = \mathcal{E}(M_0) - \mathcal{E}(M_0') = 0.5857 - 0.3603 = 0.2255$. Indeed, λ_M^+ is an upper estimation of λ_M, as shown below.

Theorem 7. *For a cleaned-up NFA M, $\lambda_M^+ \geq \lambda_M$.*

We note that in Theorem 7, the upper estimation λ_M^+ can be efficiently computed from M. Finally, we point out that the estimation is asymptotically tight; the proof will be included in the full version of the paper. We shall also point out all the results can be generalized to cost function U over $k > 2$ (k is constant) symbols instead of two symbols.

3   A Free-Energy Based Similarity Metric for Automata and Languages

We are now ready to use the theory developed so far to define the similarity metric. Let M^i ($i = 1, 2$) be an NFA with a cost function V_i assigned on edges. Notice that the two automata M^1 and M^2 work on the same input alphabet Σ. Consider a word $w = a_0 \cdots a_{n-1}$ accepted by both automata. Suppose that the following transition sequences $(q_0, a_0, q_1)(q_1, a_1, q_2) \cdots (q_{n-1}, a_{n-1}, q_n)$ in M^1 and $(p_0, a_0, p_1)(p_1, a_1, p_2) \cdots (p_{n-1}, a_{n-1}, p_n)$ in M^2 witness the acceptance. However, the total cost (i.e., the total energy or potential) on the first accepting sequence is defined by $V_1(q_0, a_0, q_1) + \cdots + V_1(q_{n-1}, a_{n-1}, q_n)$ while the total cost on the second is $V_2(p_0, a_0, p_1) + \cdots + V_2(p_{n-1}, a_{n-1}, p_n)$. The sum of the two costs shall tell the deviation of the free energy on the two sequences on the input word w. The sum can be expressed on each individual transition as $V_1(q_0, a_0, q_1) + V_2(p_0, a_0, p_1) + \cdots + V_1(q_{n-1}, a_{n-1}, q_n) + V_2(p_{n-1}, a_{n-1}, p_n)$, which again is the free energy on a properly defined (below) "shared sequence" between the two automata.

We define M to be the Cartesian product of M^1 and M^2 in a standard way. That is, $((q, p), a, (q', p'))$ is a transition in M iff (q, a, q') is a transition in M^1 and (p, a, p') is a transition in M^2, for all states p, q, p', q'. The initial state in M is the pair of initial states in M^1 and in M^2 and the accepting states in M are all the pairs of an accepting state in M^1 and an accepting state in M^2. Again, we assume that M is cleaned up. Clearly, M is an NFA accepting $L(M^1) \cap L(M^2)$.

We now define the cost functions V on the M as $V((q, p), a, (q', p')) = V_1(q, a, q') + V_2(p, a, p')$. The semantic similarity metric $\Delta(M_{V_1}^1, M_{V_2}^2)$ is defined as the free energy of M; i.e., $\mathcal{E}(M_V)$. Intuitively, this definition catches the average "shared free energy" per step on the shared accepting runs between M_1 and M_2. We have

$$0 \leq \Delta(M_{V_1}^1, M_{V_2}^2) \leq \mathcal{E}(M_{V_1}^1) + \mathcal{E}(M_{V_2}^2). \tag{13}$$

To see (13), $\Delta(M_{V_1}^1, M_{V_2}^2) = \mathcal{E}(M_V) \geq 0$ is obvious, since the LHS of (11) in Theorem 2 is nonnegative and so is the LHS of the equation in Theorem 4. To show $\Delta(M_{V_1}^1, M_{V_2}^2) \leq \mathcal{E}(M_{V_1}^1) + \mathcal{E}(M_{V_2}^2)$ in (13), we need some effort. First we assume that M is strongly connected and hence we can use Theorem 2. Observe that the term $\sum_{\tau \in A, |\tau|=n} e^{(V)(\tau)}$ in (11) in Theorem 2 satisfies, using the definition V, the following inequality

$$\sum_{\tau \in A, |\tau|=n} e^{(V)(\tau)} \leq \sum_{\tau_1 \in A_1, |\tau_1|=n} e^{(V_1)(\tau_1)} \cdot \sum_{\tau_2 \in A_2, |\tau_2|=n} e^{(V_2)(\tau_2)}$$

where A_1 and A_2 are accepting transition sequences of M_1 and M_2, respectively. Now we plug-in the RHS of the inequality into (11) and obtain

$$\limsup_{n \to \infty} \frac{1}{n} \ln \sum_{\tau \in A, |\tau|=n} e^{(V)(\tau)} \leq \limsup_{n \to \infty} \frac{1}{n} \ln \sum_{\tau_1 \in A_1, |\tau_1|=n} e^{(V_1)(\tau_1)}$$
$$+ \limsup_{n \to \infty} \frac{1}{n} \ln \sum_{\tau_2 \in A_2, |\tau_2|=n} e^{(V_2)(\tau_2)}.$$

Then, we use Theorem 4 on $M_{V_1}^1$ and $M_{V_2}^2$ and hence the RHS of the above inequality becomes

$$\limsup_{n \to \infty} \frac{1}{n} \ln \sum_{\tau \in A, |\tau|=n} e^{(V)(\tau)} \leq \mathcal{E}(M_{V_1}^1) + \mathcal{E}(M_{V_2}^2).$$

Again, using (11) in Theorem 2 on the LHS of the inequality, we have

$$\mathcal{E}(M_V) \leq \mathcal{E}(M_{V_1}^1) + \mathcal{E}(M_{V_2}^2).$$

Using Theorem 4 again on the LHS, we finally obtain for a general M (which may not be strongly connected), the above inequality still holds, which is essentially the result in (13).

In particular, the reader can easily check that the inequality in (13) is tight: when M^1 and M^2 are completely independent (i.e., $L(M^1) \cap L(M^2) = \emptyset$), the similarity metric $\Delta(M_{V_1}^1, M_{V_2}^2) = 0$. However, when M^1 and M^2 are the same and the V_1 and V_2 are also the same, we have the similarity metric $\Delta(M_{V_1}^1, M_{V_2}^2)$ reaches the maximum $\mathcal{E}(M_{V_1}^1) + \mathcal{E}(M_{V_2}^2)$. Intuitively, the metric $\Delta(M_{V_1}^1, M_{V_2}^2)$ characterizes the "shared free energy" between the two finite state programs $M_{V_1}^1$ and $M_{V_2}^2$ as follows. Imagine that the two programs (automata) are two moving gas molecules. At each step of observation, the molecules can be highly random and hence it makes little sense to compare every step of the observations to figure out the similarity between the two molecules. The approach we take in defining the metric $\Delta(M_{V_1}^1, M_{V_2}^2)$ is to "create" a third molecule M_V that, at each step, possesses the potential as the sum of the first two molecules (this is a reward) whenever the first two molecule share the same orbit (i.e., the same input symbol) – otherwise when the orbit are different, the potential of the third molecule is $-\infty$ (this is a penalty). Clearly, the dynamics of the third molecule would be very different from the first two. However, as we have shown above, the long term characteristic of free energy of the third molecule reflects the shared free energy between the first two molecules when they follow the same orbit.

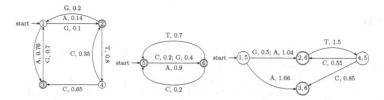

Fig. 2. The first figure is an NFA $M_{V_1}^1$ with cost function V_1; the second figure is an NFA $M_{V_2}^2$ with cost function V_2; the third figure is the Cartesian product M_V.

We now look at an example of two NFAs $M_{V_1}^1$ and $M_{V_2}^2$ where the cost functions V_1 and V_2 are labeled in the Fig. 2. One can think that the two automata try to specify a genome pattern on the four nucleotides (G, A, C, T) in DNA while the costs could be interpreted as probabilities, weights of choices, etc. Notice that the semantics of the automata are the nucleic acid sequences that the automata accept, and those sequences are associated with a cost on each nucleotide. For instance, the following sequence $(A, 0.14)(T, 0.8)(C, 0.65)$ is accepted by $M_{V_1}^1$. Hence, the semantic similarity between the two automata shall concern the similarity between the sequences (with costs) accepted by the two automata instead of the naive similarity on the appearances of the two graphs themselves. Again, we imagine each such sequence accepted as a molecule moving along the orbit specified on the sequence and use the shared free energy between the molecule specified by $M_{V_1}^1$ and the molecule specified by

$M_{V_2}^2$ to measure the semantic similarity $\Delta(M_{V_1}^1, M_{V_2}^2)$. We compute, using the algorithms in Theorem 4 and the definition of the Cartesian product M_V shown earlier, the results $\mathcal{E}(M_{V_1}^1) = 0.3500, \mathcal{E}(M_{V_2}^2) = 1.4087$, and the similarity metric $\Delta(M_{V_1}^1, M_{V_2}^2) = 1.025$, which indeed satisfies (13).

Notice that $\Delta(M_{V_1}^1, M_{V_2}^2)$ can be computed efficiently (which involves only Cartesian product of the two automata, and largest eigenvalues of the Gurevich matrices in Theorems 2 and 4).

Let L_1 and L_2 be two regular languages associated with two cost functions U_1 and U_2, respectively, According to Theorem 5(2), we can construct DFAs $M_{V_1}^1$ and $M_{V_2}^2$ and use $\Delta(M_{V_1}^1, M_{V_2}^2)$ to serve as the semantic similarity metric for the two regular languages. However, it does not seem that $\Delta(M_{V_1}^1, M_{V_2}^2)$ can be efficiently computed from the regular languages since the known construction from regular languages to deterministic finite automata involves exponential blowup on the state space. There might be other alternative definitions on the metric over regular languages (such as using the estimation of nondeterminism in an NFA shown earlier in the paper) such that the metric can be efficiently computed. We leave this for future work.

Acknowledgements. We would like to thank Jean-Charles Delvenne, David Koslicki, Daniel J. Thompson, Eric Wang, William J. Hutton III, and Ali Saberi for discussions. We would also like to thank the seven referees for suggestions and comments that have improved the presentation of our results.

References

1. Chartrand, G., Kubicki, G., Schultz, M.: Graph similarity, distance in graphs. Aequationes Math. **55**(1), 129–145 (1998)
2. Chomsky, N., Miller, G.A.: Finite state languages. Inf. Control **1**(2), 91–112 (1958)
3. Cui, C., Dang, Z., Fischer, T.R.: Typical paths of a graph. Fundam. Inform. **110**(1-4), 95–109 (2011)
4. Cui, C., Dang, Z., Fischer, T.R., Ibarra, O.H.: Similarity in languages and programs. Theor. Comput. Sci. **498**, 58–75 (2013)
5. Cui, C., Dang, Z., Fischer, T.R., Ibarra, O.H.: Information rate of some classes of non-regular languages: an automata-theoretic approach. In: Csuhaj-Varjú, E., Dietzfelbinger, M., Ésik, Z. (eds.) MFCS 2014. LNCS, vol. 8634, pp. 232–243. Springer, Heidelberg (2014)
6. Cui, C., Dang, Z., Fischer, T.R., Ibarra, O.H.: Execution information rate for some classes of automata. Inf. Comput. **246**, 20–29 (2016)
7. Dang, Z., Dementyev, D., Fischer, T.R., Hutton III, W.J.: Security of numerical sensors in automata. In: Drewes, F. (ed.) CIAA 2015. LNCS, vol. 9223, pp. 76–88. Springer, Heidelberg (2015)
8. Dehmer, M., Emmert-Streib, F., Kilian, J.: A similarity measure for graphs with low computational complexity. Appl. Math. Comput. **182**(1), 447–459 (2006)
9. Delvenne, J.-C., Libert, A.-S.: Centrality measures and thermodynamic formalism for complex networks. Phys. Rev. E **83**, 046117 (2011)
10. ElGhawalby, H., Hancock, E.R.: Measuring graph similarity using spectral geometry. In: Campilho, A., Kamel, M.S. (eds.) ICIAR 2008. LNCS, vol. 5112, pp. 517–526. Springer, Heidelberg (2008)

11. Gurevich, B.M.: A variational characterization of one-dimensional countable state gibbs random fields. Zeitschrift für Wahrscheinlichkeitstheorie und Verwandte Gebiete **68**(2), 205–242 (1984)
12. Harel, D.: Statecharts: a visual formalism for complex systems. Sci. Comput. Program. **8**(3), 231–274 (1987)
13. Ibarra, O.H., Cui, C., Dang, Z., Fischer, T.R.: Lossiness of communication channels modeled by transducers. In: Beckmann, A., Csuhaj-Varjú, E., Meer, K. (eds.) CiE 2014. LNCS, vol. 8493, pp. 224–233. Springer, Heidelberg (2014)
14. Koslicki, D.: Topological entropy of DNA sequences. Bioinformatics **27**(8), 1061–1067 (2011)
15. Koslicki, D., Thompson, D.J.: Coding sequence density estimation via topological pressure. J. Math. Biol. **70**(1), 45–69 (2014)
16. Li, Q., Dang, Z.: Sampling automata and programs. Theor. Comput. Sci. **577**, 125–140 (2015)
17. Naval, S., Laxmi, V., Rajarajan, M., Gaur, M.S., Conti, M.: Employing program semantics for malware detection. IEEE Trans. Inf. Forensics Secur. **10**(12), 2591–2604 (2015)
18. Ruelle, D.: Thermodynamic Formalism: The Mathematical Structure of Equilibrium Statistical Mechanics. Cambridge University Press/Cambridge Mathematical Library, Cambridge (2004)
19. Sarig, O.M.: Thermodynamic formalism for countable Markov shifts. Ergodic Theor. Dyn. Syst. **19**, 1565–1593 (1999)
20. Sarig, O.M.: Lecture notes on thermodynamic formalism for topological Markov shifts (2009)
21. Shannon, C.E., Weaver, W.: The Mathematical Theory of Communication. University of Illinois Press, Urbana (1949)
22. Sokolsky, O., Kannan, S., Lee, I.: Simulation-based graph similarity. In: Hermanns, H., Palsberg, J. (eds.) TACAS 2006. LNCS, vol. 3920, pp. 426–440. Springer, Heidelberg (2006)
23. Walters, P.: An Introduction to Ergodic Theory. Graduate Texts in Mathematics. Springer, New York (1982)
24. Zager, L.A., Verghese, G.C.: Graph Similarity Scoring and Matching. Appl. Math. Lett. **21**(1), 86–94 (2008)

Metric and Permutation-Based Indexing

Supermetric Search with the Four-Point Property

Richard Connor[1]([✉]), Lucia Vadicamo[2], Franco Alberto Cardillo[3], and Fausto Rabitti[2]

[1] Department of Computer and Information Sciences,
University of Strathclyde, Glasgow G1 1XH, UK
`richard.connor@strath.ac.uk`
[2] Institute of Information Science and Technologies (ISTI),
CNR, Via Moruzzi 1, 56124 Pisa, Italy
`{lucia.vadicamo,fausto.rabitti}@isti.cnr.it`
[3] Institute of Computational Linguistics (ILC), CNR,
Via Moruzzi 1, 56124 Pisa, Italy
`francoalberto.cardillo@ilc.cnr.it`

Abstract. Metric indexing research is concerned with the efficient evaluation of queries in metric spaces. In general, a large space of objects is arranged in such a way that, when a further object is presented as a query, those objects most similar to the query can be efficiently found. Most such mechanisms rely upon the triangle inequality property of the metric governing the space. The triangle inequality property is equivalent to a finite embedding property, which states that any three points of the space can be isometrically embedded in two-dimensional Euclidean space. In this paper, we examine a class of semimetric space which is finitely 4-embeddable in three-dimensional Euclidean space. In mathematics this property has been extensively studied and is generally known as the *four-point* property. All spaces with the four-point property are metric spaces, but they also have some stronger geometric guarantees. We coin the term *supermetric space* as, in terms of metric search, they are significantly more tractable. We show some stronger geometric guarantees deriving from the four-point property which can be used in indexing to great effect, and show results for two of the SISAP benchmark searches that are substantially better than any previously published.

1 Introduction

To set the context, we are interested in searching a (large) finite set of objects S which is a subset of an infinite set U, where (U, d) is a metric space. The general requirement is to efficiently find members of S which are similar to an

---

The term *supermetric space* has previously been used in the domains of particle physics and evolutionary biology as a pseudonym for the mathematical term *ultrametric*, a concept of no interest in metric search; we believe our concept is of sufficient importance to the domain to justify its reuse with a different meaning.

© Springer International Publishing AG 2016
L. Amsaleg et al. (Eds.): SISAP 2016, LNCS 9939, pp. 51–64, 2016.
DOI: 10.1007/978-3-319-46759-7_4

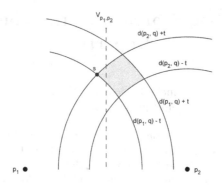

Fig. 1. In any metric space, two pivot points and any solution to a query can be isometrically embedded in ℓ_2^2. The point q cannot be drawn in the same diagram. Given its distance from p_1 and p_2, any solution in the original metric space must lie in the region bounded by the four arcs shown. If the point s lies to the right of V_{p_1,p_2}, there is therefore no requirement to search to the left of the hyperplane in the original space.

arbitrary member of U, where the distance function d gives the only way by which any two objects may be compared. There are many important practical examples captured by this mathematical framework, see for example [3,8]. Such spaces are typically searched with reference to a query object $q \in U$. A threshold search for some threshold t, based on a query $q \in U$, has the solution set $\{s \in S$ such that $d(q, s) \leq t\}$.

1.1   Metric Spaces and Finite Isometric Embeddings

An isometric embedding of one metric space (V, d_v) in another (W, d_w) can be achieved when there exists a mapping function $f : V \rightarrow W$ such that $d_v(x, y) = d_w(f(x), f(y))$, for all $x, y \in V$. A finite isometric embedding occurs whenever this property is true for any finite selection of n points from V, in which case the terminology used is that V is isometrically n-embeddable in W.

The first observation to be made in this context is that any metric space is isometrically 3-embeddable in ℓ_2^2. This is apparent from the triangle inequality property of a proper metric. In fact the two properties are equivalent: for any semi-metric space (V, d_v) which is isometrically 3-embeddable in ℓ_2^2, triangle inequality also holds. It is interesting to consider the standard exclusion mechanisms of pivot-based exclusion and hyperplane-based exclusion in the light of an isometric 3-embedding in ℓ_2^2; Fig. 1 for example shows a basis for hyperplane exclusion using only this property rather than triangle inequality explicitly.

1.2   Supermetric Spaces: Isometric 4-Embedding in ℓ_2^3

It turns out that many useful metric spaces have a stronger property: they are isometrically 4-embeddable in ℓ_2^3. In the mathematical literature, this has been referred to as the *four-point property*. We have studied such spaces in the context of metric indexing in [4], where we develop in detail the following outcomes:

1. Any metric space which is isometrically embeddable in a Hilbert space has the four-point property.
2. Important spaces with the property include, for any dimension, spaces with the following metrics: Euclidean, Jensen-Shannon, Triangular, and (a variant of) Cosine distances.
3. Important spaces which do not have the property include those with the metrics: Manhattan, Chebyshev, and Levenshtein distances.
4. However, for any metric space (U, d), the space $(U, \sqrt{d})$ does have the four-point property.

In terms of practical impact on metric search, in [4] we show only how the four-point property can be used to improve standard hyperplane partitioning. We consider a situation where a subspace is divided according to which of two selected reference points p_1 and p_2 is the closer. When relying only on triangle inequality, that is in a metric space without the four-point property, then for a query q and a query threshold t, the subspace associated with p_1 can be excluded from the search only if $d(q, p_1) - d(q, p_2) > 2t$. As the region defined by this condition when projected onto the plane is a hyperbola (see Fig. 1), we name this Hyperbolic Exclusion.

If the space in question has the four-point property, however, we show that, for the same subspaces, there is no requirement to search that associated with p_1 whenever $\frac{d(q,p_1)^2 - d(q,p_2)^2}{d(p_1,p_2)} > 2t$; this is a weaker condition and therefore allows, in general, more exclusion. We name this condition Hilbert Exclusion.

In this paper, we examine a more general consequence of four-point embeddable spaces and show some interim results including new best-performance search of SISAP data sets.

2   Tetrahedral Projection onto a Plane

In a supermetric space, any two reference points p_1 and p_2, and query point q, *and* any solution to that query s where $d(q, s) \leq t$, can *all* be embedded in 3D Euclidean space. As such, they can be used to form the vertices of a tetrahedron. It seems that, while simple metric search is based around the properties of a triangle, there should be corresponding tetrahedral properties which give a new, stronger, set of guarantees.

Assume that for some search context, points $p_1, p_2 \in U$ are somehow selected and a data structure is built for a finite set $S \subset U$ where, for $s \in S$, the three distances $d(p_1, p_2), d(s, p_1)$ and $d(s, p_2)$ are calculated during the build process and used to guide the structuring of the data. At query time, for a query q, the two distances $d(q, p_1)$ and $d(q, p_2)$ are calculated and may be used to make some deduction relating to this structure.

This situation gives knowledge of two adjacent faces of the tetrahedron which can be formed in three dimensions. Five of the six edge lengths have been measured, and the final edge is upper-bounded by the value of t. Therefore, for a

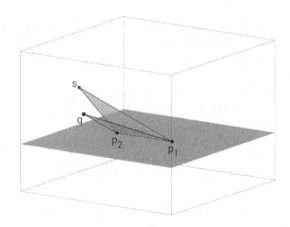

Fig. 2. Two triangles in 3D space

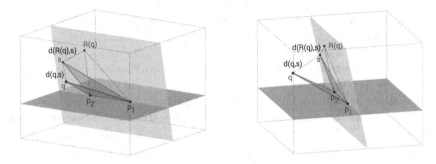

Fig. 3. Projection of the two triangles onto the same plane by rotation around p_1p_2. Note that $\ell_2^2(R(q), s) \leq \ell_2^3(q, s)$

point s to be a solution to the query, it must be possible to form a tetrahedron with the five measured edge lengths, and a last edge of length t.

Figure 2 shows a situation where five edge lengths have been embedded in 3D space. The edge p_1p_2 is shared between the two facial triangles depicted. However the distance $d(s, q)$ is not known, and therefore neither is the angle between these triangles. The observation which gives rise to the results presented here is that, if both triangles are now projected onto the same plane, which can be achieved by rotating one of them around the line p_1p_2 until it is coplanar with the other, then for any case where the final edge of the tetrahedron (qs) is less than the length t, then the length of this side in the resulting planar tetrahedron is upper bounded by t, as illustrated in Fig. 3.

Many such coplanar triangles can be depicted, representing many points in a single space, in a single scatter plot as in Fig. 4. This shows a set of 500 points, drawn from randomly generated 8-dimensional Euclidean space, and plotted with respect to their distances from two fixed reference points p_1 and p_2. The distance between the reference points is measured, and the reference points are plotted

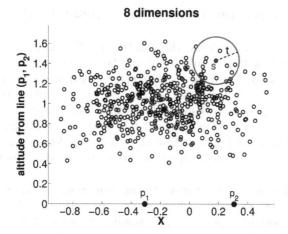

Fig. 4. Scatter diagram for 8-dimensional Euclidean Space. The distance δ between two selected reference points p_1 and p_2 is measured, and an embedding function is chosen which maps these to $(0, -\delta/2)$ and $(0, \delta/2)$ respectively. Other points s_i in the space are then plotted to preserve the distances $d(s_i, p_1)$ and $d(s_i, p_2)$. For metric spaces with the four-point property, the ℓ_2 distance between the corresponding points in this diagram is a lower bound on $d(s_i, s_j)$ in the original space. Hence, any point within t of a point s in the original space cannot lie outside the circle of radius t centered around s in the scatter plot.

on the X-axis symmetrically either side of the origin. For each point in the rest of the set, the distances $d(s, p_1)$ and $d(s, p_2)$ are calculated, and used to plot the unique corresponding point in a triangle above the X-axis, according to these edge lengths. In this figure, in consideration with our observations over Fig. 3, it can be seen that, if any two points are separated by less that some constant t in the original space, and thus also in the 3D embedding, then they are also within t of each other in this scatter plot.

It is important to be aware, in this and the following figures, of the importance of the four-point property. The same diagram can of course be plotted for a simple metric space, but in this case no spatial relationship is implied between any two points plotted: no matter how close two points are in the plot, there is no implication for the distance between them in the original space. However if the diagram is plotted for a metric with the four-point property, then the distance between any two points on the plane is a lower bound on their distance in the original space; two points that are further than t on the plot cannot be within t of each other in the original space. This observation leads to an arbitrarily large number of ways of partitioning the space and allowing these partitions to excluded based on a query position, and has many potential uses in metric indexing.

3   Indexes Based on Tetrahedral/Planar Projection

During construction of an index, the constructed 2D space can be arbitrarily partitioned according to any rule based on the geometry of this plane, calculated with respect to the distances $d(s_i, p_1), d(s_i, p_2)$ and $d(p_1, p_2)$. At query time, if the query falls in any region of the plane that is further than the query threshold t from any such partition, points within that partition cannot contain any solution to the query. Since, as will be shown, different spaces give quite different distributions of points within the plane, build-time partitions can be chosen according to this distribution, rather than as a fixed attribute of an index mechanism.

There is much potential for investigating partitions of this plane, and our work is ongoing. The simplest such mechanism to consider is the application of this concept to normal hyperplane partitioning. Suppose that a data set S is simply divided according to which of the points p_1 and p_2 is the closer, which corresponds in the scatter diagram to a split over the Y axis. Then at query time, if the corresponding plot position for the query is further than t from the Y axis, no solutions can exist in the subset closer to the opposing reference point. Figure 5 shows the same points, but now highlighted according to this distinction. Those drawn in solid, either side of the Y-axis, are guaranteed to be on the same side of the corresponding hyperplane partition in the original space; therefore, if they were query points, the opposing semi-space would not require to be searched. If the same diagram is drawn for a simple metric space, a query point can be used to exclude the opposing semi-space only according to a condition algebraically derived from triangle inequality: $|d(q, p_1) - d(q, p_2)| > 2t$, which describes a hyperbola with foci at the reference points and semi-major axis of the search threshold. For the same data and search threshold, the difference in exclusion capability is shown in Fig. 5; of the 500 randomly selected queries, only 160 fail to exclude the opposing semi-space, whereas with normal hyperbolic exclusion, the number is 421. The query threshold illustrated, 0.145, is chosen to retrieve around one millionth of the space and is not therefore artificially large.

As stated, this particular situation has been extensively investigated and is fully reported in [4]. Here we will concentrate further on other properties of the planar projection, of which the derivation of Hilbert exclusion turns out to be a special case.

4   Partitions of the 2D Plane

For the purposes of this analysis only, for reasons of simplicity, we seek to divide a data set into precisely two partitions. This is without reference to details of any indexing structure which may use the concepts, although in all cases by implication there exists a simple binary partition tree structure corresponding to the partitioning. In all cases the partition is defined in terms of the 2D plane onto which all points are projected as described above. A few points are important to note for such structures:

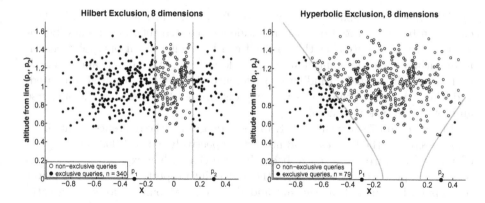

Fig. 5. Scatter diagram for 8-dimensional Euclidean Space. The data is divided into two subsets according to which side of Y-axis they lie; if the solidly-coloured points represent queries, the data on the opposing side cannot contain a solution. The left-hand side illustrates use of tetrahedral/planar projection, the right hand side illustrates use of the normal hyperbolic condition.

1. For any such strategy, other more complex ways of indexing the data exist; by analogy, for example, various forms of SAT [2], GNAT [1], M-Index [7] etc. will exist. Mechanisms normally associated with single reference point pivoting may also have equivalents. In our initial analysis we do not have time or space to investigate all of these forms.
2. There is an apparent disadvantage for any of these techniques when compared with any technique based on single-point pivoting, which is that for any conceptual tree node, two distances need to be calculated as against one. This is not the case in fact, as it is always possible to re-use one reference point from the node directly above, without significantly affecting any spatial properties of the distribution, using a technique first proposed for the monotonous bisector tree [6].
3. Furthermore, any such mechanism has a further advantage, as whenever the space is partitioned, it is also possible to store internal and external radii for the partitions, from both of the reference points, which allow further exclusions to be made at effectively no extra cost.

4.1   Reference Point Separation

An important observation is that the shape of the 2D "point cloud", upon which effective exclusion depends, is not greatly affected by the choice of reference points. In comparison with normal Hyperbolic exclusion this is a huge advantage. The hyperbola which bounds the effective queries, i.e. those which can be used to exclude the opposing semispace, is defined only by the (fixed) query radius, and the distance between the reference points, where the larger the separation of the reference points, the better the exclusion. In the extreme case where the separation is no larger than twice the query radius, which can readily occur in high-dimensional space, it is impossible for any exclusions to be made. This effect can be ameliorated by choosing widely separated reference points, but in an unevenly distributed set this in itself can be dangerous: if one point chosen is an outlier, then the point cloud will lie close to the other point, and again no exclusions will be made. Finding two reference points which are well separated, and where the rest of the points is evenly distributed between them, is of course an intractable task in general.

Figures 6 and 7 show this effect. In these diagrams, the reference points have been selected as the furthest, and nearest, respectively out of 1,000 sample pairs of points drawn from the space. It can be seen that, when exclusion is based on tetrahedral properties allowed from the four-point property, the exclusive power remains fairly constant, as the size and shape of the point cloud is not greatly affected. However, when the hyperbolic condition is used, the exclusive power is hugely affected; in this case the query threshold is only slightly less than half the separation of the reference points, and the resulting hyperbola diverges so rapidly from the separating hyperplane that no exclusions are made from the sample queries. From Fig. 6 it should also be noted that, no matter how far the reference points are separated, the four-point property always gives more exclusions; in this case, although the separating lines do not appear visually to be very different, the implied probability of exclusion in for the four-point property is 0.66, against 0.58.

To allow most partition structures to perform well, a very large part of the build cost is typically spent in the selection of good reference points and this cost is largely avoidable with any such four-point strategy.

4.2   Arbitrary Partitions

Again we stress the fact that, given the strong lower bound condition on the projected 2D plane, we can choose arbitrary geometric partitions of this plane

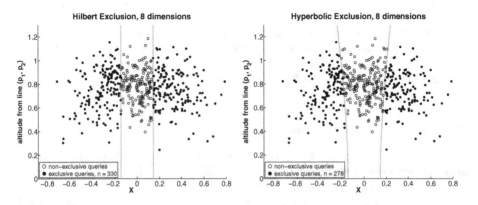

Fig. 6. Scatter diagram for 8-dimensional Euclidean Space with widely separated reference points. (The distance between reference points is such that the reference points themselves do not appear on the plot.)

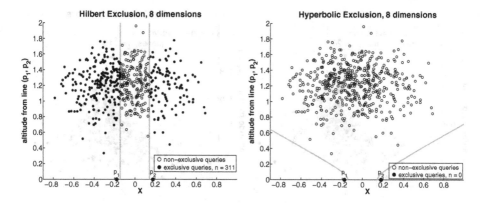

Fig. 7. Scatter diagram for 8-dimensional Euclidean Space with close reference points. Note from comparison of the left-hand graphs of this figure with Fig. 6 that the separation of the reference points has no apparent effect on the power of the four-point exclusion, whereas normal metric exclusion becomes completely useless.

to structure the data. For randomly generated, evenly distributed points there seems to be little to choose. However it is often the case that "real world" data sets do not show the same properties as generated sets; in particular, they tend to be much less evenly distributed, with significant numbers of clusters and outliers. These factors can significantly affect the performance of indexing mechanisms. Figures 8 and 9 show a sample taken from the SISAP *colors* data set with Euclidean distance applied, showing four different partitions. Four different partitions of the plane have been arbitrarily selected and applied. The query threshold illustrated is 0.052 corresponding to a query returning 0.001 % of the data.

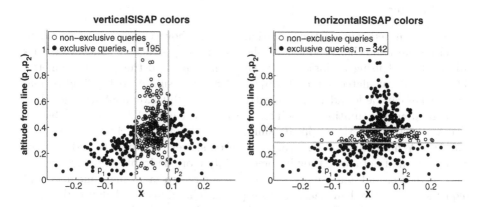

Fig. 8. Scatter diagrams dividing the plane equally in X and Y dimension, either can be used for partitioning a hyperplane tree structure. We show results for the "horizontal" pattern in Fig. 12 where it is the best available partitioning.

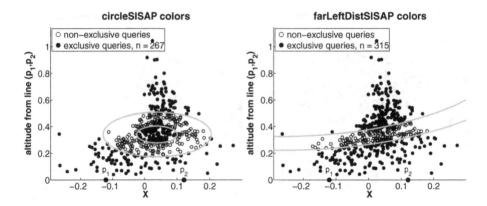

Fig. 9. Two more binary partitions, based now on median distance from arbitrary points in the plane (centre and top-left respectively); we have not yet found a use for these but include the diagrams to make the point that any such partition may be used.

In all cases, it can be noted that the partitions are even, leading to balanced indexing structures. It is very likely that skewed partitions may perform better, an aspect we have not yet investigated. However one important balanced partition is illustrated on the left hand side of Fig. 8, implying that a balanced hyperplane tree can be efficiently constructed.

It can be seen that, in this case, partitioning the plane according to the height of individual points above the X-axis is the most effective strategy. The disadvantage with this is that a little more calculation is required to plot the height of the point, rather than its offset from the Y-axis; however this is a very minor effect when significantly more distance calculations can be avoided.

4.3 Balance

As already noted, any of the partitions shown above can be simply used to bisect the data and thus produced a balanced indexing structure. These examples are all defined using a single real value with respect to the planar geometry. This can be calculated for each object within the subset to be divided, and the median can be found very efficiently using the QuickMedianSort algorithm; for a random distribution of points, the practical cost of balancing a binary tree at construction time appears similar to performing QuickSort once on all the data. While balanced structures are often slower than unbalanced ones for relatively small data sets, they become rapidly more desirable as the size of the data increases, and again more so if it is too large to fit in main memory and requires to be stored in backing store pages. The ability to balance the data without reducing the effectiveness of the exclusion mechanism therefore seems important. One further area of investigation, not yet performed, would be the effect of controlling the balance, which once again is arbitrarily possible simply by selecting different offset values. In general this will increase the probability of exclusion at cost of excluding smaller subsets of the data, and the effectiveness will depend on the individual distributions of the different strategies.

5 Experiments and Results

To illustrate the effects discussed, we have implemented a generic partition tree which can be specialised according to a number of criteria. All of the core code executed is the same[1], allowing fair comparisons to be made for both distance measurement counts and elapsed time. The generic partition tree can be parameterised according to the following criteria:

Hilbert or Hyperbolic: the essence of our investigation.

Hilbert partition type: horizontal or vertical; we have not yet experimented with any other partition of the plane.

Balanced, or Unbalanced: as explained in the text, all the partitioned spaces can be balanced, in these tests we choose an even left/right split. Unbalanced spaces tested are split according only to which reference point is closer; both Hilbert and Hyperbolic exclusion are tested for these.

Reference point selection: Three different strategies are tested. The first reference point is arbitrarily selected, and the second is one of: the closest (non-identical) value within the subset; a randomly-selected value from the subset; and the furthest value from within the subset.

In all cases, with each partition two pivot values are kept and used in conjunction with each other exclusion policy: a cover radius is stored for the respective left/right reference points, and also the minimum radius between each reference point and the closest point in the opposing semi-space.

All tests are performed over SISAP *colors* and *nasa* benchmark data sets [5], using Euclidean distance, taking 10 % of the set to act as queries over the remainder and measuring only the number of distance calculations performed per query ($n = 101.5\,\mathrm{k}$, $36.1\,\mathrm{k}$ respectively.) In the remaining text we highlight some of the more interesting results.

5.1 Results

The smallest number of distances required for indexing was achieved by the unbalanced monotonous tree using Hilbert exclusion, with the reference points separated as far as possible. Figure 10 shows these results, in each case the bottom line on the graph indicating the best performance of our Hilbert exclusion mechanism in this context[2]. For comparison, the DiSAT [2] results with the two exclusion mechanisms are shown with grey lines; DiSAT/Hyperbolic, the top line on this chart, at the time of its publication was the best known general-purpose indexing mechanism.

---

[1] All of the (Java) code for these experiments can be downloaded from https:// bitbucket.org/richardconnor/metric-space-framework/.

[2] Which we therefore believe makes the best performance yet published for these metric/dataset combinations.

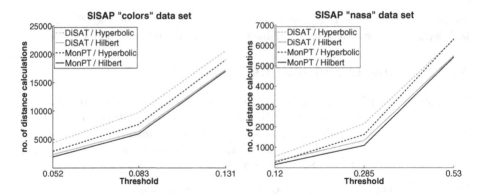

Fig. 10. Number of distance calculations per query for two SISAP benchmark sets. The best case for each data is Hilbert exclusion with a monotonous partition tree.

Figure 11 shows the relative effect of reference point separation when using Hilbert and Hyperbolic exclusion. Clearly, the furthest separation works best as would always be expected. The point here to note is the relative disadvantage suffered by the four-point metric with a cheaper choice of reference point. As collection size increases, the selection of multiple good reference points becomes relatively more expensive; with the four-point properties, building a high-performance index is much, much cheaper as the choice of reference point is much less significant.

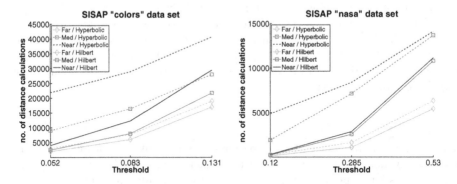

Fig. 11. Unbalanced hyperplane tree, different reference point separations. Choice of reference point is far less important for Hilbert exclusion, potentially allowing dramatic reductions to build time performance.

Finally, we can show one of our other partitions in action: the "horizontal" partition shown on the right-hand side of Fig. 8. Figure 12 shows this partition in comparison with the vertical and unbalanced Hilbert partitions when the close reference points are selected for the *colors* data set. Although, as can be seen in comparison with other graphs, this is not the best way we have found of searching this particular data set, the graph is included as a demonstration that a completely novel partitioning technique can be the best with some selections of data sets and reference points; there is much work still to do here. In fact, if partition exclusion alone is used this technique is the best available, but the way the space is partitions means many less cover radius exclusions are made; definitely the subject of further work.

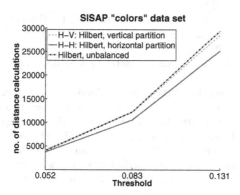

Fig. 12. Hilbert-Horizontal with close reference points; H-H is the bottom line, compared with H-V and unbalanced Hilbert. Hyperbolic exclusion with these reference points is shown as the top line of Fig. 11.

6 Conclusions

We have presented here a novel observation based on the four-point property that is possessed by many useful distance metrics. We have shown how, if it is guaranteed that any four points from the original space may be embedded in ℓ_2^3 as a tetrahedron, some much tighter geometric properties exist, in particular we have shown a lower-bound distance that can be calculated from knowledge of the sides of two tetrahedral faces. We have shown a few examples of how metric indexes can be constructed from this property and, although at an early stage of investigation, we have already shown a new best-performance for Euclidean distance search over two of the SISAP benchmark datasets. We believe a step change in improvement for exact search is possible; already our improved distance counts represent 29 % and 44 % of the previously published best results for *nasa* and *colors* respectively, using a structure which is much simpler and has a much smaller build time; we think much greater improvement is yet possible.

Acknowledgements. We would like to thank the anonymous referees for helpful comments on an earlier version of this paper. Richard Connor would like to acknowledge support by the National Research Council of Italy (CNR) for a Short-term Mobility Fellowship (STM) in June 2015, which funded a stay at ISTI-CNR in Pisa during which much of this work was conceived.

References

1. Brin, S.: Near neighbor search in large metric spaces. In 21th International Conference on Very Large Data Bases (VLDB 1995) (1995)
2. Chávez, E., Ludueña, V., Reyes, N., Roggero, P.: Faster proximity searching with the distal SAT. Inf. Syst. **59**, 15–47 (2016)
3. Chávez, E., Navarro, G.: Metric databases. In: Rivero, L.C., Doorn, J.H., Ferraggine, V.E. (eds.) Encyclopedia of Database Technologies and Applications, pp. 366–371. Idea Group, Hershey (2005)
4. Connor, R., Cardillo, F.A., Vadicamo, L., Rabitti, F.: Hilbert exclusion: improved metric search through finite isometric embeddings. ArXiv e-prints (accepted for publication ACM TOIS, July 2016), April 2016
5. Figueroa, K., Navarro, G., Chávez, E.: Metric spaces library. www.sisap.org/library/manual.pdf
6. Noltemeier, H., Verbarg, K., Zirkelbach, C.: Monotonous Bisector* Trees — a tool for efficient partitioning of complex scenes of geometric objects. In: Monien, B., Ottmann, Th (eds.) Data Structures and Efficient Algorithms. LNCS, vol. 594, pp. 186–203. Springer, Heidelberg (1992). doi:10.1007/3-540-55488-2_27
7. Novak, D., Batko, M., Zezula, P.: Metric index: an efficient and scalable solution for precise and approximate similarity search. Inf. Syst. **36**(4), 721–733 (2011). Selected Papers from the 2nd International Workshop on Similarity Search and Applications SISAP (2009)
8. Zezula, P., Amato, G., Dohnal, V., Batko, M.: Similarity Search: The Metric Space Approach. Advances in Database Systems, vol. 32. Springer, New York (2006)

Reference Point Hyperplane Trees

Richard Connor[✉]

Department of Computer and Information Sciences,
University of Strathclyde, Glasgow G1 1XH, UK
richard.connor@strath.ac.uk

Abstract. Our context of interest is tree-structured exact search in metric spaces. We make the simple observation that, the deeper a data item is within the tree, the higher the probability of that item being excluded from a search. Assuming a fixed and independent probability p of any subtree being excluded at query time, the probability of an individual data item being accessed is $(1-p)^d$ for a node at depth d. In a balanced binary tree half of the data will be at the maximum depth of the tree so this effect should be significant and observable. We test this hypothesis with two experiments on partition trees. First, we force a balance by adjusting the partition/exclusion criteria, and compare this with unbalanced trees where the mean data depth is greater. Second, we compare a generic hyperplane tree with a monotone hyperplane tree, where also the mean depth is greater. In both cases the tree with the greater mean data depth performs better in high-dimensional spaces. We then experiment with increasing the mean depth of nodes by using a small, fixed set of reference points to make exclusion decisions over the whole tree, so that almost all of the data resides at the maximum depth. Again this can be seen to reduce the overall cost of indexing. Furthermore, we observe that having already calculated reference point distances for all data, a final filtering can be applied if the distance table is retained. This reduces further the number of distance calculations required, whilst retaining scalability. The final structure can in fact be viewed as a hybrid between a generic hyperplane tree and a LAESA search structure.

1 Introduction and Background

Sections 1.1 and 1.2 set the context of metric search in very brief detail; much more comprehensive explanations are to be found in [4,12]. Readers familiar with metric search can skim these sections, although some notation used in the rest of the article is introduced.

1.1 Notation and Basic Indexing Principles

To set the context, we are interested in querying a large finite metric space (S, d) which is a subset of an infinite space (U, d). The most general form of query is a *threshold* query, where a query $q \in U$ is presented along with a threshold t, the required solution being the set $\{s \leftarrow S \mid d(q, s) \leq t\}$. In general $|S|$ is too large

© Springer International Publishing AG 2016
L. Amsaleg et al. (Eds.): SISAP 2016, LNCS 9939, pp. 65–78, 2016.
DOI: 10.1007/978-3-319-46759-7_5

for an exhaustive search to be tractable, in which case the metric properties of d require to be used to optimise the search.

In metric indexing, S is arranged in a data structure which allows exclusion of subspaces according to one or more of the exclusion conditions deriving from the triangle inequality property of the metric. As we refer to these later, we summarise them as:

pivot exclusion (a). For a reference point $p \in U$ and any real value μ, if $d(q,p) > \mu + t$, then no element of $\{s \in S \mid d(s,p) \leq \mu\}$ can be a solution to the query

pivot exclusion (b). For a reference point $p \in U$ and any real value μ, if $d(q,p) \leq \mu - t$, then no element of $\{s \in S \mid d(s,p) > \mu\}$ can be a solution to the query

hyperplane exclusion. For reference points $p_1, p_2 \in U$, if $d(q,p_1) - d(q,p_2) > 2t$, then no element of $\{s \in S \mid d(s,p_1) \leq d(s,p_2)\}$ can be a solution to the query.

1.2 Partition Trees

By "partition tree" we refer to any tree-structured metric indexing mechanism which recursively divides a finite search space into a tree structure, so that queries can subsequently be optimised using one or more of the above exclusion conditions. These structure data either by distance from a single point, such as the Vantage Point Tree, by relative distance from two points, for example the Generic Hyperplane Tree or Bisector Tree. Many such structures are documented in [4,12]. In our context we are interested only in the latter category as will become clear.

As there are many variants of both structures, we restrict our description to the simplest form of binary metric search tree in each category. The concepts extend to more complex and efficient indexes such as the GNAT [1], MIndex [10] and various forms of SAT trees [2,3,7,8], here we are only concerned with the principles.

In both cases, search trees are formed from a finite set of points in a metric space by selecting two reference points, and recursively forming child nodes to store the remaining points according to which of these reference points is the closest. During query, these nodes may be excluded from a search if it can be determined that the child node cannot contain any solutions to the query. In general, the term "bisector" is used when such exclusions are based on pivot exclusion, and the term "hyperplane" is used when exclusions are based on hyperbolic exclusion. It has long been known that, given the same basic tree structure, both exclusion techniques can be used; as this always increases the degree of exclusion, thus improving efficiency, it makes no sense to do otherwise. Therefore, any sensible index using this structure will be a hybrid of these two techniques.

1.3   Balancing the Partition

To the above exclusion conditions, we add one more first identified in [6]:

hyperbola exclusion. For reference points $p_1, p_2 \in U$ and any real value δ, if $d(q, p_1) - d(q, p_2) > 2t + \delta$, then no element of $\{s \subset S, d(s, p_1) \leq d(s, p_2) + \delta\}$ can be a solution to the query

The addition of the constant δ means that for any pair of reference points, an arbitrary balance can be chosen when constructing the tree. An algebraic proof of correctness for this property follows the same lines as that for normal hyperplane exclusion.

The purpose of this is illustrated in Fig. 1. The diagrams show the two reference points p_1, p_2 plotted centrally along the X axis $d(p_1, p_2)$ apart. Each other point is uniquely positioned according to its distances from p_1 and p_2 respectively. The data shown here is drawn from the SISAP *colors* data set under Euclidean distance.

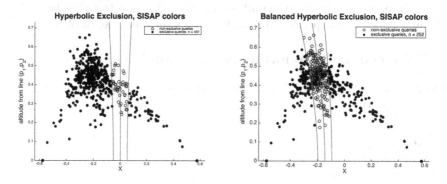

Fig. 1. Balanced hyperbolic exclusion

As we will use more such figures, it is worth explaining in a little detail what is being illustrated. The only significant geometric relationship within the scatter plot is between each point plotted and the two reference points plotted along the X axis; there is no relation between the distances of points plotted in this plane and their distance in the original space. The assumption is made however that the distribution of points in this plane is likely to be the same for both data and query as an indexing structure is being built and used; this assumption is justified by the fact that, for any metric space, any three points may be isometrically embedded in two-dimensional Euclidean space, giving a meaningful semantics to the distribution if not the individual point distances. The separation of points around the central line represents the separation of data at construction time if a structure was being built using these 500 points as data, with the two selected reference points. The effectiveness of the exclusion mechanism is illustrated by the two outer lines, which show the boundaries of queries which allow the opposing semi-space to avoid being searched for possible solutions. If the distribution is representative, it is reasonable to use the same set of example points for both purposes.

On the left-hand side of the figure, normal hyperplane exclusion is illustrated. The data is split according to which is the closer reference point, which manifests here as either side of the Y axis. The exclusion condition is the hyperbola condition, depicted by the outer (hyperbolic) curves. Any query points outside these lines do not require the opposing semi-space to be searched.

However the partition of the data is unbalanced with respect to the chosen reference points. On the right hand side of the picture, the data set is split according to the central hyperbolic curve, the value for this being chosen to achieve an arbitrary balance of the data, in this case evenly divided. From the illustration it can be seen that fewer queries will achieve the exclusion condition; however, the magnitude of the space excluded will be greater in most cases.

For our purposes here, the point is that an even balance can be achieved in all cases, for arbitrary data and any reference points. Of course, unlike in a database indexing structure, an improved balance does not imply improved performance, and our working hypothesis at this point is in fact that balancing will, on whole, degrade performance as it reduces the mean depth of the data.

2   Balanced and Monotonous Partition Trees

Algorithms 1 and 2 give the simplest algorithms for constructing, and querying, a balanced partition tree.

Data: $S_i \subset S$
Result: Node: $< p_1, p_2 : U, \delta : \mathbb{R}, \text{left}, \text{right} : \text{Node}>$
select p_1, p_2 from S_i;
if $|S_i| > 2$ **then**

> $S_i \leftarrow S_i - \{p_1, p_2\}$;
> for all $s_j \in S_i$ calculate $d(s_j, p_1) - d(s_j, p_2)$;
> find median value δ;
> create subsets S_l, S_r such that;
> $S_l = \{s \leftarrow S_i, d(s, p_1) - d(s, p_2) < \delta\}$;
> $S_r = \{s \leftarrow S_i, d(s, p_1) - d(s, p_2) \geq \delta\}$;
> left $\leftarrow$ CreateNode(S_l);
> right $\leftarrow$ CreateNode(S_r);

end

Algorithm 1. CreateNode (balanced)

This algorithm works correctly, but to work well requires the same refinements as any other hyperplane tree, as follows:

1. The reference points need to be chosen carefully to be far apart, but also not to be very close to any reference point previously used at a higher level of the tree. Otherwise, in either case, few or no exclusions will be made at the node.

Data: $q \in U, n$: Node
Result: Result: $\{s \in S, d(s, q \leq t)\}$
Result = {};
if $d(q, n.p_1) \leq t$ **then**
| Result.add(p_1)
end
if $d(q, n.p_2) \leq t$ **then**
| Result.add(p_2)
end
if $d(q, n.p_1) - d(q, n.p_2) \geq 2t + n.\delta$ **then**
| Result.add(Query(q,n.right))
end
if $d(q, n.p_2) - d(q, n.p_1) > 2t + n.\delta$ **then**
| Result.add(Query(q,n.left))
end

Algorithm 2. Query

2. As well as relying on hyperbolic exclusion, each node can also cheaply store values for use, for both partitions, with both types of pivot exclusion. For both subtrees minimum and maximum distances to either reference point can be stored and used to allow pivot exclusion for a query. Most commonly, only the cover radius is kept for the reference point closest to the subtree. The minimum distance from the opposing reference point may also be of value; an interesting observation with the balanced tree, which can be seen by studying Fig. 1, is that both these types of pivot exclusion may well function better with a higher δ value at the node.

The monotonous hyperplane tree (MHT[1]) was first described in [9] where it was described as a bisector tree using only pivot exclusion. The structure is essentially the same, but each child node of the tree shares one reference point with its parent, as shown in Algorithm 3. A significant advantage is that, for each exclusion decision required in an internal node of the tree, only a single distance needs to be calculated rather than two for the non-monotonous variant.

The query algorithm is conceptually the same, but in practice the distance value $d(q, p_1)$ is calculated in the parent node and passed through the recursion to avoid its recalculation in the child node.

The intent behind this reuse of reference points was originally geometric in origin, based on an intuition of point clustering within a relatively low dimensional space; this intuition becomes increasingly invalid as the dimensionality of the space increases. Interestingly however the monotonous tree performs substantially better that an equivalent hyperplane tree in high dimensional spaces.

---

[1] Originally named the "Monotonous Bisector* Tree".

Data: $S_i \subset S, p_1 \in S$
Result: Node: $< p_1, p_2 : U, \delta : \mathbb{R},$ left, right : Node$>$
select p_2 from S_i;
if $|S_i| > 2$ **then**
$\quad S_i \leftarrow S_i - \{p_1, p_2\}$;
$\quad$ for all $s_j \in S_i$ calculate $d(s_j, p_1) - d(s_j, p_2)$;
$\quad$ find median value δ;
$\quad$ create subsets S_l, S_r such that;
$\quad S_l = \{s \leftarrow S_i, d(s, p_1) - d(s, p_2) < \delta\}$;
$\quad S_r = \{s \leftarrow S_i, d(s, p_1) - d(s, p_2) \geq \delta\}$;
$\quad$ left $\leftarrow$ CreateNode(S_l, p_1);
$\quad$ right $\leftarrow$ CreateNode(S_r, p_2);
end

Algorithm 3. CreateNode (monotonous balanced)

3   The Effect of Depth

As balance and monotonicity are orthogonal properties of the partition tree, we have now identified four different types of tree to test in experiments. At each node of each of the four trees described, it is not unreasonable to assume that over a large range of queries the probability of being able to exclude one of the subtrees is approximately constant.

Data is embedded within the whole tree. Viewed from the perspective of an individual data item at depth d within the tree, it sits at the end of a chain of tests, each of which may result in it not being visited during a query. The probability of any data item being reached, and therefore having its distance measured, is therefore $(1 - p)^d$, where p is the probability of an exclusion being made at each higher node. It should therefore be possible to measure that (a) unbalanced trees perform better than balanced, and (b) monotonous trees perform better than non-monotonous, as in each case the mean data depth is greater.

Figure 2 shows performance for the four tree types used for Euclidean search on the SISAP *colors* and *nasa* data sets [5]. In each case ten percent of the data is used as queries over remaining 90 percent of the set, at threshold values which return 0.01, 0.1 and 1 % of the data sets respectively; results plotted are the mean number of distances required per query ($n = 101414, 36135$ respectively.) The results presented are in terms of the mean number of distance calculations required per query; for most mechanisms over similar data structures execution times are proportional to these, and are not included due to space constraints[2].

In all of these tests, a reasonable attempt to find "good" reference points is made; the selection of the first reference point is arbitrary (either randomly selected, or passed down from the construction of the parent node in the case of monotonous trees); the second reference point is selected as the point within the subset which is furthest from the first. This works reasonably well, although

---

[2] Source code to repeat these experiments, including timings, is available from https://bitbucket.org/richardconnor/metric-space-framework.

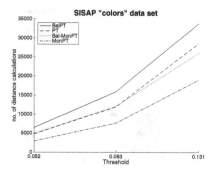 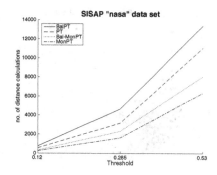

Fig. 2. Four variants of hyperplane trees (monotonous or not, balanced or not) showing number of distances performed for SISAP benchmark searches. In each case, from the bottom (best) is: monotonous unbalanced, monotonous balanced, normal unbalanced, normal balanced

performance can be improved a little by using a much more expensive algorithm at this point.

In each case, as expected, the balanced tree does not perform as well as the unbalanced tree, and the monotonous tree performs better than the non-monotonous tree.

4   Balancing and Pivot Exclusion

Before describing our proposed mechanism we briefly consider the effect on exclusion when relatively ineffective reference points are used. It should be noted that this will usually be the case towards the leaf nodes of any tree, as only a small set is available to choose from, and in fact this will affect the majority cases of any search in a high-dimensional space. In our particular context, we are going to compromise the effectiveness of the hyperplane exclusion through the tree nodes, by using relatively ineffective reference points, in exchange for placing the majority of the data at the leaf nodes.

Figure 3 shows the same data as plotted in Fig. 1, but where much less good reference points have been selected. These are not pathologically bad reference points, in that they are the worst pair tested from a randomly selected sample of only a few. The tradeoff between balanced and unbalanced exclusion is now very interesting. As can be seen in the left hand figure, the large majority (450 ex. 500 points) of queries will successfully allow exclusion of the opposing subspace; however in all but four cases the opposing subspace contains only around 2 % of the data; however, those four cases exclude 98 % of the data. On the right-hand side, only 24 ex 500 queries allow exclusion, but in each case half of the data is excluded. So for this sample of points, treated as both data and query over the same reference points, both balanced and unbalanced versions save a total of around $6k$ distance calculations out of $25k$.

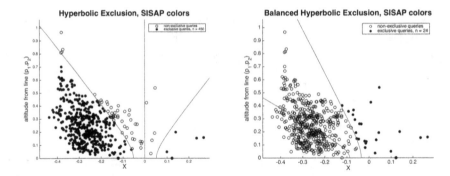

Fig. 3. The data as plotted in Fig. 1, with a much worse choice of reference points. Note that in the right-hand chart, the pivots are so skewed that the left-hand branch of the exclusion hyperbola does not exist in ℓ_2^2; the line on the left is the hyperbolic centre of the data with respect to the reference points.

However one further factor that can be noticed in general terms is that the use of pivot exclusion of both types (a cover radius can be kept for each reference point and semispace, and also the minimum distance between each reference point and the opposing semispace) may be more effective in the balanced version due to the division between the sets being skewed; it can be seen here that the left-hand cover radius in the balanced diagram is usefully smaller than the corresponding cover radius in the left-hand diagram.

This case is clearly anecdotal, and our experiments still show that on the whole the unbalanced version is more effective overall; however we believe this is because of the larger mean number of exclusion possibilities before reaching the data nodes. This is the aspect we now try to address.

5   Reference Point Hyperplane Trees

The essence of the idea presented here is to use the same partition tree structure and query strategy, but using a fixed, relatively small set of reference points to define the partitions. The underlying observation is that, given we can achieve a balanced partition of the data for any pair of reference points, we can reuse the same reference points in different sub-branches of the tree. Attempting the same tactic without the ability to control the balance degrades into, effectively, a collection of lists.

Any points from the data set not included in the reference point set will necessarily end up at the leaf nodes of the tree. Thus, although the limited set of reference points pairs may reduce the effectiveness of exclusion at each level, the mean depth traversed before another distance calculation is required will be greater.

Assuming a balanced tree is constructed as above, the binary monotonous hyperplane tree stores half of its data at the leaf nodes, which have a depth of $\log_2 n$ for n data. The non-monotonous variant has only one-third of its data in the leaf nodes, and the mean depth is corresponding smaller. The two tree types are illustrated in Fig. 4, where it is clear to see the average depth of a data item is always greater for the monotonous case. In fact empirical analysis shows that for large trees, the Reference Point tree has a weighted mean data depth of exactly one more than the Monotonous tree, which in turn has a weighted mean data depth of exactly one more than the non-monotonous tree.

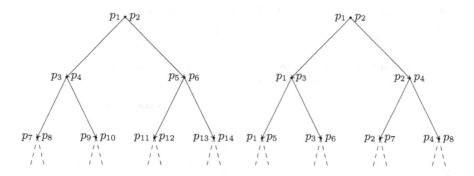

Fig. 4. Generic and monotonous hyperplane trees. Note the re-use of a single parent node for constructing the child node partition. For large trees, mean data depth is $\log_2 n - 1$ for Generic, and $\log_2 n$ for Monotonous.

To investigate this advantage further, we have considered two ways of using a small fixed set of points for the tree navigation, illustrated in Fig. 5.

5.1   Permutation Trees

The first we refer to as a *permutation tree*. The underlying observation here is that, for a fixed set of n reference points, there exist $\binom{n}{2}$ unique pairs of points that can be used to form a balanced hyperplane partition. These can be assigned a numbering, as can the internal nodes of the tree, so that a different permutation is used at each node of the tree. At construction time, the permutation used for the particular node is selected and the difference of the distances to each point are calculated, the data is then divided into two parts based on the median of these differences. At query time, the distance between the query and each of the n reference points can be pre-calculated; this gives all the information that is required to navigate the tree as far as the leaf nodes where the data is stored.

The strength of this method derives from the rate of growth of the function $\binom{n}{2}$. For n data to be resident at the leaf nodes, we require (modulo detail) around n internal nodes and therefore permutations, which in turn requires only around $\frac{\sqrt{8n}}{2}$ reference points. This equates to around 1,400 points for 1 M data, 14k for 100 M data etc.

We have built such structures and measured them; the results are shown in Fig. 6. They are encouraging; for the *colors* data set in particular, this is faster,

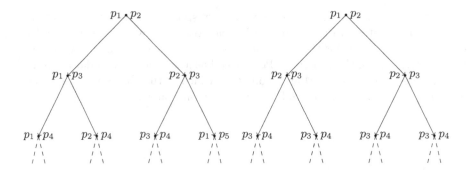

Fig. 5. Permutation and Leanest trees. In either case, on scaling, mean data depth is effectively $log_2 n + 1$ as all data is stored at the leaf nodes.

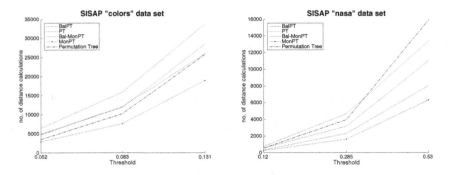

Fig. 6. Cost of Permutation Tree indexing. The costs for the two datasets are plotted in bold against the background of the costs plotted in Fig. 2 for comparison.

and requires less distance calculations, than any other balanced tree format. It seems to do relatively better at lower relative query thresholds, and for the higher-complexity data of the *colors* data set. Finally, we should note that the reference points from which the permutations are constructed are, at this point, selected randomly; we believe a significant improvement could be obtained by a better selection of these points but have not yet investigated how to achieve this.

5.2 Leanest Trees

For our other test, we have selected a strategy that we did not expect to work at all; for a set of $n + 1$ reference points, we partition each *level* of the tree with the same pair of points. That is, for the node at level 0, we use points $\{p_0, p_1\}$ to partition the space; at level two, we use the pair $\{p_1, p_2\}$, etc. For all nodes across the breadth of the tree, for depth m we use the reference pair $\{p_m, p_{m+1}\}$. This requires the selection of only $log_2 n + 1$ reference points for data of size n.

For this strategy, it is much easier to provide relatively good pairs of points for partitioning the data, as there are relatively very few of them. For the results given we used a cheap but relatively effective strategy. The first reference point is chosen at random; repeatedly, until all are selected, another is found from the data which is far apart (in these cases we only sampled 500 points), and so on until all the required points are found. One further check is required, that none of the selected points is very close (or indeed identical) to another already selected, as that would result in the whole layer of the tree performing no exclusions at all.

5.3   Leanest Trees with LAESA

We have one more important refinement. The build algorithm for the Leanest Tree, for greatest efficiency, will pre-calculate the distances from the small set of reference points to each element of the data; this data structure can then be passed into the recursive build function as a table. This table has exactly the same structure as the LAESA [11] structure.

The table has only $n \log_2 n$ entries and may therefore typically be stored along with the constructed tree. At query time, a vector of distances from the query to the same reference points is calculated before the tree traversal begins. Whenever the query evaluation reaches a leaf node of the tree, containing therefore a data node that has not been excluded during the tree traversal, a normal tree query algorithm would then calculate the distance between the query and the datum s at this node. If $d(q, s) \leq t$ then s is included in the result set.

However, before performing this calculation (these distance calculations typically representing the major cost of the search) it may be possible to avoid it, as for each p_i in the set of reference points, $d(q, p_i)$ and $d(s, p_i)$ are both available, having been previously calculated. If, for any p_i, $|d(q, p_i) - d(s, p_i)| > t$, it is not possible that $d(q, s) \leq t$ (by the principle of Pivot Exclusion (a) named in Sect. 1.1) and the datum can be discarded without its distance being calculated.

Of course this operation itself is not without cost, and should be performed only if its cost is less than that of a distance calculation. This will generally be the case at least if the size of an individual datum is greater than $\log_2 |S|$, or if a particularly expensive distance metric is being used.

6   Analysis

Figure 7 shows measurements for the Leanest Tree and its LAESA hybrid. These are plotted in bold, again set in the context of the greyed-out measurements copied from Fig. 2.

In each case, the top dotted line is for the Leanest Tree measured without using the LAESA hybrid. This is comparable to the Permutation Tree, and again it is worth noting that this is a good performance for a balanced tree; in cases where balancing is required, for example if the data does not fit in main memory and requires paging, this mechanism is worthy of consideration.

The lower dotted line is the raw number of distance measurements made by the hybrid mechanism. This is by far the best performance measured in these terms for both data sets; however, for reasons explained above, it must be noted

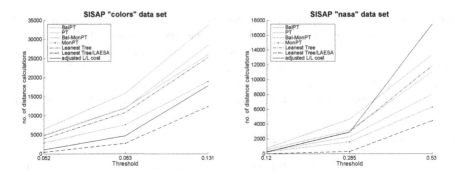

Fig. 7. Cost of Leanest Tree indexing. The costs for the two datasets are plotted in bold against the background of the costs plotted in Fig. 2 for comparison. The LAESA hybrid is the lowest line in each graph; this is not a true representation of overall cost, as explained in the text; the solid line gives a good estimate of the true cost of the hybrid mechanism.

that this does not represent actual measured performance in terms of query time, as there is a significant extra cost entailed in performing the LAESA-based filtering. In some cases however the number of distance calculations will swamp any other cost of the query and this line would be representative for the mechanism.

To give a fair comparison for these data sets and this metric, the solid black line is generated, rather than measured, to take the LAESA overhead cost into account. In fact this is done pessimistically by considering only the size of the data access required. Thus, to the raw distance count, a factor is added according to the number of reference points used, as a proportion of the size of the original data points. Thus for example for the *nasa* data set the tree depth is 15, requiring 16 reference points, and the original data is 20 dimensions. For every time the LAESA exclusion is attempted, 0.8 of a distance calculation is added to the total measured from the original space. For the *colors* data, these figures are 17 and 118 respectively, making the hybrid mechanism relatively more useful.

It may be noted that the total number of distance measurements made by this mechanism is similar (although in general smaller) to that required by the pure LAESA mechanism; however, a query using LAESA requires a linear scan of the whole LAESA table, therefore giving a cost directly proportional to the data size. In contrast our hybrid mechanism resorts to using the LAESA table only for data which has not already been excluded through the tree traversal, and therefore retains the scalability implied by recursive search structures.

The main outcome of our work is thus represented by the solid black line in the left hand figure, which gives a substantially better performance for this data set than any other we are aware of. The hybrid Leanest/LAESA mechanism appears to be very well suited for data sets which are very large and require paging, whose individual data items are very large, or whose distance metrics are very expensive.

7 Conclusions and Further Work

Having made the observation that the monotonous hyperplane tree is sub-stantially more efficient that the non-monotonous equivalent, even in high-dimensional spaces, we formed the hypothesis that this is primarily due to the longer mean search paths to each data item. We have taken this idea to its extreme, in conjunction with an ability to force balance onto a hyperplane par-tition, through the design of "permutation" and "leanest" hyperplane trees. In particular, the latter requires only $log_2 n + 1$ reference points for data of size n, therefore leaving effectively all of the data at the leaf nodes of the tree. We have tested both mechanisms against two SISAP benchmark data sets, and found good realistic performance in comparison with other structures that are bal-anced and therefore usable for very large data sets, or very large data points, which require to be paged.

Furthermore we note that the balanced tree mechanism can also be viewed as a scalable implementation of a LAESA structure, giving very good performance in particular for high-dimensional and expensive distance metrics. For very little extra cost, LAESA-style filtering can be performed on the results of the tree search, apparently giving the best of both worlds. We continue to investigate this mechanism in metric spaces more challenging that the benchmark sets reported so far.

Acknowledgements. Richard Connor would like to acknowledge support by the National Research Council of Italy (CNR) for a Short-term Mobility Fellowship (STM) in June 2015, which funded a stay at ISTI-CNR in Pisa where some of this work was done. The work has also benefitted considerably from conversations with Franco Alberto Cardillo, Lucia Vadicamo and Fausto Rabitti, as well as feedback from the anonymous referees. Thanks also to Jakub Lokoč for pointing out his earlier invention of parameterised hyperplane partitioning!

References

1. Brin, S.: Near neighbor search in large metric spaces. In: 21st International Confer-ence on Very Large Data Bases (VLDB 1995) (1995). http://ilpubs.stanford.edu:8090/113/
2. Chávez, E., Ludueña, V., Reyes, N., Roggero, P.: Faster proximity searching with the distal SAT. In: Traina, A.J.M., Traina Jr., C., Cordeiro, R.L.F. (eds.) SISAP 2014. LNCS, vol. 8821, pp. 58–69. Springer, Heidelberg (2014)
3. Chávez, E., Ludueña, V., Reyes, N., Roggero, P.: Faster proximity searching with the distal SAT. In: Traina, A.J.M., Traina, C., Cordeiro, R.L.F. (eds.) SISAP 2014. LNCS, vol. 8821, pp. 58–69. Springer, Heidelberg (2014). doi:10.1007/978-3-319-11988-5_6
4. Chávez, E., Navarro, G.: Metric databases. In: Rivero, L.C., Doorn, J.H., Ferraggine, V.E. (eds.) Encyclopedia of Database Technologies and Applications, pp. 366–371. Idea Group, Hershey (2005)
5. Figueroa, K., Navarro, G., Chávez, E.: Metric spaces library. www.sisap.org/library/manual.pdf

6. Lokoč, J., Skopal, T.: On applications of parameterized hyperplane partitioning. In: Proceedings of the Third International Conference on SImilarity Search and Applications, SISAP 2010, pp. 131–132. ACM, New York (2010). http://doi.acm.org/10.1145/1862344.1862370

7. Navarro, G.: Searching in metric spaces by spatial approximation. VLDB J. **11**(1), 28–46 (2002)

8. Navarro, G., Reyes, N.: Fully dynamic spatial approximation trees. In: Laender, A.H.F., Oliveira, A.L. (eds.) SPIRE 2002. LNCS, vol. 2476, p. 254. Springer, Heidelberg (2002)

9. Noltemeier, H., Verbarg, K., Zirkelbach, C.: Monotonous Bisector* Trees — a tool for efficient partitioning of complex scenes of geometric objects. In: Monien, B., Ottmann, T. (eds.) Data Structures and Efficient Algorithms. LNCS, vol. 594, pp. 186–203. Springer, Heidelberg (1992). doi:10.1007/3-540-55488-2_27

10. Novak, D., Batko, M., Zezula, P.: Metric index: an efficient and scalable solution for precise and approximate similarity search. Inf. Syst. **36**(4), 721–733 (2011). Selected Papers from the 2nd International Workshop on Similarity Search and Applications (SISAP) 2009

11. Ruiz, E.V.: An algorithm for finding nearest neighbours in (approximately) constant average time. Pattern Recogn. Lett. **4**(3), 145–157 (1986). http://www.sciencedirect.com/science/article/pii/0167865586900139

12. Zezula, P., Amato, G., Dohnal, V., Batko, M.: Similarity Search: The Metric Space Approach. Advances in Database Systems, vol. 32. Springer, New York (2006)

Quantifying the Invariance and Robustness of Permutation-Based Indexing Schemes

Stéphane Marchand-Maillet[1]([⊠]), Edgar Roman-Rangel[1], Hisham Mohamed[1], and Frank Nielsen[2]

[1] Department of Computer Science, University of Geneva, Geneva, Switzerland
stephane.marchand-maillet@unige.ch
[2] LIX Polytechnique, Paris, France

Abstract. Providing a fast and accurate (exact or approximate) access to large-scale multidimensional data is a ubiquitous problem and dates back to the early days of large-scale Information Systems. Similarity search, requiring to resolve nearest neighbor (NN) searches, is a fundamental tool for structuring information space. Permutation-based Indexing (PBI) is a reference-based indexing scheme that accelerates NN search by combining the use of landmark points and ranking in place of distance calculation.

In this paper, we are interested in understanding the approximation made by the PBI scheme. The aim is to understand the robustness of the scheme created by modeling and studying by quantifying its invariance properties. After discussing the geometry of PBI, in relation to the study of ranking, from empirical evidence, we make proposals to cater for the inconsistencies of this structure.

Keywords: Permutation based indexing · Ranking · Geometry

1 Introduction

Providing a fast and accurate (exact or approximate) access to large-scale multidimensional data is a ubiquitous problem and dates back to the early days of large-scale Information Systems. The approach generally taken is that of defining a structure of the space based on information similarity and to partition the information space according to this structure for quantized or hierarchical access. The most common base for structuring the space is to assume the existence of a relevant metric in the space and to base the indexing on the properties of that metric space to resolve the Nearest Neighbor (NN) search problem. From there, a large variety of indexing techniques have been defined [10,29,36,37].

In this paper, we are interested in a finer understanding of the approximations made by the PBI scheme (and, more generally, permutation-based distance measurements). In particular, the aim is to understand the robustness of the scheme created by quantifying its invariance properties. The main contributions is the definition of a formal space partitioning model for the PBI scheme, embarking power tools from geometry modeling.

© Springer International Publishing AG 2016
L. Amsaleg et al. (Eds.): SISAP 2016, LNCS 9939, pp. 79–92, 2016.
DOI: 10.1007/978-3-319-46759-7_6

We demonstrate the validity of our proposal with extensive empirical evidence. In this paper, we are interested in understanding the approximation made by the PBI scheme. The aim is to understand the robustness of the scheme created, or conversely, quantify its invariance properties. After discussing the geometry of PBI, in relation to the study of ranking, from extensive empirical evidence, we make proposals to cater for the inconsistencies of this structure.

2   Related Work

A large family of indexing techniques is that of reference-based indexing schemes, where some reference points (sometimes referred to as pivots or anchor points) are selected, based on their local or global properties and then organized for facilitating query resolution and data access. In the list of such structures, we can cite tree-based indexing that place a hierarchical structure over these pivots. These include BK-Tree [6], Vantage Point Tree [23,32] or M-Tree (Metric Tree) [11].

More recent structures such as the Fixed Query Array [9], M-Index [24] or Permutation Based Indexing [8] use pivots to partition the space and to encode the data according to the structure of the partition. These structures have a number of parameters on which their actual performance depend and their choice are generally made empirically, either based on heuristics or on the statistics of the data in question [1,3,4,7]. However, a formal modeling of the relationship between these choices and the impact on the performance, based on a sound modeling of the encoding created by the indexing scheme is still missing [2,20].

They also relate to the statistical properties of high-dimensional representation spaces within which the *curse of dimensionality* applies [5,13,33,34]. Although indexing performance decreases in such a setup and hardware advances (such as GPU computations) allow brute-force exhaustive search to be fast and robust [12,17], it is still relevant to look at indexing structures acting either within subspaces or data manifolds [35].

We have studied how PBI may be distributed over parallel architectures [19], how PBI schemes may be simplified (pruned) to scale while preserving an adequate level of approximation [20]. We have worked on large-scale data processing, including with GPU processing [14,21,22]. Here, we extend an initial modeling for the geometry of PBI [18].

3   Formal Modeling of Permutation-Based Indexing Schemes

We follow and adapt notations from [8,16]. Given $\mathcal{U} = \{o_1, \ldots, o_N\}$ a collection (universe) of N D-dimensional objects $o_i \in \mathbb{R}^D$, and given a continuous distance function $d(.,.)$ operating on objects, typically any Minkowsky distance (including the Euclidean distance d_E) or other classical distance function (including the cosine similarity distance).

We choose from $\mathcal{U}$ a set of n $(0 < n \leq N)$ *reference objects* $R = \{r_1, \ldots, r_n\}$ where, for every k, $r_k = o_i$ for some i.

Definition 1 (Ordered list). *Given $o_i \in \mathcal{U}$, we define the ordered list of object o_i as the permutation $\pi_i : [\![1, n]\!] \to [\![1, n]\!]$ such that for all $k \in [\![1, n-1]\!]$:*[1]

$$\begin{cases} d(r_{\pi_i(k)}, o_i) < d(r_{\pi_i(k+1)}, o_i) \\ \text{or } d(r_{\pi_i(k)}, o_i) = d(r_{\pi_i(k+1)}, o_i) \text{ and } \pi_i(k) < \pi_i(k+1) \end{cases}$$

We note $\boldsymbol{\pi_i} = (\pi_i(1), \ldots, \pi_i(n))$.
Given $p \in \mathbb{R}^D$, we note π_p the ordered list of any point p.

In other words, $\boldsymbol{\pi_i}$ is the list of indices of the reference objects r_k sorted in increasing distance values from o_i. To remove randomness completely from the ranking, in case of a tie on distances, the reference object of lower index appears first in the list.

Viewing the ordered list as a bijective function, we can define π_i^{-1} as its inverse function, providing the position of a reference object in the ordered list.

We also extend the notation to apply the function π_i (resp π_i^{-1}) on ordered sets. In that case, for example, $\pi_i^{-1}(J) = (\pi_i^{-1}(j_1), \ldots, \pi_i^{-1}(j_l))$, where $J = (j_1, \ldots, j_l)$.

The function π_i encodes the position of object o_i with respect to the list of reference objects R and it is the purpose of this paper to study further the properties of π_i.

Based on this position encoding, we can define a new distance approximation using any distance that can be computed between rankings (ordered lists). The Spearman Footrule Distance (SFD) based on set R or the Spearman Rho (ρ) are typically used:

$$\delta_R(o_i, o_j) = \sum_{k=1}^{n} \left| \pi_i^{-1}(k) - \pi_j^{-1}(k) \right| \tag{1}$$

$$\rho_R(o_i, o_j) = \sqrt{\sum_{k=1}^{n} \left(\pi_i^{-1}(k) - \pi_j^{-1}(k) \right)^2} \tag{2}$$

It has been shown that such distance functions can be used to resolve the k nearest neighbor problem (k-NN) since δ_R and ρ_R approximate, in terms of ranking, continuous distances for the search of k-NN [8]. In other words, for example,

$$\delta_R(o_i, o_j) \overset{\text{rank}}{\simeq} d(o_i, o_j) \qquad \forall o_i, o_j \in \mathcal{U} \tag{3}$$

Hence, *Permutation-based Indexing* (PBI) aims at facilitating and optimizing, for any query q ($q \in \mathcal{U}$ or $q \notin \mathcal{U}$) the computation of rank-based distances such as $\delta_R(q, o_i)$ for all $o_i \in \mathcal{U}$.

We will base our formal analysis on δ_R but, unless otherwise stated, any other rank-based distance function (such as ρ_R) may apply instead.

---

[1] We use the compact notation $[\![1, n]\!] = \{1, \cdots, n\}$ for sets of successive integers.

3.1   Invariance

Computing distances over ordered lists creates distance approximations, which in turn create equivalence relationship.

Definition 2 (Equivalence relationship). *Given $R \subset \mathcal{U}$ and $o_i, o_j \in \mathcal{U}$, we note $o_i \equiv o_j$ if and only if*

- $\delta_R(o_i, o_j) = 0$,
- *equivalently, $\pi_i = \pi_j$ (since δ_R is a distance function).*

Definition 3 (Equivalence class - Invariance). *The* equivalence class *of object o_i is*

$$[o_i] = \{p \in \mathbb{R}^D \text{ such that } p \equiv o_i\}$$

The quotient space $\mathcal{U}/\equiv$ is the set of all equivalence classes of δ_R from $\mathcal{U}$.

The equivalence class is the set of all positions p an object can take in the initial space without changing its encoding in the permutation space. As an immediate consequence, the value of the δ_R distance between any pair of points of respective classes does not vary. Hence, the equivalence classes show the extent of the invariance of the π_i encoding. Similarly, the equivalence classes measure the approximation made by the distance function δ_R.

We now construct a geometric structure for analyzing the PBI scheme.

3.2   Geometry

Objects o_i are points of the $\mathbb{R}^D$ space over which some geometrical properties can be inferred. We use the Euclidean distance in $\mathbb{R}^D$ but this analysis may be extended with using other metrics.

A D-dimensional space may be partitioned by $(D-1)$-dimensional hyperplanes. In our context, perpendicular bisectors are particular such hyperplanes.

Definition 4 (Perpendicular bisector). *Given $r_k, r_l \in R$, we define Δ_{kl} as the $(D-1)$-dimensional perpendicular bisector[2] of the segment $[r_k, r_l]$.*

Proposition 1. *If two given objects $o_i, o_j \in \mathcal{U}$ are separated by Δ_{kl} then*

$$(\pi_i^{-1}(k) - \pi_i^{-1}(l)).(\pi_j^{-1}(k) - \pi_j^{-1}(l)) < 0$$

If Δ_{kl} is the only bisector separating o_i and o_j, then in that case, in particular, $\delta_R(o_i, o_j) = 2$.

Proof. Traversing Δ_{kl} flips the ranking of r_k and r_l in the ordered list, while leaving other values of $\pi_i^{-1}(m)$ and $\pi_j^{-1}(m)$ unchanged for all $m \neq k, l$.

---

[2] We initially restrict ourselves to $\mathbb{R}^D$ spaces. The generalisation of these notions to generic metric spaces is left for future work.

Definition 5 (Local flip). *We call the fact of traversing a bisector Δ_{kl} a local flip, ($|\pi_i^{-1}(k) - \pi_i^{-1}(l)| = 1$).*

There is therefore a direct relationship between the geometrical organization of the points and the organization of the ordered list. More generally, neighboring relationships between objects relate to Voronoi diagrams, themselves formed out of bisectors Δ_{kl}. We define the base element of $\mathcal{V}(R)$, the classical Voronoi diagram of R, as follows.

Definition 6 (Voronoi cell). *Given $r_k \in R$, we define $V_R(r_k) \subset \mathbb{R}^D$ as the Voronoi cell of r_k with respect to R. $V_R(r_k)$ is the subset:*

$$V_R(r_k) = \left\{ p \in \mathbb{R}^D \text{ such that } d(p, r_k) \leq d(p, r_l) \ \forall r_l \in R \right\}$$

$V_R(r_k)$ is a D-dimensional simplex bounded by bisectors Δ_{kl}. r_k is then said to be a generator of $V_R(r_k)$.
$\mathcal{V}(R) = \{V_R(r_k) \ \forall r_k \in R\}$ is the Voronoi diagram of R.

Remark 1. We assume that, considering the randomness of the positions of objects in $\mathcal{U}$ (and therefore in R):

- The Voronoi diagram of R is not degenerate, i.e., no more than $D+1$ reference objects lie on the same D-dimensional hypersphere;
- no object o_i lies exactly on the boundary of two or more Voronoi cells.

A number of properties of the Voronoi diagrams help us understanding the structure of PBI. We recall the definition of the Delaunay graph.

Definition 7 (Delaunay graph). *Given R and $\mathcal{V}(R)$, we define $G = (R, E)$ the Delaunay graph with vertices $r_k \in R$ and edges E such that:*

$$(r_k, r_l) \in E \text{ if and only if } V_R(r_k) \text{ and } V_R(r_l) \text{ share a common facet.}$$

Definition 6 considers a unique object as generator for each Voronoi cell. Hence, by definition, for all objects $o_i \in V_R(r_k)$, we have $\pi_i^{-1}(k) = 1$.

Consider now r_l and $r_m \in R$, neighbors of r_k in G. Δ_{kl} and Δ_{km} support facets of $V_R(r_k)$. Suppose we extend Δ_{lm} within $V_R(r_k)$. Δ_{lm} separates objects o_i for which $\pi_i^{-1}(l) > \pi_i^{-1}(m)$ from objects o_i for which $\pi_i^{-1}(l) < \pi_i^{-1}(m)$. In particular, because r_l and $r_m \in R$ are neighbors of r_k, one may isolate a portion of $V_R(r_k)$ bounded by Δ_{lm} where, for each object o_i in that region $\pi_i^{-1}((k, l, m)) = (1, 2, 3)$. Repeating that process, leads to the construction of the *ordered order-2 Voronoi diagram*, where the generators of the cells at the ordered pairs of reference objects (Fig. 1).

Generalizing this construction, we obtain the *ordered order-k Voronoi diagram* (OOkVD).

Definition 8 (Ordered order-k Voronoi diagram). *Given $R_k = (r_{j_1}, \ldots, r_{j_k})$ an ordered subset of R, we define $V_R^k(R_k) \subset \mathbb{R}^D$ as the OOkVD cell of R_k with respect to R. $V_R^k(R_k)$ is such that:*

$$o_i \in V_R^k(R_k) \quad \Leftrightarrow \quad \pi_i^{-1}((j_1, \ldots, j_k)) = [\![1, k]\!]$$

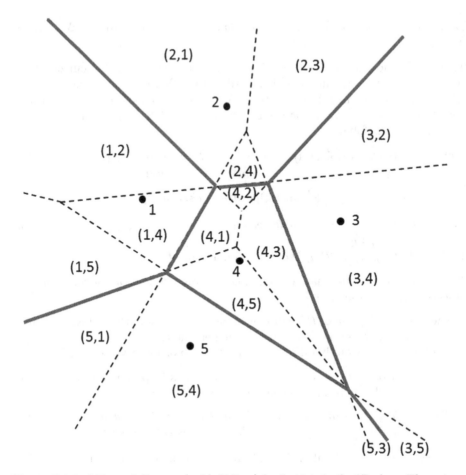

Fig. 1. Original Voronoi diagram (*red bold lines*) for 5 points in the 2D plane. The union with the order-2 Voronoi diagram (*dashed lines*) forms the ordered order-2 Voronoi partition. Cell centers act as reference points. The label for every cell is given as the permutation of the 2 closest reference points from points in the cell. Every original cell is repartitioned by the order-2 neighboring relationships (adapted from [26]) (Color figure online)

Proposition 2. *The equivalent classes of the δ_R distance $(\mathcal{U}/\equiv)$ are cells of the ordered order-$(n-1)$ Voronoi diagram of R: if $o_i \in V_R^k(R_k)$ then $[o_i] = V_R^k(R_k)$.*

Proof. By construction. Knowing that $p \in V_R^k(R_k)$ for all $k < n$ is sufficient to determine the ordered list π_p.

As noted in [2], equivalent classes are the vertices of the permutahedron of order n, the polytope whose edges are connecting all permutations differing from a local flip.

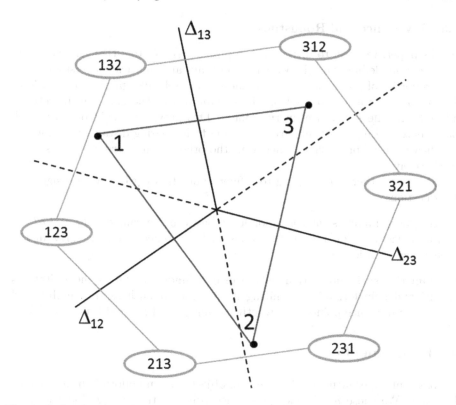

Fig. 2. Ordered order-2 Voronoi diagram of 3 points (*black lines*). Edges of the Delaunay graph (*red lines*). Edges of the order-n permutahedron mapped on the same plane (*green lines*) (Color figure online)

Proposition 3. *The edges of the order-n permutahedron form a equivalent "order-$(n-1)$ Delaunay graph" for the ordered order-$(n-1)$ Voronoi diagram. In other words, permutations differing from one local flip (connected vertices of the order-n permutahedron) relate to neighboring cells of the ordered order-$(n-1)$ Voronoi diagram.*

Proof. Direct from Definition 5 and Proposition 2.

Propositions 2 and 3 provide us with powerful geometric tools to study the performance of the δ_R distance and therefore the permutation-based encoding. For example, it is easily seen that local flips between positions k and $k+1$ in the list relate to crossing edges of the order-k Voronoi diagram (Fig. 2). Similarly, relationships between Voronoi cells, Delaunay simplices and enclosing spheres help us understanding which of the $n!$ possible permutations will actually exist in the permutation-based encoding defined by a given choice of R. Upperbounds and D-dimensional constructs that achieve these bounds are presented in [30, 31]. An empirical study on the number of Pivot Permutations prefixes is proposed in [25].

3.3 Invariance and Robustness

In this paper, we wish to investigate empirically the factors that emerged from the above modeling. Namely, we wish to obtain an empirical understanding of the properties of invariance and robustness of the scheme against perturbations. The related literature focused on the capabilities of the encoding to retrieve all and only the k-NN of a query point p. This provides insight on how much balls centered on p grow similarly according to increasing distances d_E and δ_R, which we use as prototypical metric in the original and permutations spaces, respectively.

Here, we rather aim at going to a finer understanding by giving insights on the questions:

- How much unique is the correspondence between the values of d_E and δ?
- How much position information does each reference point r_k carry in the encoding of object o_i?

We think that such information will advance the understanding of the limitations of PBI and help formally optimizing its parameters such as the number and position of reference points, and whether using partial ordered lists is useful.

4 Experiments

We base our experiments on dense sets of objects drawn uniformly from the unit $\mathbb{R}^D$ cube. We chose n reference points according to the greedy global locality approach [18].

4.1 Original Versus Permutation-Based Distances

We first investigate the match between distance values in the original space and the permutation space. Ideally, for every original distance value, we should find a corresponding permutation-based distance value. However, due to rank approximation and invariance, this is not the case. To measure this invariance, for every value of the permutation-based distance[3], we gather the corresponding histogram of the original distance values. The less peaked the histogram, the more invariance, and the more confusion in discriminating objects.

As can be seen from Fig. 3, both original and permutation-based distance functions show a decent correlation (dark diagonal corresponding to the peak value of the histograms). δ_R and ρ_R behave similarly. However, the higher the value of the permutation-based distance, the more spread the original corresponding distance values are. This can be interpreted as the fact that the ball of the permutation distance will grow more and more with irregular borders. In other words, there is more and more uncertainty in the match between original distance values and permutation-based distance values.

---

[3] We use $\delta_R(o_i, o_j) = \frac{1}{2} \sum_{k=1}^{n} \left| \pi_i^{-1}(k) - \pi_j^{-1}(k) \right|$, to avoid systematically empty odd bins.

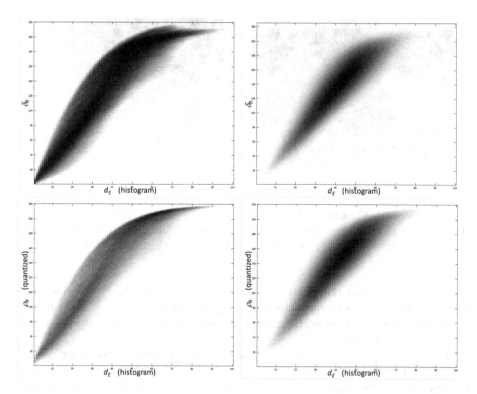

Fig. 3. Collection of histograms (horizontal lines of the images - the darker the higher the value) of Euclidean distance values for every value of the permutation-based distance (vertical value). From left to right, top to bottom: (a) Uniform distribution of 2D objects with δ_R based on 30 reference points. (b) Uniform distribution of 4D objects with δ_R based on 30 reference points. (c) Uniform distribution of 2D objects with ρ_R (values quantized) based on 30 reference points. (d) Uniform distribution of 4D objects with ρ_R (values quantized) based on 30 reference points.

4.2   Local Invariance Properties

We now wish to get a more detailed understanding of how permutation-based distance work. From their definition, these distance functions (e.g. δ_R or ρ_R) essentially count the discrepancy between the ordered list, without accounting for the position in the lists at which this difference arises. For example, if $\delta_R(o_i, o_j) = 1$, the corresponding ordered lists differs from only one local flip. However this local flip may indifferently be between elements at the beginning of the list (e.g. changing cell of the order-1 Voronoi diagramà) or at the end of the list (crossing Δ_{kl} where r_k and r_l are far from o_i and o_j).

Definition 9 (Activation). *We say that a reference object r_k is activated in the computation of $\delta_R(o_i, o_j)$ if $\pi_i^{-1}(k) \neq \pi_j^{-1}(k)$.*

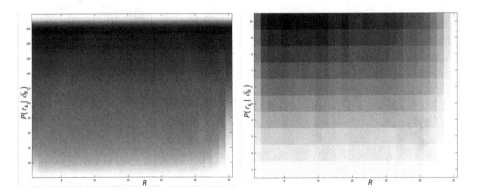

Fig. 4. Collection of histograms (horizontal lines of the images - the darker the higher the value) for the activation (see text) of each reference object depending of the value of δ_R ($D = 2$, $n = 30$). (left) full statistic. (right) zoom on low values of δ_R.

Ideally, we would like the position of an object be encoded mostly by its local reference objects. This corresponds to making the position encoding independent of far structures. As a result, this would support the use of local criteria for the choice of reference objects.

In that case, when computing permutation-based distance values for neighboring objects, local reference objects would be activated. Conversely, low values of permutation-based distance should be due to the activation of local reference objects. This would for example justify formally that ordered list pruning is a sound operation.

We plot in Fig. 4 the statistics of activation of reference objects ($n = 30$) for each value of the permutation-based distance.

We read a rather uniform distribution of activation, which counters to the idea of local encoding. This may be understood by looking at Fig. 1. One can see that bisectors resulting from the high order Voronoi partition splits cells into a fine grain partition. Hence, pairs of distance reference objects do participate in the determination of the fine sensitivity of the encoding. This is rather undesirable and motivates the use of weighted permutation-based distance functions such as that proposed in [15] to enforce a local penalty on distance measurements.

4.3   Real Data

We now study a real use case where indexing invariance is desirable and should be adapted to the data. We study Maya hieroglyphs images. A part of Maya writing consists into *glyphs* (base signs) combined into *glyph blocks*. Each glyph can be referenced by a Thompson code (T-code, e.g. T0168) and glyphblocks can be therefore described by the combination of the T-codes of the glyphs that compose the block, which we call a T-string (see Fig. 5).

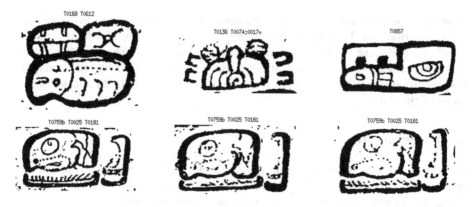

Fig. 5. Maya glyphblocks annotated with their corresponding T-strings. (1st row) Glyphblocks with different number of composing glyphs. (2nd row) Glyphblocks of the same class, illustrating the visual variability of the classes

Table 1. Average equivalence class population and precision with respect to the number of reference objects. Values in brackets indicate the standard deviation

n	6	7	8	9	10	20
Pop.	233.56 (187.76)	99.31 (99.14)	43.77 (57.46)	19.74 (35.58)	7.20 (11.87)	1.06 (1.44)
Prec.	0.05 (0.09)	0.10 (0.17)	0.19 (0.25)	0.36 (0.34)	0.54 (0.37)	99.84 (0.03)

It is interesting to study visual similarity of Maya hieroglyphs [28]. Figure 5 (2nd row) motivates the fact that the indexing scheme should absorb the visual variation of the symbols. Integrating this with our earlier discussion, the question is how to map similarity-based classification onto the notion of invariance of the indexing scheme. Here, "invariance" is understood as "invariance to writing style". In other words the challenge is to tune the parameters of the indexing scheme to align with the semantics of the data.

Here, we use a set of 15,500 annotated glyphblocks in 155 classes (same Tstring) of 100 individuals. We extract features from an autoencoder. We use the values of the L most activated neurons on the encoding layer (joining the encoder and the decoder architectures). Our initial experiments show that this encoding captures relevant visual features[4]. We extract the $L = 30$ values of the most activated neurons of the encoding layer as features and use the Euclidean distance to measure similarity. Here, we adapt the number n of reference objects using the greedy global locality approach.

The above numbers in Table 1 illustrate the reduction in size of the partition cells, leading to a reduction of the size of their population. In that particular application, standard deviation figure on the cell population show that the choice of reference objects is not adapted to the data since there is a large variation in the number of items per cell. A higher number of reference objects creates a

---

[4] The details of this study may be found in [27].

finer partition. As a result, the precision inside the equivalence class mechanically increases. However, here again, the figures show the need for an adapted choice of reference objects to align the equivalence classes (cells of the partition) with the semantic classes of the data. It is therefore a critical challenge to formulate the optimisation of the choice of reference objects according to the semantic value of the data.

5   Conclusion

Permutation-based indexing schemes have shown to be effective to support the resolution of kNN queries. Their main parameters are the number and location of reference objects and the permutation-based distance used.

In this paper, the main contribution is a formal modeling of the mechanics of PBI schemes, helped by ranking theory and computational geometry. This base model provides insights and powerful tools for the fine study of properties of permutation-based geometry. Here, we focus on invariance, which relates to robustness to data variation (e.g. due to noise). We motivate such a study by the use of PBI in applications where items may be grouped by classes with internal variation. In that case, kNN queries may be resolved directly using the space partition thus created.

Our initial experiments following our model reveal an adequate transfer of neighboring information from the original feature space onto the permutation-based representation space. However, the analysis also demonstrates that permutation-based distances such as δ_R or ρ_R do not localize the measurements, as it would be desirable. The use of adapted permutation-based distance functions (such as weighted by rank position [15]) may be beneficial here.

This paper opens many avenues for deeper studies on PBI. We plan to extend our formal model in the direction of a better understanding of the geometry of PBI and the design or choice of adapted parameters such as permutation-based distance incorporating pruning or weighting. Getting deeper insights on the geometry of the partition will also be a way to optimize the use of reference objects and therefore their location and number.

Acknowledgments. This work has been partly supported by the Swiss National Science Foundation under project MAAYA (SNF Grant number 144238).

Dr. Hisham Mohamed is now with Sensirion AG, Staefa, Switzerland.

References

1. Amato, G., Esuli, A., Falchi, F.: Pivot selection strategies for permutation-based similarity search. In: Brisaboa, N., Pedreira, O., Zezula, P. (eds.) SISAP 2013. LNCS, vol. 8199, pp. 91–102. Springer, Heidelberg (2013)
2. Amato, G., Falchi, F., Rabitti, F., Vadicamo, L.: Some theoretical and experimental observations on permutation spaces and similarity search. In: Traina, A.J.M., Traina Jr., C., Cordeiro, R.L.F. (eds.) SISAP 2014. LNCS, vol. 8821, pp. 37–49. Springer, Heidelberg (2014)

3. Amato, G., Rabitti, F., Savino, P., Zezula, P.: Region proximity in metric spaces and its use for approximate similarity search. ACM Trans. Inf. Syst. **21**(2), 192–227 (2003)
4. Ares, L.G., Brisaboa, N.R., Esteller, M.F., Pedreira, O., Places, A.S.: Optimal pivots to minimize the index size for metric access methods. In: Proceedings of the 2009 Second International Workshop on Similarity Search and Applications, SISAP 2009, pp. 74–80. IEEE Computer Society, Washington, DC (2009)
5. Beyer, K., Goldstein, J., Ramakrishnan, R., Shaft, U.: When is "nearest neighbor" meaningful? In: International Conference on Database Theory, pp. 217–235 (1999)
6. Burkhard, W.A., Keller, R.M.: Some approaches to best-match file searching. Commun. ACM **16**(4), 230–236 (1973)
7. Bustos, B., Navarro, G., Chávez, E.: Pivot selection techniques for proximity searching in metric spaces. Pattern Recogn. Lett. **24**(14), 2357–2366 (2003)
8. Chavez, E., Figueroa, K., Navarro, G.: Effective proximity retrieval by ordering permutations. IEEE Trans. Pattern Anal. Mach. Intell. **30**(9), 1647–1658 (2008)
9. Chávez, E., Marroquín, J.L., Navarro, G.: Fixed queries array: a fast and economical data structure for proximity searching. Multimed. Tools Appl. **14**(2), 113–135 (2001)
10. Chávez, E., Navarro, G., Baeza-Yates, R.A., Marroquín, J.L.: Searching in metric spaces. ACM Comput. Surv. **33**(3), 273–321 (2001)
11. Ciaccia, P., Patella, M., Zezula, P.: M-tree: an efficient access method for similarity search in metric spaces. In: Proceedings of the 23rd International Conference on Very Large Data Bases, VLDB 1997, San Francisco, CA, USA, pp. 426–435 (1997)
12. Garcia, V., Debreuve, E., Nielsen, F., Barlaud, M.: K-nearest neighbor search: fast GPU-based implementations and application to high-dimensional feature matching. In: 2010 17th IEEE International Conference on Image Processing (ICIP), pp. 3757–3760. IEEE (2010)
13. Hinneburg, A., Aggarwal, C.C., Keim, D.A.: What is the nearest neighbor in high dimensional spaces? In: Proceedings of the 26th International Conference on Very Large Data Bases, VLDB 2000, pp. 506–515. Morgan Kaufmann Publishers Inc., San Francisco (2000)
14. Kruliš, M., Osipyan, H., Marchand-Maillet, S.: Optimizing sorting and top-k selection steps in permutation based indexing on GPUs. In: Morzy, T., Valduriez, P., Bellatreche, L. (eds.) ADBIS 2015. CCIS, vol. 539, pp. 305–317. Springer, Heidelberg (2015)
15. Kumar, R., Vassilvitskii, S.: Generalized distances between rankings. In: Proceedings of the 19th International Conference on World Wide Web, WWW 2010, New York, NY, USA, pp. 571–580 (2010)
16. Lebanon, G., Lafferty, J.D.: Cranking: combining rankings using conditional probability models on permutations. In: Proceedings of the Nineteenth International Conference on Machine Learning, ICML 2002, pp. 363–370. Morgan Kaufmann Publishers Inc., San Francisco (2002)
17. Li, S., Amenta, N.: Brute-force k-nearest neighbors search on the GPU. In: Amato, G., et al. (eds.) SISAP 2015. LNCS, vol. 9371, pp. 259–270. Springer, Heidelberg (2015). doi:10.1007/978-3-319-25087-8_25
18. Mohamed, H.: Scalable approximate k-NN in multidimensional Big Data (in particular, Chap. 3). Ph.D. thesis, Viper Group, CS Department, University of Geneva, August 2014
19. Mohamed, H., Marchand-Maillet, S.: Distributed media indexing based on MPI and mapreduce. Multimed. Tools Appl. **69**(2), 513–537 (2014)

20. Mohamed, H., Marchand-Maillet, S.: Quantized ranking for permutation-based indexing. Inf. Syst. **52**, 163–175 (2015)
21. Mohamed, H., Osipyan, H., Marchand-Maillet, S.: Multi-core (CPU and GPU) for permutation-based indexing. In: Traina, A.J.M., Traina Jr., C., Cordeiro, R.L.F. (eds.) SISAP 2014. LNCS, vol. 8821, pp. 277–288. Springer, Heidelberg (2014)
22. Mohammed, H., Marchand-Maillet, S.: Scalable indexing for big data processing. In: Li, K.-C., Jiang, H., Yang, L.T., Cuzzocrea, A. (eds.) Big Data: Algorithms, Analytics, and Applications. Chapman & Hall, Boca Raton (2015)
23. Nielsen, F., Piro, P., Barlaud, M.: Bregman vantage point trees for efficient nearest neighbor queries. In: IEEE International Conference on Multimedia and Expo, 2009, ICME 2009, pp. 878–881. IEEE (2009)
24. Novak, D., Batko, M., Zezula, P.: Metric index: an efficient and scalable solution for precise and approximate similarity search. Inf. Syst. **36**(4), 721–733 (2011)
25. Novak, D., Zezula, P.: Performance study of independent anchor spaces for similarity searching. Comput. J. **57**(11), 1741–1755 (2014)
26. Okabe, A., Boots, B., Sugihara, K., Chui, S.N.: Spatial Tessellations: Concepts and Applications of Voronoi Diagrams, 2nd edn. Wiley, New York (2000)
27. Roman-Rangel, E., Marchand-Maillet, S.: Indexing Mayan hieroglyphs with neural codes. In: International Conference on Pattern Recognition (ICPR 2016), Cancun, Mexico (2016)
28. Roman-Rangel, E., Wang, C., Marchand-Maillet, S.: Simmap: similarity maps for scale invariant local shape descriptors. Neurocomputing (Part B) **175**, 888–898 (2016)
29. Samet, H.: Foundations of Multidimensional and Metric Data Structures. The Morgan Kaufmann Series in Computer Graphics and Geometric Modeling. Elsevier/Morgan Kaufmann, California (2006)
30. Skala, M.: Counting distance permutations. In: IEEE 24th International Conference on Data Engineering Workshop, 2008, ICDEW 2008, pp. 362–369, April 2008
31. Skala, M.: Aspects of metric spaces in computation. Ph.D. thesis, University of Waterloo (2008)
32. Uhlmann, J.K.: Satisfying general proximity/similarity queries with metric trees. Inf. Process. Lett. **40**(4), 175–179 (1991)
33. Volnyansky, I., Pestov, V.: Curse of dimensionality in pivot based indexes. In: Second International Workshop on Similarity Search and Applications, 2009, SISAP 2009, pp. 39–46, August 2009
34. Weber, R., Schek, H.J., Blott, S.: A quantitative analysis and performance study for similarity-search methods in high-dimensional spaces. In: Proceedings of the 24rd International Conference on Very Large Data Bases, VLDB 1998, pp. 194–205. Morgan Kaufmann Publishers Inc., San Francisco (1998)
35. Weinberger, K.Q., Saul, L.K.: Distance metric learning for large margin nearest neighbor classification. J. Mach. Learn. Res. **10**, 207–244 (2009)
36. Yianilos, P.N.: Data structures and algorithms for nearest neighbor search in general metric spaces. In: Proceedings of the Fourth Annual ACM-SIAM Symposium on Discrete Algorithms, SODA 1993, Philadelphia, PA, USA, pp. 311–321 (1993)
37. Zezula, P., Amato, G., Dohnal, V., Batko, M.: Similarity Search: The Metric Space Approach. Advances in Database Systems, vol. 32. Springer, New York (2006)

Deep Permutations: Deep Convolutional Neural Networks and Permutation-Based Indexing

Giuseppe Amato$^{(\boxtimes)}$, Fabrizio Falchi$^{(\boxtimes)}$, Claudio Gennaro$^{(\boxtimes)}$, and Lucia Vadicamo$^{(\boxtimes)}$

ISTI-CNR, via G. Moruzzi 1, 56124 Pisa, Italy
{giuseppe.amato,fabrizio.falchi,claudio.gennaro,
lucia.vadicamo}@isti.cnr.it

Abstract. The activation of the Deep Convolutional Neural Networks hidden layers can be successfully used as features, often referred as Deep Features, in generic visual similarity search tasks.

Recently scientists have shown that permutation-based methods offer very good performance in indexing and supporting approximate similarity search on large database of objects. Permutation-based approaches represent metric objects as sequences (permutations) of reference objects, chosen from a predefined set of data. However, associating objects with permutations might have a high cost due to the distance calculation between the data objects and the reference objects.

In this work, we propose a new approach to generate permutations at a very low computational cost, when objects to be indexed are Deep Features. We show that the permutations generated using the proposed method are more effective than those obtained using pivot selection criteria specifically developed for permutation-based methods.

Keywords: Similarity search · Permutation-based indexing · Deep convolutional neural network

1 Introduction

The activation of the Deep Convolutional Neural Networks (DCNNs) hidden layers has been used in the context of transfer learning and content-based image retrieval [10,23]. In fact, Deep Learning methods are "representation-learning methods with multiple levels of representation, obtained by composing simple but non-linear modules that each transform the representation at one level (starting with the raw input) into a representation at a higher, slightly more abstract level" [19]. These representations can be successfully used as features in generic recognition or visual similarity search tasks. The first layers are typically useful in recognizing low-level characteristics of images such as edges and blobs, while higher levels have demonstrated to be more suitable for semantic similarity search.

However, DCNN features are typically of high dimensionality. For instance, in the well-known AlexNet architecture [18] the output of the sixth layer (fc6)

© Springer International Publishing AG 2016
L. Amsaleg et al. (Eds.): SISAP 2016, LNCS 9939, pp. 93–106, 2016.
DOI: 10.1007/978-3-319-46759-7_7

has 4,096 dimensions, while the fifth layer (pool5) has 9,216 dimensions. This represents a major obstacle to the use of DCNN features on large scale, due to the well-known dimensionality curse [13].

An effective approach to tackle the dimensionality curse problem is the application of approximate access methods. Permutation-based approaches [4,9,11,22] are promising access methods for approximate similarity search. They represent metric objects as sequences (permutations) of reference objects, chosen from a predefined set of objects. Similarity queries are executed by searching for data objects whose permutation representations are similar to the query permutation representation. Each permutation is generated by sorting the entire set of reference objects according to their distances from the object to be represented.

The total number of reference objects, to be used for building permutations, depends on the size of the dataset to be indexed, and can amount to tens of thousands [4]. In these cases, both indexing time and searching time is affected by the cost of generating permutations for objects being inserted, or for the queries.

In this paper, we propose an approach to generate permutations for Deep Features at a very low computational cost since it does not require the distance calculation between the reference objects and the objects to be represented. Moreover, we show that the permutations generated using the proposed method are more effective than those obtained using pivot selection criteria specifically developed for permutation-based methods.

The rest of the paper is organized as follows. In Sect. 2, we briefly describe related work. Section 3 provides background for the reader. In Sect. 4, we introduce our approach to generate permutations for Deep Features. Section 5 presents some experimental results using real-life datasets. Section 6 concludes the paper.

2   Related Work

Pivot selection strategies for permutation-based methods were discussed in [2]. In the paper the Farthest-First Traversal (FFT) technique was identified as the one providing a set of reference objects such that the sorting performed with similarity computed among the permutations was the most correlated to sorting performed using the original distance. We will see that the techniques proposed here for Deep Features outperform also the FFT technique.

The permutation-based approach was used in PPP-Codes index [21] to index a collection of 20 million images processed by a deep convolutional neural network. However, no special techniques was used to generate permutations for Deep Features.

Some recent works try to treat the features in a convolutional layer as local features [5,25]. This way, a single forward pass of the entire image through the DCNN is enough to obtain the activation of its local patches, which are then encoded using *Vector of Locally Aggregated Descriptors* (VLAD). A similar approach uses *Bag of Words* (BoW) encoding instead of VLAD to take advantage of sparse representations for fast retrieval in large-scale databases. However, although authors claim that their approach is very scalable in terms of search time, they did not report any efficiency measurements and experiments have been carried out on datasets of limited size.

Liu et al. [20] proposed a framework that adapts Bag-of-Word model and inverted table to DCNN feature indexing, which is similar to the one we propose. However, for large-scale datasets, Liu et al. have to build a large-scale visual dictionary that employs the product quantization method to learn a large-scale visual dictionary from a training set of global DCNN features. In any case, using this approach the authors reported a search time that is one order higher than in our case for the same dataset.

An approach, called *LuQ* and introduced in [1], exploits the quantization of the vector components of the DCNN features that allows one to use a text retrieval engine to perform image similarity search. In LuQ, each real-valued vector component x_i of the deep feature is transformed in a natural numbers n_i given by $\lfloor Qx_i \rfloor$; where $\lfloor \rfloor$ denotes the floor function and Q is a multiplication factor >1 that works as a *quantization factor*. n_i are then used as term frequencies for the "term-components" of the text documents representing the feature vectors.

3   Background

In the following we introduce the needed notions of permutation-based similarity search approach and Deep Features.

3.1   Permutation-Based Indexing

Given a domain $\mathcal{D}$, a *distance function* $d : \mathcal{D} \times \mathcal{D} \to \mathbb{R}$, and a fixed set of reference objects $P = \{p_1 \dots p_n\} \subset \mathcal{D}$ that we call *pivots* or *reference objects*, we define a permutation-based representation Π_o (briefly *permutation*) of an object $o \in \mathcal{D}$ as the sequence of pivots identifiers sorted in ascending order by their distance from o [4,9,11,22].

Formally, the permutation-based representation $\Pi_o = (\Pi_o(1), \dots, \Pi_o(n))$ lists the pivot identifiers in an order such that $\forall j \in \{1, \dots, n-1\}$, $d(o, p_{\Pi_o(j)}) \leq d(o, p_{\Pi_o(j+1)})$, where $p_{\Pi_o(j)}$ indicates the pivot at position j in the permutation associated with object o.

If we denote as $\Pi_o^{-1}(i)$ the position of a pivot p_i, in the permutation of an object $o \in \mathcal{D}$, so that $\Pi_o(\Pi_o^{-1}(i)) = i$, we obtain the equivalent *inverted* representation of permutations Π_o^{-1}:

$$\Pi_o^{-1} = (\Pi_o^{-1}(1), \dots, \Pi_o^{-1}(n)).$$

In Π_o the value in each position of the sequence is the identifier of the pivot in that position. In the inverted representation Π_o^{-1}, each position corresponds to a pivot and the value in each position corresponds to the rank of the corresponding pivot. The inverted representation of permutations Π_o^{-1} allows us to easily define most of the distance functions between permutations.

Permutations are generally compared using Spearman rho, Kendall Tau, or Spearman Footrule distances. As an example given two permutations Π_x and Π_y, Spearman rho distance is defined as:

$$S_\rho(\Pi_x, \Pi_y) = \sqrt{\sum_{1 \leq i \leq n} (\Pi_x^{-1}(i) - \Pi_y^{-1}(i))^2}$$

Following the intuition that the most relevant information of the permutation Π_o is in the very first, i.e. nearest, pivots [4], the Spearman rho distance with location parameter $S_{\rho,l}$ is a generalization intended to compare top-l lists (i.e., truncated permutations). It was defined in [12] as:

$$S_{\rho,l}(\Pi_x, \Pi_y) = \sqrt{\sum_{1 \leq i \leq n} (\tilde{\Pi}_{x,l}^{-1}(i) - \tilde{\Pi}_{y,l}^{-1}(i))^2}$$

$S_{\rho,l}$ differs from S_ρ for the use of an inverted top-l permutation $\tilde{\Pi}_{o,l}^{-1}$, which assumes that pivots further than $p_{\Pi_o(l)}$ from o are assigned to position $l+1$. Formally, $\tilde{\Pi}_{o,l}^{-1}(i) = \Pi_o^{-1}(i)$ if $\Pi_o^{-1}(i) \leq l$ and $\tilde{\Pi}_{o,l}^{-1}(i) = l+1$ otherwise.

It is worth noting that only the first l elements of the permutation Π_o are used, in order to compare any two objects with the $S_{\rho,l}$.

3.2   Deep Features

Recently, a new class of image descriptor, built upon Deep Convolutional Neural Networks, have been used as effective alternative to descriptors built using local features such as SIFT, SURF, ORB, BRIEF, etc. DCNNs have attracted enormous interest within the Computer Vision community because of the state-of-the-art results [18] achieved in challenging image classification challenges such as ImageNet Large Scale Visual Recognition Challenge (ILSVRC). In computer vision, DCNN have been used to perform several tasks, including not only image classification, but also image retrieval [7,10] and object detection [14], to cite some. In particular, it has been proved that the multiple levels of representation, which are learned by DCNN on specific task (typically supervised) can be used to transfer learning across tasks [10,23]. The activation of neurons of a specific layers, in particular the last ones, can be used as features for describing the visual content.

In order to extract Deep Features, we used a trained model publicly available for the popular Caffe framework [17]. Many deep neural network models, in particular trained models, are available for this framework[1]. Among them, we chose the HybridNet for several reasons: first, its architecture is the very same of the famous AlexNet [18]; second, the HybridNet has been trained not only on the ImageNet subset used for ILSVRC competitions (as many others), but

---

[1] https://github.com/BVLC/caffe/wiki/Model-Zoo.

also on the Places Database [26]; last, but not least, experiments conducted on various datasets demonstrate the good transferability of the learning [6,8,26]. We decided to use the activation of the first fully connected layer, the fc6 layer, given the results reported on [7,8,10].

The activations at the fc6 layer is a vector of 4,096 of floats. Generally, the rectified linear unit (ReLU) is used to bring to zero all negative activation values. In this way, feature vectors contain only values greater or equal to zero. Feature vectors are sparse, so that in average about 75 % of elements are zero.

4   Permutation Representation for Deep Features

As introduced in Sect. 3.1 the basic idea of permutation-based indexing techniques is to represent data objects with permutations built using a set of reference object identifiers as permutants. Given an object o, its permutation-based representation Π_o is the list of reference object identifiers, sorted in ascending order with respect to the distance between o and the various reference objects.

Using the permutation-based representation, similarity between two objects is estimated computing the similarity between the two corresponding permutations, rather than using the original distance function. The rationale behind this is that, when permutations are built using this strategy, objects that are very close one to the other, have similar permutation representations as well. In other words, if two objects are very close one to the other, they will sort the set of reference objects in a very similar way.

Notice however that, the relevant aspect, when building permutations, is the capability of generating sequences of identifiers (permutations) in such a way that similar objects have similar permutations as well. Sorting a set of reference objects, according to their distance with the object to be represented is just one, yet effective, approach.

Here, we propose an approach to generate sequence of identifiers, not necessarily associated with reference objects, when objects to be indexed are Deep Features. The basic idea is as follows. Permutants are the indexes of elements of the deep feature vectors. Given a deep feature vector, the corresponding permutation is obtained by sorting the indexes of the elements of the vector, in descending order with respect to the values of the corresponding elements. Suppose for instance the feature vector is $fv = [0.1, 0.3, 0.4, 0, 0.2]^2$. The permutation-based representation of fv is $\Pi_{fv} = (3, 2, 5, 1, 4)$, that is permutant (index) 3 is in position 1, permutant 2 is in position 2, permutant 5 is in position 3, etc. The inverted representation, introduced in Sect. 3.1 is $\Pi_{fv}^{-1} = (4, 2, 1, 5, 3)$, that is permutant (index) 1 is in position 4, permutant 2 is in position 2, permutant 3 is in position 1, etc.

The intuition behind this is that features in the high levels of the neural network carry-out some sort of high-level visual information. We can imagine that individual dimensions of the deep feature vectors represent some sort of

---

^{2} In reality, the number of dimensions is 4,096 or more.

visual concept, and that the value of each dimension specifies the importance of that visual concept in the image. Similar deep feature vectors sort the visual concepts (the dimensions) in the same way, according to the activation values.

More formally, let $fv = [v_1, \ldots, v_n]$ be a deep feature vector (where $n = 4,096$, in our case). The corresponding permutation is $\Pi_{fv} = (\Pi_{fv}(1), \ldots, \Pi_{fv}(n))$ such that $\forall i \in \{1, \ldots, n-1\}, fv[\Pi_{fv}(i)] \geq fv[\Pi_{fv}(i+1)]$. Using the inverted representation, introduced in Sect. 3.1, we have that $\Pi_{fv}^{-1} = (\Pi_{fv}^{-1}(1), \ldots, \Pi_{fv}^{-1}(n))$ such that if $\Pi_{fv}^{-1}(i) \leq \Pi_{fv}^{-1}(j)$ then $fv[i] \geq fv[j]$, that is if index i of the vector appears before index j in the permutation, then the value of element i of the vector is greater than that of element j.

Let us discuss more in details the process of creating permutations with the activations of the deep neural network. Note that when two elements of a deep feature vectors have the same value, their position in the permutation cannot be uniquely assigned. This is very rare, with elements having a value different than zero, since we are using real values. However, as we said in Sect. 3.2, on average, 75 % of the dimension have value equal to zero. This means that order of these elements is not unique.

In order to face this problem, we define and compare two different strategies:

– The first strategy, which we call *zeros-to-l*, assigns all elements having value equal to zero to position $l + 1$, where l is the location parameters.
– The second strategy, which we call *no-ReLU* does not use the ReLU (See Sect. 3.2) so that negative values are not flattened to 0 and vector components with the same activation values occur very rarely.

If we restrict to the case of Spearman rho distance and considering deep feature vectors L2-normalized to the unit length, it easy to see that our strategy of generating permutations is equivalent to the following permutation generation strategy.

– Create a set of 4,096 reference vectors (pivots) such that the i-th reference object has 1 in dimension i and 0 in all other elements of the vectors.
– Given an object o sort all reference objects in ascending order to their distance from o, as described in Sect. 3.1.

A question that might arise after this description is what is the benefit of this approach given that vector of permutations are of the same dimension of DCNN vectors. The advantage of the proposed approach is that permutation vectors can be easily encoded into an inverted index, which exhibits high efficiency as shown in [3].

5   Experiments

The similarity search paradigm employs a similarity (or a distance) function to retrieve objects similar to a query. The similarity function and the object representations are chosen so that they reflect the user (or the application) requirements for the task being executed. However, generally, the similarity function

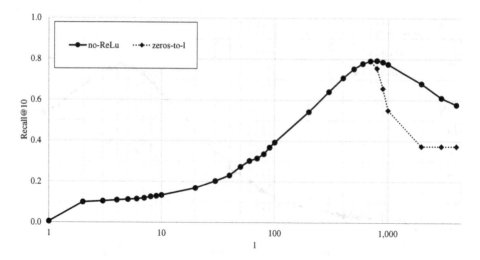

Fig. 1. Comparison between the *no-ReLU* and the *zeros-to-l* techniques, varying the location parameter l (length of the truncated permutations).

does not capture precisely the semantic of the indexed objects, and some errors occur in the similarity search results.

In our case, we are testing an approximate similarity search algorithm. That is, an algorithm that returns a result that is approximate with respect to the exact similarity search result, which in turns tries to satisfy the user retrieval requirements. The assumption is that, although the approximate similarity search result is an approximation of the exact similarity search results, the user does not notice the possible degradation of accuracy, given that also the exact similarity search algorithm is already an approximation of his/her intuition of similarity.

In this respect, we performed two type of experiments. We first evaluated the performance of the proposed technique in a pure similarity search task, where we use an exact similarity search ground-truth to assess the quality of the approximate similarity search, obtained with the permutation-based approach. Then, we evaluated the performance in a multimedia information retrieval task. Here, the ground-truth was manually generated associating each query with a set of results pertinent to the query. In this way, we were able to evaluate both the approximation introduced with respect to the exact similarity search algorithm, and the impact of this approximation with respect to the user perception of the retrieval task.

5.1   Experimental Settings

For assessing the proposed technique in a pure similarity search task, we used the Deep Features extracted as discussed in Sect. 3.2 from the *Yahoo Flickr Creative Commons 100 Million dataset (YFCC100M)* [24].

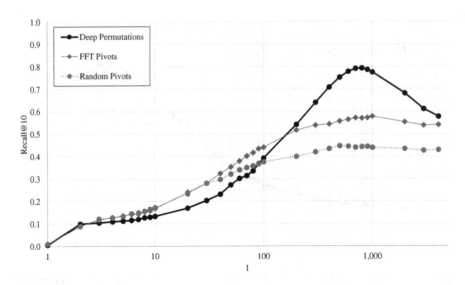

Fig. 2. Comparisons of the proposed Deep Permutation approach, with standard permutation-based methods using random selection of pivots, and Farthest-First Traversal (FFT) pivot selection strategy.

The assessment of the proposed algorithm in a multimedia information retrieval task was performed using the Deep Features extracted from the *INRIA Holidays dataset* [16].

The YFCC100M dataset [24] contains almost 100M of the images, all uploaded to Flickr between 2004 and 2014 and published under a CC commercial or noncommercial license. The ground-truth was built selecting 1,000 different queries and executing an exact similarity search on these queries using the euclidean distance to compare Deep Features.

INRIA Holidays [16] is a collection of 1,491 images, which mainly contains personal holidays photos. The images are of high resolution and represent a large variety of scene type (natural, man-made, water, fire effects, etc.). The authors selected 500 queries and manually identified a list of qualified results for each of them. As in [15], we merged the Holidays dataset with the distraction dataset MIRFlickr including 1M images[3].

5.2   Evaluation in a Similarity Search Task

In order to assess quality of search results of our approach, we use the measure called *recall@k*, which determines the ratio of correct results for a given query in the top-k results returned. Let $ER_Q(k)$ and $AR_Q(k)$ be the top-k sets of results returned by exact and approximate similarity search, respectively. The recall@k is the ratio between the number of correct results in the approximate result set

---

[3] http://press.liacs.nl/mirflickr/.

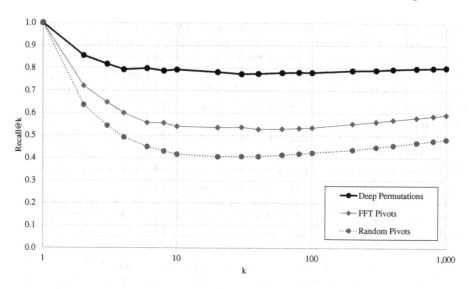

Fig. 3. *Recall@k* varying k for our approach with $l = 800$ and 4,096 random and FFT pivots.

and the number of correct results that should have been retrieved:

$$recall@k = \frac{|AR_Q(k) \cap ER_Q(k)|}{|ER_Q(k)|}.$$

Where $|\cdot|$ denote the size of a set.

We first discuss the comparison of the two approaches that we defined for handling elements of the vectors having zero as value: *no-ReLU* and the *zeros-to-l*. These tests were executed on a subset of the YFCC100M dataset of size 1M and the results are shown in Fig. 1. In experiments, we vary the location parameter l, that is the length of the truncated top-l permutation, and we compute the recall@10.

The figure shows that the plots corresponding to the two strategies are overlapped until $l = 700$. Them, the *zeros-to-l* degrades with respect to the other. At $l = 900$ also the *no-ReLU* starts degrading, remaining always higher than the other. This behavior is due to the presence of elements with value equal to zero. As we said in Sect. 4, it is not possible to distinguish and to sort the elements having value equal to zero and, on average, about the 75 % of elements of fc6 vectors are zeros. This means that when l approaches to 1,000, there are no more elements with non-zero values, which up to now were correctly sorted, and we encounter elements having value equal to zero. These elements are all assigned to position $l+1$ in the *zeros-to-l* approach, and are replaced by the negative activation values, seen before applying the ReLU, in the *no-ReLU* approach. The graphs show that using negative value for sorting these elements helps, until a certain degree. For larger values of l the quality degrades. It is worth mentioning that when the neural network was trained, the ReLU was used. Therefore, negative values were never seen at the output and they were always flattened to zero. Therefore, the negative values were not subject of fine tuning during the learning phase, and were always all treated as zeros. This is the reason why we see a degradation also using negative values. It would be interesting comparing with a network trained without using ReLU. However, this was out of the scope of this paper, where we wanted to use a standard DCNN, and we leave it to future investigation.

Figure 2 compares the proposed approach, with the *no-ReLU* strategy using a standard permutation-based approach, where pivots where both selected randomly and using Farthest-First Traversal (FFT), which in [2] was shown to be the best pivot selection method for permutation-based searching. Random selection and FFT offer better performance for values of l up to 200. Then the Deep Permutation approach is much better, reaching a recall of 80%. The FFT approach is always lower than 60% and the random approach reaches just 45% recall.

Figure 3 compares our approach against random selection and FFT, computing the *recall@k* for k ranging from 1 to 1,000. Also in this case, we can see that the new proposed approach outperforms the others and remains practically stable for all ks.

Tests discussed above were executed on a subset of size 1M of the entire YFCC100M dataset. Figure 4 shows the performance of the Deep Permutations approach increasing the size of the indexed dataset up to 100M. Here, we compute recall for k equal to 10, 100, and 1,000. Also in this case we do not see significant differences for different values of k, and the recall remains also stable around 80% for the various tested sizes of the dataset.

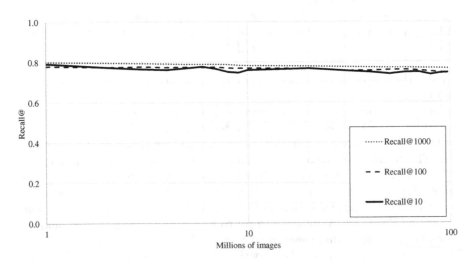

Fig. 4. *Recall@k* for various k varying dataset size (expressed in millions) obtained by the proposed approach for $l = 800$.

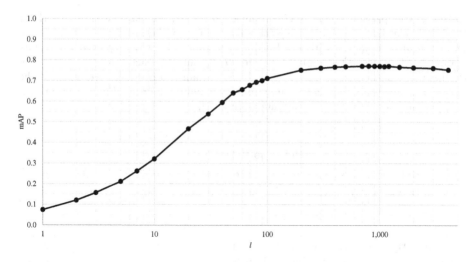

Fig. 5. mAP obtained on INRIA Holidays varying l

5.3   Evaluation in a Multimedia Information Retrieval Task

In this set of experiments, we only test the *no-ReLU* approach. Figure 5 shows the graph of the mean average precision (mAP) varying the location parameter l, that is the length of the truncated permutation. We can see that the mAP improves rapidly until l is 100, then remains stable slightly below 0.8. The maximum is reached when $l = 800$, where the mAP is 0.77.

These values of mAP are rather surprising and competing with state of the art methods tested on the INRIA Holidays dataset. To further investigate this we have compared the obtained results with the direct use of the Deep Features, using the Euclidean distance ($L2$) as distance, and with the LuQ method [1]. The comparison was performed on the INRIA Holidays dataset alone and together with the MIRFlickr dataset.

Results are reported in Table 1. The direct use of the Deep Features on the INRIA Holidays dataset, using the $L2$ distance, exhibits a mAP of 0.75 with *ReLU*, 0.76 without *ReLU*. On the INRIA Holidays dataset with the MIRFlickr distraction set, it exhibits a mAP of 0.69 with *ReLU*, 0.62 without *ReLU*.

Our approach on the INRIA Holidays dataset, exhibits a mAP of 0.75 with full permutations, 0.77 with $l = 800$. On the INRIA Holidays dataset with MIRFlickr distraction set, we obtain a mAP of 0.60 with full permutations, 0.62 with $l = 800$. The results obtained using $l = 800$ are always greater or equal to the one obtained directly using the Deep Features, and equal to the results obtained by LuQ.

Looking at these results we can make an additional observation. Deep Features are generally compared using the $L2$ distance. However, results above suggest that possibly this is not the best distance function to be used. In fact, transforming Deep Features into permutations and comparing them using the

Table 1. Comparison of the *mAP* obtained on INRIA Holidays (with and without the MIRFlickr distraction set) using the following approaches: a) the direct use of the Deep Features compared with the Euclidean distance (L2). We reported the results obtained before and after applying the *ReLu*; b) our approach based on the use of the Deep Permutations compared with the Spearman rho distance. We reported the results obtained using the full length permutation and the truncated permutation with location parameter $l = 800$; c) the LuQ method [1].

	L2		Deep permutations		LuQ [1]
	ReLu	no-ReLu	full	$l = 800$	
Holidays	0.75	0.76	0.75	0.77	0.77
Holidays+MIRFlickr	0.60	0.62	0.60	0.62	0.62

Spearman rho distance has slightly better performance. Thus, investigations on better distance functions to be used with Deep Features is worth being considered.

6   Conclusion

In this paper, we presented an approach for representing and fast indexing Deep Convolutional Neural Network Features as permutations. Compared to the classical approach based on permutation, this technique does not need computing distances between pivots and data objects but uses the same activation values of the neural network as a source for associating Deep Feature vectors with permutations.

The proposed technique when evaluated in a pure similarity search task offers a recall up to 80 %, much higher than other permutation-based methods. We also evaluated this technique in a multimedia information retrieval context. Here, surprisingly, the proposed technique offers a mean average precision of 0.77, slightly higher than the direct use of the Deep Features with the L2 distance.

This suggests that probably the L2 is not the most effective distance function to be used with Deep Features, given that permutation representation together with the Spearman rho distance provide better performance.

Note also that, probably, the same approach can be applied to any feature represented as a vector, not just DCNN features, provided its dimensionality is high. We are going to investigate how this idea generalizes to other features and to other distance functions as future work.

Acknowledgments. This work was partially founded by: EAGLE, Europeana network of Ancient Greek and Latin Epigraphy, co-founded by the European Commission, CIP-ICT-PSP.2012.2.1 - Europeana and creativity, Grant Agreement no. 325122; and Smart News, Social sensing for breakingnews, co-founded by the Tuscany region under the FAR-FAS 2014 program, CUP CIPE D58C15000270008.

References

1. Amato, G., Debole, F., Falchi, F., Gennaro, C., Rabitti, F.: Large scale indexing and searching deep convolutional neural network features. In: Madria, S., Hara, T. (eds.) DaWaK 2016. LNCS, vol. 9829, pp. 213–224. Springer, Heidelberg (2016). doi:10.1007/978-3-319-43946-4_14

2. Amato, G., Esuli, A., Falchi, F.: Pivot selection strategies for permutation-based similarity search. In: Brisaboa, N., Pedreira, O., Zezula, P. (eds.) SISAP 2013. LNCS, vol. 8199, pp. 91–102. Springer, Heidelberg (2013). doi:10.1007/978-3-642-41062-8_10

3. Amato, G., Gennaro, C., Savino, P.: MI-File: using inverted files for scalable approximatesimilarity search. Multimedia Tools Appl. 1–30 (2012)

4. Amato, G., Gennaro, C., Savino, P.: MI-File: using inverted files for scalable approximate similarity search. Multimedia Tools Appl. **71**(3), 1333–1362 (2014). doi:10.1007/s11042-012-1271-1

5. Arandjelović, R., Gronat, P., Torii, A., Pajdla, T., Sivic, J.: NetVLAD: CNN architecture for weakly supervised place recognition. arXiv preprint arXiv:1511.07247 (2015)

6. Azizpour, H., Razavian, A., Sullivan, J., Maki, A., Carlsson, S.: From generic to specific deep representations for visual recognition. In: Proceedings of the IEEE Conference on Computer Vision and Pattern Recognition Workshops, pp. 36–45 (2015)

7. Babenko, A., Slesarev, A., Chigorin, A., Lempitsky, V.: Neural codes for image retrieval. In: Fleet, D., Pajdla, T., Schiele, B., Tuytelaars, T. (eds.) ECCV 2014, Part I. LNCS, vol. 8689, pp. 584–599. Springer, Heidelberg (2014). doi:10.1007/978-3-319-10590-1_38

8. Chandrasekhar, V., Lin, J., Morère, O., Goh, H., Veillard, A.: A practical guide to CNNs and fisher vectors for image instance retrieval. arXiv preprint arXiv:1508.02496 (2015)

9. Chávez, E., Figueroa, K., Navarro, G.: Effective proximity retrieval by ordering permutations. IEEE Trans. Pattern Anal. Mach. Intell. **30**(9), 1647–1658 (2008)

10. Donahue, J., Jia, Y., Vinyals, O., Hoffman, J., Zhang, N., Tzeng, E., Darrell, T.: Decaf: a deep convolutional activation feature for generic visual recognition. arXiv preprint arXiv:1310.1531 (2013)

11. Esuli, A.: Use of permutation prefixes for efficient and scalable approximate similarity search. Inf. Process. Manag. **48**(5), 889–902 (2012)

12. Fagin, R., Kumar, R., Sivakumar, D.: Comparing top k lists. In: Proceedings of the Fourteenth Annual ACM-SIAM Symposium on Discrete Algorithms, SODA 2003, pp. 28–36. Society for Industrial and Applied Mathematics (2003)

13. Ge, Z., McCool, C., Sanderson, C., Corke, P.: Modelling local deep convolutional neural network features to improve fine-grained image classification. In: 2015 IEEE International Conference on Image Processing (ICIP), pp. 4112–4116. IEEE (2015)

14. Girshick, R., Donahue, J., Darrell, T., Malik, J.: Rich feature hierarchies for accurate object detection and semantic segmentation. In: Proceedings of the IEEE Conference on Computer Vision and Pattern Recognition, pp. 580–587 (2014)

15. Jégou, H., Douze, M., Schmid, C.: Packing bag-of-features. In: IEEE 12th International Conference on Computer Vision, 29 November 2009–2 October 2009, pp. 2357–2364 (2009)

16. Jegou, H., Douze, M., Schmid, C.: Hamming embedding and weak geometric consistency for large scale image search. In: Forsyth, D., Torr, P., Zisserman, A. (eds.) ECCV 2008, Part I. LNCS, vol. 5302, pp. 304–317. Springer, Heidelberg (2008). doi:10.1007/978-3-540-88682-2_24

17. Jia, Y., Shelhamer, E., Donahue, J., Karayev, S., Long, J., Girshick, R., Guadarrama, S., Darrell, T.: Caffe: convolutional architecture for fast feature embedding. arXiv preprint arXiv:1408.5093 (2014)

18. Krizhevsky, A., Sutskever, I., Hinton, G.E.: Imagenet classification with deep convolutional neural networks. In: Advances in Neural Information Processing Systems, pp. 1097–1105 (2012)

19. LeCun, Y., Bengio, Y., Hinton, G.: Deep learning. Nature **521**(7553), 436–444 (2015)

20. Liu, R., Zhao, Y., Wei, S., Zhu, Z., Liao, L., Qiu, S.: Indexing of CNN features for large scale image search. arXiv preprint arXiv:1508.00217 (2015)

21. Novak, D., Batko, M., Zezula, P.: Large-scale image retrieval using neural net descriptors. In: Proceedings of the 38th International ACM SIGIR Conference on Research and Development in Information Retrieval, pp. 1039–1040. ACM (2015)

22. Novak, D., Kyselak, M., Zezula, P.: On locality-sensitive indexing in generic metric spaces. In: Proceedings of the Third International Conference on Similarity Search and Applications, SISAP 2010, pp. 59–66. ACM (2010)

23. Razavian, A.S., Azizpour, H., Sullivan, J., Carlsson, S.: CNN features off-the-shelf: an astounding baseline for recognition. In: 2014 IEEE Conference on Computer Vision and Pattern Recognition Workshops (CVPRW), pp. 512–519. IEEE (2014)

24. Thomee, B., Elizalde, B., Shamma, D.A., Ni, K., Friedland, G., Poland, D., Borth, D., Li, L.J.: YFCC100M: the new data in multimedia research. Commun. ACM **59**(2), 64–73 (2016)

25. Yue-Hei Ng, J., Yang, F., Davis, L.S.: Exploiting local features from deep networks for image retrieval. In: Proceedings of the IEEE Conference on Computer Vision and Pattern Recognition Workshops, pp. 53–61 (2015)

26. Zhou, B., Lapedriza, A., Xiao, J., Torralba, A., Oliva, A.: Learning deep features for scene recognition using places database. In: Advances in Neural Information Processing Systems, pp. 487–495 (2014)

Multimedia

Patch Matching with Polynomial Exponential Families and Projective Divergences

Frank Nielsen[1,2(✉)] and Richard Nock[3,4]

[1] École Polytechnique, Palaiseau, France
Frank.Nielsen@acm.org
[2] Sony Computer Science Laboratories Inc., Tokyo, Japan
[3] Data61, Sydney ATP, Sydney, Australia
[4] The Australian National University, Sydney, Australia

Abstract. Given a query patch image, patch matching consists in finding similar patches in a target image. In pattern recognition, patch matching is a fundamental task that is time consuming, specially when zoom factors and symmetries are handled. The matching results heavily depend on the underlying notion of distances, or similarities, between patches. We present a method that consists in modeling patches by flexible statistical parametric distributions called polynomial exponential families (PEFs). PEFs model universally arbitrary smooth distributions, and yield a compact patch representation of complexity independent of the patch sizes. Since PEFs have computationally intractable normalization terms, we estimate PEFs with score matching, and consider a projective distance: the symmetrized γ-divergence. We demonstrate experimentally the performance of our patch matching system.

1 Introduction and Prior Work

Given a query patch image I_s of dimension (w_s, h_s), patch matching asks to find "similar" patches in a target image I_t of dimension (w_t, h_t). Patch matching find countless applications [1–3] in image processing. A basic dissimilarity measure of patches I_s and sub-image patch $I_t(x_0)$ of dimension (w_s, h_s) anchored at location x_0 is the Sum of Square Differences (SSDs) of the pixel intensities: $D(I_s, I_t(x_0)) = \sum_{x \in [1,w_s] \times [1,h_s]} (I_s(x) - I_t(x + x_0))^2$, that can be interpreted as the squared Euclidean distance on the vectorized patch intensities. Thus finding similar patches amount to find *close(st) neighbors* in $\mathbb{R}^{w_s \times h_s}$. This basic SSD distance may be further extended to color or multi-channel images (like hyperspectral images) either by taking the average or the maximum of the SSDs for all channels. A naïve brute-force baseline patch matching algorithm computes a matching score for each potential pixel position $x_0 \in [1, w_t] \times [1, h_t]$ at the target image in time $O(w_t h_t w_s h_s)$. This is prohibitively too expensive in practice. When dealing with pure translations, the Fourier phase correlation method [4,5] can be used to speed up the alignment of images within subpixel precision in $O(wh \log(wh))$ time using the Fast Fourier Transform (FFT) with $w = \max(w_s, w_t)$ and $h = \max(h_s, h_t)$. To factorize and speed up the distance

© Springer International Publishing AG 2016
L. Amsaleg et al. (Eds.): SISAP 2016, LNCS 9939, pp. 109–116, 2016.
DOI: 10.1007/978-3-319-46759-7_8

calculation between the source patch and the target patch when scanning the target image, a general Patch Matching (PM) method [1,2] has been designed that computes a Nearest Neighbor Field (NNF). However, those methods are too time consuming when dealing with large patches, and are not robust to smooth patch deformations, symmetry detections (like reflections [6,7]) of patches, and zooming factors (requiring guessing and rescaling the source patch accordingly).

We propose a fast statistical framework to match patches: Our algorithm models (potentially large) patches by statistical parametric distributions estimated from patch pixels, and define the distance between patches by a corresponding statistical distance between those compact patch representations. In order to handle a flexible faithful modeling of any arbitrary smooth probability density, we consider Polynomial Exponential Families [8,9] (PEFs) that have intractable normalizing constants. Since we cannot computationally normalize those PEF distributions, we bypass this problem by considering statistical projective distances [10] that ensures that $D(\lambda q, \lambda' q') = D(q, q')$ for any $\lambda, \lambda' > 0$, where q and q' are the unnormalized distributions of patches.

The paper is organized as follows: Sect. 2 presents our patch statistical representation, the chosen distance function, and the fast batched patch parameter estimation procedure using integral images. Section 3 reports on our experiments. Section 4 concludes this work by hinting at further perspectives.

2   Patch Matching with Polynomial Exponential Families

2.1   Polynomial Exponential Families: Definition and Estimation

We consider the *univariate* parametric probability distributions with the following probability density:

$$p(x; \theta) = \exp(\langle \theta, t(x) \rangle - F(\theta)),$$

where $t(x) = (x^1, \ldots, x^D)$ denotes the sufficient statistics [10,11], and θ the natural parameters. $\langle x, y \rangle = x^\top y \in \mathbb{R}$ denotes the Euclidean inner product. Let $\Theta \subseteq \mathbb{R}^D$ be the natural parameter space so that $\int \exp(\langle \theta, t(x) \rangle) \mathrm{d}x < \infty$.

The function $F(\theta) = \log \int \exp(\langle \theta, t(x) \rangle) \mathrm{d}x$ is called the log-normalizer or partition function in statistical physics [10,11]. The order of this exponential family is the dimension of the natural parameter space, D. Since $t_i(x) = x^i$ is the monomial of degree i for $i \in [D] = \{1, \ldots, D\}$, this family of distribution is called a *polynomial exponential family* (PEF). PEFs are *universal* density estimators that allow one to model arbitrarily finely any smooth multimodal density [8]. This can be easily seen by considering the log-density that is a polynomial function, and polynomial functions are well-known to model any smooth function.

For an exponential family, the maximum likelihood estimator $\hat{\theta}$ from a set of independently and identically distributed (iid) scalar observations $x_1, \ldots, x_n$ (the patch pixel intensities with $n = w_s h_s$) satisfies the following identity equation: $\nabla F(\hat{\theta}) = \frac{1}{n} \sum_{i=1}^{n} t(x_i)$.

Since $F(\theta)$ is *not* known in closed form for PEFs, one cannot compute its gradient $\nabla F(\theta)$, and we need a different method to estimate $\hat{\theta}$. Let $q(x; \theta) = \exp(\langle \theta, t(x) \rangle)$ be the unnormalized model, a positive probability measure (and $p(x; \theta) = \frac{q(x;\theta)}{e^{F(\theta)}} \propto q(x; \theta)$). We use the *Score Matching* Estimator [12,13] (SME[1]):

$$\hat{\theta} = - \left(\sum_{i=1}^{n} D(x_i)^{\top} D(x_i) \right)^{-1} \left(\sum_{i=1}^{n} \Delta t(x_i) \right), \qquad (1)$$

where $D(x) = \nabla t(x) = (t_1'(x), \ldots, t_D'(x))$ is the vector of derivatives of $t(x)$ (term by term), and $\Delta t(x)$ is the vector of the Laplacian operators of the $t_i(x)$'s, (computed from the second derivatives, term by term). We have $t_i'(x) = ix^{i-1}$ when $i \geq 1$ (and 0 otherwise) and $\Delta_i t(x) = t_i''(x) = i(i-1)x^{i-2}$ when $i \geq 2$ (and 0 otherwise). Notice that SME is not efficient when $p(x; \theta)$ is not Gaussian [10] (but MLE is efficient). See also [8] for alternative estimation recursion moment method of PEFs. Here, we consider $\mathcal{X} = \mathbb{R}^+$ the support of all PEFs (although intensity values are clamped to 255 for fully saturated pixels).

Thus a patch of size (w_s, h_s) is represented *compactly* by a natural parameter of a PEF of order $D \ll w_s \times h_s$, and is *independent* of the patch resolutions.

2.2   Statistical Projective Divergences

In order to measure the (dis)similarity between two patches described by their natural parameters θ_s and θ_t, we need a proper *statistical distance* [14–16] like the relative entropy also called the Kullback-Leibler (KL) divergence: $KL(p(x; \theta_s), p(x; \theta_t)) = \int_{x \in \mathbb{R}^+} p(x; \theta_s) \log \frac{p(x;\theta_s)}{p(x;\theta_t)} dx$. It is well-known that the KL divergence amounts to a Bregman divergence on swapped natural parameters when the distributions come from the same exponential family [17]. However, we point out that we do not have the log-normalizer $F(\theta)$ in closed form for PEFs. Hence, we consider a *projective divergence* that ensures that $D(\lambda q, \lambda' q') = D(q, q')$ for any $\lambda, \lambda' > 0$. For PEFs, we need to consider a statistical projective distance, and we choose the γ-divergence [18–20] (for $\gamma > 0$) between two distributions p and q:

$$D_\gamma(p, q) = \frac{1}{\gamma(1 + \gamma)} \log I_\gamma(p, p) - \frac{1}{\gamma} \log I_\gamma(p, q) + \frac{1}{1 + \gamma} \log I_\gamma(q, q), \qquad (2)$$

where

$$I_\gamma(p, q) = \int_{x \in \mathcal{X}} p(x) q(x)^\gamma dx. \qquad (3)$$

When $\gamma \to 0$, $D_\gamma(p, q) \to KL(p, q)$. For our patch matching application, we consider the *symmetrized γ-divergence*: $S_\gamma(p, q) = \frac{1}{2}(D_\gamma(p, q) + D_\gamma(q, p))$. Although $D_\gamma(p, q)$ can be applied to unnormalized densities, its value depend on the log-normalizer F. Indeed, the term $I_\gamma(p : q)$ admits a closed-form solution provided that $\gamma \theta_p + \theta_q \in \Theta$:

$$I_\gamma(\theta_p, \theta_q) = \exp\left(F(\theta_p + \gamma \theta_q) - F(\theta_p) - \gamma F(\theta_q) \right). \qquad (4)$$

---

[1] http://user2015.math.aau.dk/presentations/invited_steffen_lauritzen.pdf.

Proof. We have $I_\gamma(\theta_p, \theta_q) = \int \exp(\langle t(x), \theta_p + \gamma\theta_q \rangle - F(\theta_p) - \gamma F(\theta_q))\mathrm{d}x$. Expanding the right-hand side, we get $\exp(F(\theta_p + \gamma\theta_q) - F(\theta_p) - \gamma F(\theta_q)) \int \exp(\langle t(x), \theta_p + \gamma\theta_q \rangle - F(\theta_p + \gamma\theta_q))\mathrm{d}x$. By definition, when $\gamma\theta_p + \theta_q \in \Theta$, we have $\int \exp(\langle t(x), \theta_p + \gamma\theta_q \rangle - F(\theta_p + \gamma\theta_q))\mathrm{d}x = 1$, and the result follows. The $\gamma\theta_p + \theta_q \in \Theta$ condition is always satisfied when the natural parameter space [14–16] is a *cone* (since $\gamma > 0$), like the multivariate Gaussians distributions, the multinomial distributions, and the Wishart distributions, just to name a few.

One can check that by taking a Taylor expansion on the γ-divergence expressed using the closed-form expression of Eq. 4 for exponential families with conic natural space, we obtain the Bregman divergence [17] when $\gamma \to 0$.

To fix ideas, we shall consider $\gamma = 0.1$ in the remainder. Since the support is univariate, we may approximate the γ-divergence by discretizing the integral of Eq. 3 with a *Riemann sum* (discretization). Another method that works also for arbitrary multivariate distributions, is to use *stochastic integration* by sampling $x_1, \ldots, x_m$ following distribution p (importance sampling). Then we have:

$$I_\gamma(p, q) = \int_{x \in \mathcal{X}} p(x)q(x)^\gamma \mathrm{d}x \simeq \frac{1}{m} \sum_{i=1}^{m} q(x_i)^\gamma.$$

In practice, we set $m = 100000$ for importance sampling.

Notice that some statistical divergences are only *one-sided* projective divergence. For example, Hyvärinen divergence [21]:

$$D_H(p, q) = \frac{1}{2} \int p(x) \|\nabla_x \log p(x) - \nabla_x \log q(x)\|^2 \mathrm{d}x.$$

2.3 Fast PEF Estimations Using Summed Area Tables

Recall that to compute the score matching estimator of Eq. 1 we need to compute both sums for *all* $w_t \times h_t$ locii of the patches. In order to estimate the PEF parameters in constant time (for a prescribed order D) instead of time proportional to the source patch size, we use *Summed Area Tables* [22] (SATs also called integral images, interestingly also used to detect mirror symmetry of patches in [6]). For every pixel of position (x, y), the value of the summed area table $F(x, y)$ is $\sum_{x' \leq x, y' \leq y} f(x', y')$, where f is the function that we want to sum up. In our setting, we need two SATs (cumulative sum arrays) per channel for $D^\top D$ and Δt evaluated for the intensity (or red, green, blue values when considering color images). The value of the table at position (x, y) can be computed using previously computed values when filling the SAT,

$$F(x, y) = f(x, y) + F(x - 1, y) + F(x, y - 1) - F(x - 1, y - 1)$$

(when $x - 1$ or $y - 1$ is zero, the value is just zero). Once the table is constructed, for the score matching method, when we want to compute the sums in the equation for a patch (a rectangle with bottom-left corner of position(x_0, y_0) and top-right corner of position(x_1, y_1)), and we compute it in $O(1)$ as: $\sum_{x_0 \leq x \leq x_1, y_0 \leq y \leq y_1} f(x, y) = F(x_1, y_1) + F(x_0, y_0) - F(x_0, y_1) - F(x_1, y_0)$.

aligned pixel-based (SSD)   PEF ($D = 4$) with S_γ

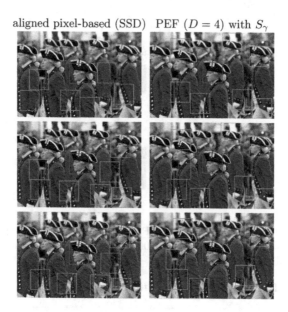

Fig. 1. Comparison of PEFPM with the baseline SSD patch matching (patch size 150×150 and image size 960×640): Observe that due to its flexibility the statistical modeling got more face hits (4 faces) than the pixel-aligned SSD method (one face). The upper row is on pixel intensities, the middle row on sum of RGB channel dissimilarities, and the last row on the maximum of RGB channel dissimilarity. (Color figure online)

3   Experiments

All our experiments were performed in Java^{TM} on a HP Elitebook 840 G1 (i7-4600U CPU 2.1 GHz with 8 GB RAM). First, let us compare our PEFPM method (with $\gamma = 0.1$) with the baseline SSD method: Fig. 1 displays the results obtained when considering intensity values, sum of distances (sum of SSDs or sum of S_γ divergences), or max of distances for the dissimilarity between patches. Observe that our statistical method successfully detected 4 visually similar patches (human faces) while the aligned pixel-based distance detected only one face.

Next, we study the impact of the PEF order D on the visual patch search in Fig. 2. We notice that results depend on the chosen order, and that the most visually similar patches are found for $D = 4$ and $D = 5$. This raises the problem of *model selection* for future research.

Table 1 reports the computational times breaked down into (i) the SAT construction stage, (ii) the PEF estimations, and (iii) the S_γ approximation according to the order of the PEF. Clearly, the higher the order the more costly. The PEF estimation stage scales linearly with the order with slope 1, while the SAT construction and S_γ search have slopes <1.

We then tested the *stability* of our statistical PEFPM method by adding some gaussian noise (Fig. 3). Corrupting the pixel channels with a Gaussian

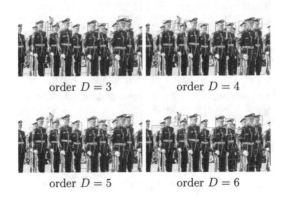

order $D = 3$ order $D = 4$

order $D = 5$ order $D = 6$

Fig. 2. Impact of the polynomial degree of PEFs, the order D of the exponential family (image size 1280×853, patch size 100×100). Here, order $D = 4$ and $D = 5$ find the most visually similar patches.

Table 1. Computational time in seconds for PEFs.

	Degree of PEF				
	2	3	4	5	6
SAT construction	5.99	6.37	8.04	9.56	11.86
PEF estimation	5.24	7.44	8.96	12.84	14.82
S_γ search		7.02	8.86	9.70	10.30	10.52

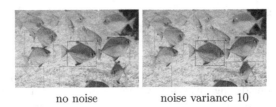

no noise   noise variance 10

Fig. 3. Effect of Gaussian noise on PEFs patch matching (patch size 300×250).

noise change the estimated distributions, but the distance evaluations with the gamma divergence provably attenuate the distortions [18], and we can check that we obtained the same matching patches.

Finally, one big advantage of our statistical modeling other pixel-based patch distance is to allow to consider different zoom values and symmetries. Indeed, geometric symmetries do not change (much) the estimated distribution in a patch. For example, see Fig. 4 that illustrates this property: PEFPM nicely detected the two butterflies even if one looks like the reflected copy of the other. Here, the patch sizes are large (about $1/3$ of the image dimension) and SSD-based method will be very costly and inefficient.

Fig. 4. Patch matching with symmetry (reflection) detected by PEFPM of order 6 (patch siwe 250×250, image dimension $(960, 640)$).

4   Concluding Remarks

We proposed a novel statistical flexible modeling of image patches for fast pattern recognition. Our approach is particularly well-suited for handling large patches and accounting for patch symmetries [7] or other deformations in target images. This work offers many avenues for future research: (1) *model selection* of the polynomial exponential family according to the query patch, (2) *foreground/background* detection in patches (say, using Grabcut [23]) and matching only the foreground statistical distributions to improve the accuracy of patch matching, (3) *multivariate PEFs* to bypass sum/max of univariate PEF distances, etc.

Acknowledgments. We gratefully thank Quei-An Chen (École Polytechnique, France) for implementing our patch matching system and performing various experiments.

References

1. Barnes, C., Shechtman, E., Finkelstein, A., Goldman, D.B.: PatchMatch: a randomized correspondence algorithm for structural image editing. ACM Trans. Graph. (TOG) **28**(3), 24 (2009)
2. Barnes, C., Shechtman, E., Goldman, D.B., Finkelstein, A.: The generalized PatchMatch correspondence algorithm. In: Daniilidis, K., Maragos, P., Paragios, N. (eds.) ECCV 2010. LNCS, vol. 6313, pp. 29–43. Springer, Heidelberg (2010). doi:10.1007/978-3-642-15558-1_3
3. Li, Y., Dong, W., Shi, G., Xie, X.: Learning parametric distributions for image super-resolution: where patch matching meets sparse coding. In: 2015 IEEE International Conference on Computer Vision (ICCV), pp. 450–458, December 2015
4. Kuglin, C.: The phase correlation image alignment method. In: Proceedings of the International Conference on Cybernetics and Society, pp. 163–165 (1975)
5. Foroosh, H., Zerubia, J.B., Berthod, M.: Extension of phase correlation to subpixel registration. IEEE Trans. Image Process. **11**(3), 188–200 (2002)
6. Patraucean, V., Gioi, R., Ovsjanikov, M.: Detection of mirror-symmetric image patches. In: Proceedings of the IEEE Conference on Computer Vision and Pattern Recognition Workshops, pp. 211–216 (2013)
7. Wang, Z., Tang, Z., Zhang, X.: Reflection symmetry detection using locally affine invariant edge correspondence. IEEE Trans. Image Process. **24**(4), 1297–1301 (2015)

8. Cobb, L., Koppstein, P., Chen, N.H.: Estimation and moment recursion relations for multimodal distributions of the exponential family. J. Am. Stat. Assoc. **78**(381), 124–130 (1983)

9. Rohde, D., Corcoran, J.: MCMC methods for univariate exponential family models with intractable normalization constants. In: 2014 IEEE Workshop on Statistical Signal Processing (SSP), pp. 356–359, June 2014

10. Amari, S.: Information Geometry and Its Applications. Applied Mathematical Sciences. Springer, Tokyo (2016)

11. Brown, L.D.: Fundamentals of Statistical Exponential Families: With Applications in Statistical Decision Theory, vol. 9. Institute of Mathematical Statistics, Hayward (1983). 283 pages. ISSN 0749-2170. https://projecteuclid.org/euclid.lnms/1215466757

12. Hyvärinen, A.: Estimation of non-normalized statistical models by score matching. J. Mach. Learn. Res. **6**, 695–709 (2005)

13. Forbes, P.G., Lauritzen, S.: Linear estimating equations for exponential families with application to Gaussian linear concentration models. Linear Algebra Its Appl. **473**, 261–283 (2015). Special Issue on Statistics

14. Nielsen, F., Nock, R.: A closed-form expression for the Sharma-Mittal entropy of exponential families. J. Phys. A Math. Theor. **45**(3), 032003 (2011)

15. Nielsen, F.: Closed-form information-theoretic divergences for statistical mixtures. In: 21st International Conference on Pattern Recognition (ICPR), pp. 1723–1726. IEEE (2012)

16. Nielsen, F., Nock, R.: On the chi square and higher-order chi distances for approximating f-divergences. IEEE Sig. Process. Lett. **1**(21), 10–13 (2014)

17. Banerjee, A., Merugu, S., Dhillon, I.S., Ghosh, J.: Clustering with Bregman divergences. J. Mach. Learn. Res. **6**, 1705–1749 (2005)

18. Fujisawa, H., Eguchi, S.: Robust parameter estimation with a small bias against heavy contamination. J. Multivariate Anal. **99**(9), 2053–2081 (2008)

19. Notsu, A., Komori, O., Eguchi, S.: Spontaneous clustering via minimum gamma-divergence. Neural Comput. **26**(2), 421–448 (2014)

20. Chen, T.L., Hsieh, D.N., Hung, H., Tu, I.P., Wu, P.S., Wu, Y.M., Chang, W.H., Huang, S.Y., et al.: γ-SUP: a clustering algorithm for cryo-electron microscopy images of asymmetric particles. Ann. Appl. Stat. **8**(1), 259–285 (2014)

21. Ehm, W.: Unbiased risk estimation and scoring rules. Comptes Rendus Mathématiques **349**(11), 699–702 (2011)

22. Crow, F.C.: Summed-area tables for texture mapping. ACM SIGGRAPH Comput. Graph. **18**(3), 207–212 (1984)

23. Rother, C., Kolmogorov, V., Blake, A.: GrabCut: Interactive foreground extraction using iterated graph cuts. ACM Trans. Graph. (TOG) **23**(3), 309–314 (2004)

Known-Item Search in Video Databases with Textual Queries

Adam Blažek, David Kuboň[(✉)], and Jakub Lokoč

SIRET Research Group, Faculty of Mathematics and Physics,
Department of Software Engineering, Charles University in Prague,
Prague, Czech Republic
blazekada@gmail.com, kubondavid@seznam.cz, lokoc@ksi.mff.cuni.cz

Abstract. In this paper, we present two approaches for known-item search in video databases with textual queries. In the first approach, we require the database objects to be labeled with an arbitrary ImageNet classification model. During the search, the set of query words is expanded with synonyms and hypernyms until we encounter words present in the database which are consequently searched for. In the second approach, we delegate the query to an independent database such as Google Images and let the user pick a suitable result for query-by-example search. Furthermore, the effectiveness of the proposed approaches is evaluated in a user study.

1 Introduction

With the renaissance of artificial neural networks trained by deep learning algorithms [5,16], query-by-example similarity search has become an effective retrieval scenario in many domains. The retrieval scenario requires one strong assumption: the user has an example query object to start the search. Unfortunately, this assumption is not always satisfied. The search intents can be rather abstract (e.g., find an animal) or the searched object can be present only in the mind of a user (e.g., face of a person). In both cases, the user is not able to find one ideal query object that perfectly fits the search intents, directly express the query by keywords, or draw a sketch. Such retrieval scenarios are denoted as a *known-item search* (or *mental query retrieval*). In order to solve known-item search tasks, retrieval systems [13] put the user into the center of the retrieval process and provide support tools to minimize the number of necessary interactions with the user.

Given a known-item search system, it is difficult to evaluate its effectiveness automatically. Therefore, competitions like the Video Browser Showdown (VBS) [12] are organized[1], where many different systems and their operators compete in predefined known-item search tasks. At the VBS, there are two types of tasks for large video collection: visual search tasks where one finds a short

---

[1] VBS was set up as a follow up of known-item search TRECVID [15] task organized from 2010 to 2012. Since 2016, VBS has become a part of TRECVID.

L. Amsaleg et al. (Eds.): SISAP 2016, LNCS 9939, pp. 117–124, 2016.
DOI: 10.1007/978-3-319-46759-7_9

presented video clip (recording is not allowed) and textual search tasks where the requested scene is described only by a short text. Although the tasks only simulate real complex retrieval scenarios, the results of the competition point to promising approaches and open problems. Based on the results of the last VBS (searching in 300 h of video), we may conclude that visual search tasks can be effectively solved by many traditional approaches [1] and, to some extent, even with a sequential scan of the database [4]. On the other hand, the textual search tasks represent an open challenge. In this case, it is significantly harder to initialize the search with an appropriate query-example or sketch.

In this paper, we focus on textual known-item search task initialization. We compare two orthogonal approaches to find an initial query object using keyword search. The first approach employs an external image retrieval system designed for effective keyword search (e.g., Google Images). In such system, users can usually materialize their ideas and use some of the returned image to initialize similarity search. The problem of this approach is that the first idea of the user does not have to correspond (in the terms of similarity) with the contents of the dataset. For this reason, the second approach is dataset-oriented, where automatic annotations are created for the searched video. Unlike other state-of-the-art approaches, our approach is not restricted to speech transcripts only [17] nor do they require manual pre-annotation [8]. Given the annotations, user-defined keyword query is matched directly to the contents of the dataset. Such an approach has been used in several known-item search tools [9]. We enhance it with text processing and exploration of semantic relations between words. Since the user is part of the search process, only a comparative user study can decide which of the two approaches is more effective given specific retrieval tasks.

The paper is organized as follows: Sect. 2 presents an overview of a reference known-item search system successful in visual search tasks, which we want to improve for textual search tasks by two keyword search approaches presented in Sect. 3. In Sect. 4, the results of a comparative user study are presented and Sect. 5 concludes the paper.

2   Reference Known-Item Search Tool

Although this paper focuses on effective processing of textual queries, the benefits of other approaches used for known-item search should not be omitted. Therefore, we implement the textual search option into a well-established system from the Video Browser Showdown competition. More specifically, we extend the Multi-sketch Semantic Video Browser [7] that won the competition twice in 2014 and 2015, while in 2016 in took the third place, still in the cluster of the most successful tools. Whereas the novel textual-query approaches are presented in the following section, in this section we briefly summarize the key features of the tool.

The crucial feature of this tool is the multi-modal sketching canvas (Fig. 1 - right), where users can define the most significant visual clues of the searched scene. The modalities are colors, more specifically feature signatures [6], and

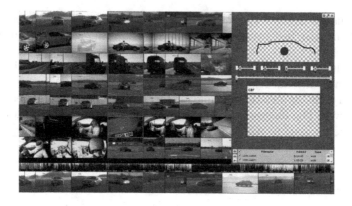

Fig. 1. Multi-sketch semantic video browser

edges [10]. The tool also enables the user to define two consecutive sketches and include temporal information in the query. However, most of the area is dedicated for presenting and browsing the top matching results. Any displayed key-frame might be selected for query-by-example search with DeCAF features [2].

So far, the presented features were successful in handling visual search tasks. In the past, the tool outperformed even tools with advanced concept detectors and filtering approaches. However, if the user is unaware of the colors and edges in search scenes, sketch-based techniques often fail to initiate a successful search. For this reason, we introduce two types of textual queries and analyze their performance in connection with the described visual search approaches.

3   Textual Queries for Known-Item Search in Video

Given a textual description of the searched scene, it is difficult to initialize the search in a content-based way. If the textual description does not contain any clues for color or edge/shape, but only a concept label, the keyword search becomes the preferred filtering choice. Even if the search item is known visually, it might be easier, especially for novice users, to describe it textually. For these reasons we are introducing and evaluating two orthogonal approaches for textual search for known-item in video.

3.1   External Image Search Engine

The first approach we investigate is to use the text query in an external keyword-based image search engine to get sample images and use these to initialize the search. Such an engine can be, for example, well-established Google Images search. However, this approach has two significant limitations. First, a sufficiently fast Internet connection is necessary and the engine has to be available. Furthermore, as the selected image has to be preprocessed for the video retrieval system, all the feature extraction techniques (or services) have to be available for instant usage. Second, the selected image does not have to correspond to the content of the searched scene, while the materialized image may be distracting for the search.

The Multi-sketch Semantic Video Browser can be extended easily just by providing an additional text search field in the application. When user issues a textual query, the application downloads top k images from the external image retrieval server. The images are displayed to the user and the user selects a candidate image for search. Appropriate features are then extracted (e.g., DeCAF descriptors) and used to initialize similarity search browsing.

3.2   ImageNet Labels

In the second approach investigated in this paper, we consider a search method based on labels automatically assigned to key-frames using an arbitrary ImageNet [11] classification model; in our case, Deep Neural Network [14]. While the model provides highly specific labels from the ImageNet set of 1000 labels, users tend to form more elaborate yet generic queries or even whole sentences. For example, we would like to match a key-frame labeled as 'golden retriever' with a query 'A dog is playing with a ball'. To interconnect the worlds of ImageNet labels and user queries, we introduce two different processing pipelines: key-frames labeling and query processing and matching.

Key-Frames Labeling. During the video preprocessing, the top five labels for each key-frame are extracted together with their probabilities. Every ImageNet label is in fact a synset (a group of elements, which are semantically equivalent) from the WordNet [3] lexical database. Of the five extracted labels, a tree of their hypernyms is built, such as the one in Fig. 2. The nodes represent synsets and edges represent WordNet hypernym–hyponym relations between them. A probability of a newly inferred synset is a sum of probabilities of its children. Using this tree, we can deduce probability values even for labels not available in the ImageNet labels set. Both extracted and inferred labels and their probabilities are stored in an inverted index for efficient retrieval.

Query Processing and Matching. The processing of a user query follows a simple procedure. First, stop-words (such as 'a', 'been' or 'your') are removed for higher efficiency while the remaining words are transformed into their basic forms. Since only nouns (objects) are present in ImageNet, all other parts of speech are also removed. Consequently, the noun's meanings are explored using WordNet and a set of synsets is assigned to each of the nouns, which usually contains several different meanings of the word (e.g. a noun 'horse' might be an animal as well as a gymnastic equipment).

Now, we iteratively generalize each query synset (exchange it for its hypernym) until it is present in the database. For example, a query synset 'horse' (not present) might be exchanged after few iterations for synset 'animal' (present). This way, each query synset would yield some results — in the worst case, we

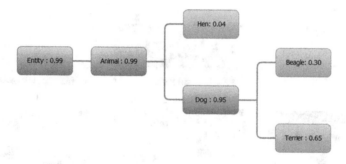

Fig. 2. A synset tree of one particular key-frame. Each node represents one synset together with the probability of its presence in the key-frame. The tree includes both the synsets extracted by the model (green) and the inferred ones (blue). (Color figure online)

would search for the most general synset – 'entity' which is present in every key-frame. In other words, we are going to search for the most specific synsets we can find in the dataset.

For each query synset, we rank all the key-frames with the probabilities of occurrence of the synset in the key-frames. The rankings of query synset s are weighted by $1 - avg_p(s)$ where $avg_p(s)$ stands for average probability of synset s over all the dataset key-frames. In other words, we prefer more specific synsets (rarely present) over generic synsets, which have a weaker filtering power. This scheme follows the same idea as the term frequency weighing technique utilized in general textual search.

User Feedback Loop. Since all the word meanings are extracted from the query, it is vital for the user to be able to overlook and control the actual searched labels. We provide a checklist of them and only a subset might be selected. The effects of this procedure affect the results immediately. Experiments reveal that this provides a highly used and convenient way to further specify the search, now based on information actually contained in the dataset.

4 User Study

We carried out a simple user study in order to determine which of the proposed approaches is more effective. A total of 21 novice participants were briefly introduced to the tool and asked to find 6 different video segments presented via playback with a limit of 3 min for each task. The database contained almost 30 h of diverse video content including, TV shows, sports, indoor and outdoor activities, etc. Example key-frames from the searched video segments are displayed in Fig. 3.

The two textual search approaches presented in this paper are referred to as Google (Sect. 3.1) and ImageNet (Sect. 3.2). For each task, participants were

Task 1   Task 2   Task 3

Task 4   Task 5   Task 6

Fig. 3. Example key-frames from the searched video segments.

randomly divided in three groups, where each group was enabled to use either Google, ImageNet or both approaches.

Out of 102 individual searches, 96 included a textual query, and 33 of them consecutively lead to a success (34.4 %). The numbers of successful searches are captured in Fig. 4. In 13 out of the successful 33 searches, participants did not use any feature other than textual search. The data collected do not demonstrate statistically significant differences between success rates — for Google vs. ImageNet p-value is 0.731014, for Google vs. both p-value is 0.291958 and for ImageNet vs. both p-value 0.261518. In the case of the group enabled to use both techniques, total number of Google queries was 108, as opposed to 72 ImageNet

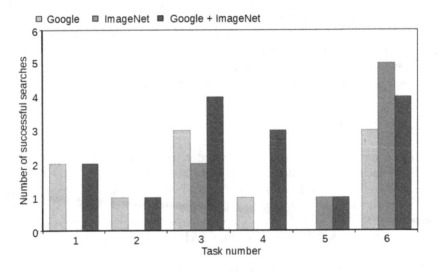

Fig. 4. Numbers of successful searches for each task.

queries. However, if we restrict this to the successful tasks, there were 17 Google queries and 20 ImageNet queries.

ImageNet turned to be effective search initialization for scenes, that contain a particular, easily identifiable object, e.g. a *harvester* in Task 3 or a *golf cart* in Task 6. Regarding complex scenes containing plurality of concepts (Tasks 2, 4 and 5), users struggled to select the one actually detected by ImageNet.

Arguably, the main limitation of Google approach is that the retrieved images frequently do not fit users expectations. Although they may contain the searched object the context happen to be quite dissimilar to the searched scene. We observed that inexperienced users were using these unfitting images instead of refining the query. On the other hand, we are able to retrieve suitable images with seemingly unrelated query independently on the content of our database.

Apparently, Tasks 6 and 3 were rather easy as both approaches lead to success almost instantaneously. We contribute this to the fact that distinct keywords were available to begin the search with (e.g. *golf cart, harvester*). Task 1 could have been solved just by using the sketch-based techniques. Task 4 was very hard for users limited to ImageNet as none of the obvious concepts (*pumpkin, goat*) were actually detected. Task 5 required to follow the obvious text query *football* with additional browsing techniques. Task 2 was rather confusing as some of the apparent search words such as *military* or *kitchen* provided misleading results.

5 Conclusion

In this paper, we presented two orthogonal approaches for known-item search in video with textual queries. The approaches were evaluated in a user study which revealed that both approaches provide a viable way to search for a known-item in certain scenarios. We may also conclude that the textual queries are preferred by novice users, as a third of the successful searches was carried out without any other modalities, such as color and edge sketches.

Acknowledgements. This research was supported by Charles University in Prague Grant Agency – GAUK project no. 1134316. Furthermore, we are grateful to Mr. Jan Pavlovsky for his help with the user study.

References

1. Barthel, K.U., Hezel, N., Mackowiak, R.: Navigating a graph of scenes for exploring large video collections. In: Tian, Q., Sebe, N., Qi, G.-J., Huet, B., Hong, R., Liu, X. (eds.) MMM 2016. LNCS, vol. 9517, pp. 418–423. Springer, Heidelberg (2016). doi:10.1007/978-3-319-27674-8_43
2. Donahue, J., Jia, Y., Vinyals, O., et al.: DeCAF: a deep convolutional activation feature for generic visual recognition. CoRR, abs/1310.1531 (2013)
3. Fellbaum, C.: WordNet: An Electronic Lexical Database. Bradford Books, Bradford (1998)

4. Hürst, W., van de Werken, R., Hoet, M.: A storyboard-based interface for mobile video browsing. In: He, X., Luo, S., Tao, D., Xu, C., Yang, J., Hasan, M.A. (eds.) MMM 2015, Part II. LNCS, vol. 8936, pp. 261–265. Springer, Heidelberg (2015). doi:10.1007/978-3-319-14442-9_25

5. Krizhevsky, A., Sutskever, I., Hinton, G.E.: Imagenet classification with deep convolutional neural networks. In: Pereira, F., Burges, C.J.C., Bottou, L., Weinberger, K.Q. (eds.) Advances in Neural Information Processing Systems, vol. 25, pp. 1097–1105. Curran Associates Inc. (2012)

6. Kruliš, M., Lokoč, J., Skopal, T.: Efficient extraction of clustering-based feature signatures using GPU architectures. Multimedia Tools Appl. 1–33 (2015)

7. Kuboň, D., Blažek, A., Lokoč, J., Skopal, T.: Multi-sketch semantic video browser. In: Tian, Q., Sebe, N., Qi, G.-J., Huet, B., Hong, R., Liu, X. (eds.) MMM 2016. LNCS, vol. 9517, pp. 406–411. Springer, Heidelberg (2016). doi:10.1007/978-3-319-27674-8_41

8. Lin, D., Fidler, S., Kong, C., Urtasun, R.: Visual semantic search: retrieving videos via complex textual queries. In: 2014 IEEE Conference on Computer Vision and Pattern Recognition, pp. 2657–2664, June 2014

9. Moumtzidou, A., et al.: VERGE: an interactive search engine for browsing video collections. In: Gurrin, C., Hopfgartner, F., Hurst, W., Johansen, H., Lee, H., O'Connor, N. (eds.) MMM 2014, Part II. LNCS, vol. 8326, pp. 411–414. Springer, Heidelberg (2014). doi:10.1007/978-3-319-04117-9_48

10. Park, D.K., Jeon, Y.S., Won, C.S.: Efficient use of local edge histogram descriptor. In: Proceedings of the 2000 ACM Workshops on Multimedia, MULTIMEDIA 2000, pp. 51–54. ACM, New York (2000)

11. Russakovsky, O., Deng, J., Hao, S., et al.: ImageNet large scale visual recognition challenge. Int. J. Comput. Vis. (IJCV) **115**(3), 211–252 (2015)

12. Schoeffmann, K.: A user-centric media retrieval competition: the video browser showdown 2012–2014. IEEE Multimedia **21**(4), 8–13 (2014)

13. Schoeffmann, K., Hudelist, M.A., Huber, J.: Video interaction tools: a survey of recent work. ACM Comput. Surv. **48**(1), 14:1–14:34 (2015)

14. Simonyan, K., Zisserman, A.: Very deep convolutional networks for large-scale image recognition. CoRR, abs/1409.1556 (2014)

15. Smeaton, A.F., Over, P., Kraaij, W.: Evaluation campaigns and trecvid. In: Proceedings of the 8th ACM International Workshop on Multimedia Information Retrieval, MIR 2006, pp. 321–330, ACM, New York (2006)

16. Szegedy, C., Liu, W., Jia, Y., et al.: Going deeper with convolutions. CoRR, abs/1409.4842 (2014)

17. Volkmer, T., Natsev, A.: Exploring automatic query refinement for text-based video retrieval. In: 2006 IEEE International Conference on Multimedia and Expo, pp. 765–768, July 2006

Combustion Quality Estimation in Carbonization Furnace Using Flame Similarity Measure

Fredy Martínez[1(✉)], Angelica Rendón[1], and Pedro Guevara[2]

[1] District University Francisco José de Caldas, Bogotá D.C., Colombia
fhmartinezs@udistrital.edu.co, avrendonc@correo.udistrital.edu.co
[2] Tecsol Industries Limited, Bogotá D.C., Colombia
pedro_guevarap@yahoo.com
http://www.udistrital.edu.co

Abstract. Similarity distance measures are used to study the similarity between patterns. We propose the use of similarity measures between images to estimate the quality of combustion in a furnace designed for carbonization processes in the production of activated carbon. Broadly speaking, the production of activated carbon requires two thermal processes: carbonization and activation. One of the most sensitive variables in both processes is the level of oxygen. For carbonization, the process involves thermal decomposition of vegetal material in the absence of air. For activation, the gasification of the material at high temperature is required, and one of the oxidizing agents used is oxygen. Given the complexity of measuring the oxygen level because of the functional characteristics of the furnaces, we propose a strategy for estimating the quality of combustion, which is directly related to the oxygen level, based on similarity measures between reference photographs and the flame states. This strategy corresponds to the instrumentalization of methods used by operators in manual control of the furnaces. Our algorithm is tested with reference photos taken at the production plant, and the experimental results prove the efficiency of the proposed technique.

Keywords: Activated carbon · Carbonization · Distance · Flame · Similarity

1 Introduction

Similarity is the measure of how alike two data objects are, and it is a particularly powerful tool to discover patterns in large data sets, as is the case of color images [1]. Normally it is described as a distance between features of an object. If this distance is small, there will be high degree of similarity. And if instead the distance is large, there will be low degree of similarity.

Similarity measure is highly dependent on the domain and application. Two objects can be similar because of their size, shape or color, which is why care

© Springer International Publishing AG 2016
L. Amsaleg et al. (Eds.): SISAP 2016, LNCS 9939, pp. 125–133, 2016.
DOI: 10.1007/978-3-319-46759-7_10

should be taken to avoid calculating distances with features that are not related. The process is called Content-Based Image Retrieval (CBIR), also known as Query By Image Content (QBIC) [2]. This seeks to relate images according to their content, as opposed to based on metadata. A CBIR system performs using features like color, size, texture and shape computed from images.

Similarity comparison between two images can be made using different metrics. Each metric corresponds to a strategy that produces a quantitative assessment of similarity. A large number of image similarity metrics have been proposed, however, there is no a right image similarity metric, but a set of metrics that are appropriated for a particular application. Different metrics can be classified into three categories:

– Image based
– Histrogram based
– Based on similarity of objects contained in images.

The histogram involves the construction of a representation of the color characteristics of each image [4–6]. It corresponds to the definition of a function on each pixel of the image, but the value is stored for each pixel. For the case of a grayscale image, if (Eq. 1):

$$f : [1, n] \times [1, m] \longrightarrow [0, 255] \tag{1}$$

is a gray value image, then (Eq. 2):

$$H(f) : [0, 255] \longrightarrow [0, n \times m] \tag{2}$$

is its histogram. In general, similar images have similar histograms.

In this paper we propose and evaluate the use of the histogram. While is true that different images can have similar histograms, when the designer aims to compare images that knows in advance that are similar (for example, with different states of the same system), this strategy turns out to be extremely useful. Besides, the histogram is invariant to rotation, translation, and scaling of the image.

The similarity of two images described by colour histograms is also measured by a distance between the histograms in the colour space. A very simple and fairly used distance is the histogram intersection proposed by Swain-Ballard [8] (Eq. 3):

$$d(f, g) = 1 - \sum_{i=1}^{n} \sum_{j=1}^{m} min \left[f(i, j), g(i, j) \right] \tag{3}$$

We propose the use of histogram-based similarity measures on images of the flame in a carbonization furnace, with respect to reference images, as a strategy to estimate the amount of oxygen in the process, and therefore, the quality of combustion [3,7]. To do this, we evaluate four different similarity metrics, selecting the one with best performing in order to integrate it to the sensing loop of the control system.

The prototype hardware is comprised of a small digital camera (5 megapixel native resolution, sensor-capable of 2592 × 1944 pixel static images). The control board uses a Broadcom system on a chip (SoC), which includes an ARM CPU (900 MHz 32-bit quad-core Cortex-A7, with 256 KB shared L2 cache) and an on chip graphics processing unit (GPU, a VideoCore IV).

The hardware uses a variant of Debian Linux as operating system. The digital image processing (filters, scaling, histograms and calculation of distances) is done in Python code. The system has WiFi connectivity, this allows us to remotely monitor images, histograms and other data. The platform is installed directly in the furnace on the opposite end of the burner, and in front of it.

The paper is organized as follows. In Sect. 2 presents a description of the problem of determining the quality of combustion to control the carbonization process. Section 3 describes the strategies used to estimate the content of oxygen in the combustion by flame similarity measure. Section 4 introduces some results obtained with the proposed strategy. Finally, conclusion and discussion are presented in Sect. 5.

2   Problem Statement

Carbonization is a process of thermal decomposition of organic waste in the absence of air. The idea is to remove from the vegetable material hydrogen, oxygen and nitrogen to increase the proportion of carbon. In production plant, this process is performed in a furnace, on which must be controlled: (1) Oxygen level, (2) Heating rate, (3) Final temperature, and (4) Residence time.

Unlike temperature, due to the configuration of the furnace it is not possible the use of an oxygen sensor to feedback the state of this variable to the control system. Our experiments show that it is possible to control the content of oxygen in combustion if we regulate the entry of air into the furnace burner. The proposed similarity measure considers this fact to compare the state of the flame in the furnace with four reference images corresponding to four states under which the oxygen percentage is known.

The four reference images correspond to the flame in four operating conditions for which the oxygen level is known. These conditions are, according to the opening of the air valve:

- Flama with 0 % air.
- Flama with 40 % air.
- Flama with 80 % air.
- Flama with 100 % air.

The fuel used is natural gas, and the system can independently control the supply of fuel and air.

To solve the problem of sensing oxygen into the furnace, an embedded system with a camera to take pictures of the flame and estimate the quality of combustion through similarity measures with reference images is proposed. In the case of carbonization (the production plant also includes thermal activation processes), the reference will be the image with 0 % air content.

3   Methodology

Choosing which histogram comparison function to use is normally dependent on:

- The size of the dataset.
- The quality of the images in the dataset.
- The set of relevant characteristics for comparison.

In principle, to compare the histograms we selected four distances because they are the most used in these types of applications:

- Correlation: Computes the correlation between the two histograms. The distance between the H_1 and H_2 histograms is calculated as follows (Eq. 4):

$$d\left(H_1, H_2\right) = \frac{\sum\limits_I \left(H_1\left(I\right) - \overline{H_1}\right)\left(H_2\left(I\right) - \overline{H_2}\right)}{\sqrt{\sum\limits_I \left(H_1\left(I\right) - \overline{H_1}\right)^2 \left(H_2\left(I\right) - \overline{H_2}\right)^2}} \tag{4}$$

where (Eq. 5):

$$\overline{H_k} = \frac{1}{N}\sum_J H_k\left(J\right) \tag{5}$$

and N is the total number of histogram bins.

- Chi-Squared: Applies the Chi-Squared distance to the histograms. The distance between the H_1 and H_2 histograms is calculated as follows (Eq. 6):

$$d\left(H_1, H_2\right) = \sum_I \frac{\left(H_1\left(I\right) - H_2\left(I\right)\right)^2}{H_1\left(I\right)} \tag{6}$$

- Intersection: Calculates the intersection between the two histograms. The distance between the H_1 and H_2 histograms is calculated as follows (Eq. 7):

$$d\left(H_1, H_2\right) = \sum_I min\left(H_1\left(I\right), H_2\left(I\right)\right) \tag{7}$$

- Bhattacharyya: Bhattacharyya distance, used to measure the *overlap* between the two histograms. The distance between the H_1 and H_2 histograms is calculated as follows (Eq. 8):

$$d\left(H_1, H_2\right) = \sqrt{1 - \frac{1}{\sqrt{H_1 H_2 N^2}}\sum_I \sqrt{H_1\left(I\right) \cdot H_2\left(I\right)}} \tag{8}$$

As a first evaluation strategy of the four selected distances, we compared the four reference images and organize them according to their similarity with respect to the reference image for 0 % air content, this thinking about the carbonization process. The expected result is that the metrics organize the four images from smaller air content to higher air content (ordered by oxygen level). The result is shown in Fig. 1.

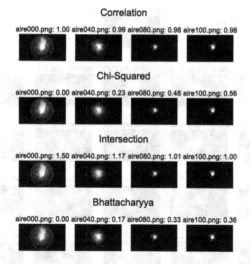

Fig. 1. Comparison between the four metrics with respect to the four reference images. (a) Correlation: Scale [0.9, 1], values close to one indicate too little oxygen. (b) Chi-Squared: Scale [0, 0.6], values close to zero indicate too little oxygen. (c) Intersection: Scale [1, 1.5], values close to 1.5 indicate too little oxygen. (d) Bhattacharyya: Scale [0, 0.4], values close to zero indicate too little oxygen.

The four metrics managed to organize the reference images as expected. However, we select among them the Chi-Squared distance and the Bhattacharyya distance because they produce a higher resolution in the differentiation of each reference state. The final selection of the metric to use in the control system was performed according to the behavior with images of the furnace operation.

4   Results

We have tested our proposed algorithm on a database generated in the Tecsol Industries Limited production plant (25 images of different operating conditions). The images were taken facing the flame, from the opposite side of the carbonization furnace. Images are captured in color with a quality of 1280 × 720 pixels. The processing system maintains this quality for calculating the histogram. The illumination inside the furnace is completely dependent on the flame.

We first compared the database with respect to the reference image for 0 % air content using the Chi-Squared distance. Then, we repeat the procedure for the Bhattacharyya distance. The results are shown in Figs. 2 and 3.

Figure 2 shows that the Chi-Squared distance has trouble for properly categorize images in which there is very low level of oxygen, below the reference to 0 % (018, 002 and 001). These images were taken greatly reducing the flow of fuel, and so we force a much wider and orange flame. This distance puts

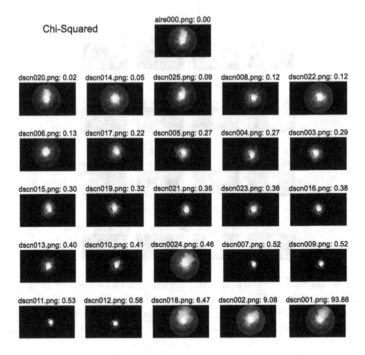

Fig. 2. Oxygen estimation in the furnace using the Chi-Squared distance. (Color figure online)

these three cases at the end, after the values corresponding to 100 % air, which is clearly wrong. Another problem is the increase in the scale for the distance value, which due to these three cases rises to almost 100. This impairs normal readings (distances with values from 0 to 1) by reducing the resolution of the sensor.

Figure 3 shows that the Bhattacharyya distance is also struggling with the same three images, but they are best handled by assigning values below the extreme case (011 and 012 cases, where the air valve is opened almost completely). The behavior is the expected, and the scale allows good resolution in the identification of system status, this is the reason why this distance is selected for the implementation of smart sensor.

Table 1 shows a comparison of errors for each of these two metrics. All scales are normalized in the range [0, 1], and the error for each image is calculated at full scale. In general, the distance Bhattacharyya has a greater error, however, statistically the error is similar to the distance Chi-Squared error. In addition, the three images out of scale do not alter the behavior of the scale.

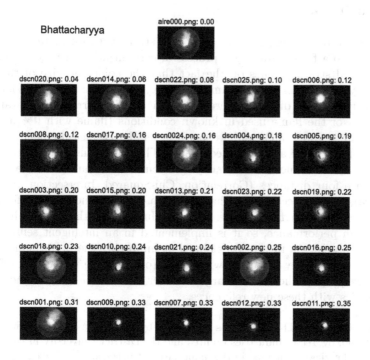

Fig. 3. Oxygen estimation in the furnace using the Bhattacharyya distance.

Table 1. Percentage error of each metric for each image at full scale

Image	Expected distance	Chi-Squared distance	Error	Bhattacharyya distance	Error
dscn001	0.00	93.88	**15646.7%**	0.31	77.5%
dscn002	0.00	9.08	1513.3%	0.25	62.5%
dscn003	0.48	0.29	-0.1%	0.20	1.6%
dscn004	0.44	0.27	1.3%	0.18	1.2%
dscn005	0.47	0.27	-1.9%	0.19	0.6%
dscn006	0.20	0.13	1.4%	0.12	9.7%
dscn007	0.97	0.52	-10.2%	0.33	-14.4%
dscn008	0.22	0.12	-1.9%	0.12	8.1%
dscn009	0.95	0.52	-8.6%	0.33	-12.8%
dscn010	0.75	0.41	-6.7%	0.24	-15.0%
dscn011	1.00	0.53	-11.7%	0.35	-12.5%
dscn012	0.98	0.58	-1.8%	0.33	-15.9%
dscn013	0.59	0.40	7.3%	0.21	-6.9%
dscn014	0.08	0.05	0.5%	0.06	7.2%
dscn015	0.56	0.30	-6.2%	0.20	-6.3%
dscn016	0.81	0.38	-17.9%	0.25	-18.8%
dscn017	0.34	0.22	2.3%	0.16	5.6%
dscn018	0.00	6.47	1078.3%	0.23	57.5%
dscn019	0.64	0.32	-10.7%	0.22	-9.1%
dscn020	0.00	0.02	3.3%	0.04	10.0%
dscn021	0.77	0.35	-18.2%	0.24	-16.6%
dscn022	0.08	0.12	12.2%	0.08	12.2%
dscn023	0.63	0.36	-2.5%	0.22	-7.5%
dscn024	0.06	0.46	70.4%	0.16	33.8%
dscn025	0.14	0.09	0.9%	0.10	10.9%
	Scale: [0,1]	Scale: [0,0.6]		Scale: [0,4]	

5   Conclusions

The objective of this paper is to propose a strategy to estimate the amount of oxygen into a furnace during a process of carbonization of vegetal material. The proposed strategy relies on the fact that the color and size of the flame are directly proportional to the amount of oxygen on the furnace, a fact supported by tests under different operating conditions. To carry out the evaluation four images of the flame used in known conditions (flame with 0 % air, 40 % air, 80 % air y 100 % air), and we perform similarity measurements between the state of the flame and reference images. The scheme uses the histogram of the image, and examines the similarity by calculating distance. We evaluate the performance four distances (Correlation, Chi-Squared, Intersection and Bhattacharyya), finally selecting the Bhattacharyya distance due to its performance. Over several tests on the furnace, it is concluded that the scheme has more than enough performance, so it is implemented in an intelligent sensor on an embedded system. As future work aimed at improving the performance of the algorithm, particularly for unknown images (out of scale formed by reference images), we propose the use of a larger number of reference images, and the use of all metrics with these new references.

Acknowledgments. This work was supported by Colciencias through the project 622470149090, by Tecsol Industries Limited and the District University Francisco José de Caldas. The views expressed in this paper are not necessarily endorsed by Colciencias, Tecsol or District University. The authors thank the research group ARMOS for the evaluation carried out on prototypes of ideas and strategies.

References

1. Barrero, A., Robayo, M., Jacinto, E.: Algoritmo de navegación a bordo en ambientes controlados a partir de procesamiento de imágenes. Tekhnê **12**(2), 23–34 (2015). ISSN 1692–8407
2. Beecks, C., Kirchhoff, S., Seidl, T.: On stability of signature-based similarity measures for content-based image retrieval. Multimedia Tools Appl. **71**(1), 349–362 (2014). ISSN 1380–7501
3. Beecks, C., Zimmer, A., Seidl, T., Martin, D., Pischke, P., Kneer, R.: Applying similarity search for the investigation of the fuel injection process. In: 4th International Conference on Similarity Search and Applications (SISAP 2011), pp. 117–118 (2011)
4. Chen, S., Li, F.: Color image retrieval based on vector quantization. In: International Conference on Electrical and Control Engineering (ICECE 2010), pp. 5092–5095 (2010)
5. Pang, Y., Shi, X., Jia, B., Blasch, E., Sheaff, C., Pham, K., Chen, G., Ling, H.: Multiway histogram intersection for multi-target tracking. In: 18th International Conference on Information Fusion (Fusion 2015), pp. 1938–1945 (2015)
6. Rajalakshmi, T., Minu, R.: Improving relevance feedback for content based medical image retrieval. In: International Conference on Information Communication and Embedded Systems (ICICES 2014), pp. 1–5 (2014)

7. Sabeti, L., Wu, J.: New similarity measure for illumination invariant content-based image retrieval. In: International Conference on Automation and Logistics, pp. 279–283 (2008)
8. Swain, M., Ballard, D.: Indexing via color histograms. In: Third International Conference on Computer Vision, pp. 390–393 (1990)

Text and Document Similarity

Bit-Vector Search Filtering with Application to a Kanji Dictionary

Matthew Skala[✉]

IT University of Copenhagen, Copenhagen, Denmark
mskala@ansuz.sooke.bc.ca

Abstract. Database query problems can be categorized by the expressiveness of their query languages, and data structure bounds are better for less expressive languages. Highly expressive languages, such as those permitting Boolean operations, lead to difficult query problems with poor bounds, and high dimensionality in geometric problems also causes their query languages to become expressive and inefficient. The IDSgrep *kanji* dictionary software approaches a highly expressive tree-matching query problem with a filtering technique set in 128-bit Hamming space. It can be a model for other highly expressive query languages. We suggest improvements to bit vector filtering of general applicability, and evaluate them in the context of IDSgrep.

1 Introduction

Many data structure problems of interest are specializations of the following general database query problem.

Problem 1. Given a universe U, preprocess a database $S \subseteq U$ into a data structure, to efficiently answer queries of the form "Find $Q \cap S$ for a given query Q." The database will be given by explicitly listing its n elements, but the query Q will be specified in some much more concise query language, which will not necessarily permit all arbitrarily-chosen subsets of U to be queries.

For example, in one kind of *similarity search* the query is described by a single element $q \in U$; the set Q to be intersected with S is then all elements "similar" to q, for some definition of similarity. In the present work, we are interested in database queries for which each element of S is or is not part of the query result independently—excluding such things as *nearest neighbour* queries, where the presence of one element in the database can affect whether some other element should or should not be returned.

The range of different query sets Q, determined by the expressive power of the query language, affects how this problem can be solved. A query language with very little expressive power, for instance consisting only of intervals of permitted values in a single numeric attribute, permits the use of tree-based data structures with typical $O(\log n)$ query time. An even less expressive language, such as one expressing only singleton sets (thus, membership queries) would

© Springer International Publishing AG 2016
L. Amsaleg et al. (Eds.): SISAP 2016, LNCS 9939, pp. 137–150, 2016.
DOI: 10.1007/978-3-319-46759-7_11

be naturally solved using hash tables for constant query time. But when Q is specified using an *advanced* query language, loosely defined as one with great expressive power, it becomes more difficult to apply data structure techniques, and the data structures give less useful guarantees.

The naive solution to Problem 1 for an advanced query language would examine every element of S in $\Omega(n)$ time. Similarity queries of a linear-space data structure based on distance in a metric space also quickly approach $\Omega(n)$ number of elements examined when the dimensionality is large [5]; and because some problems arising from similarity search become $\mathcal{NP}$-hard in arbitrary dimension [21,26], and others have disappointingly large polynomial lower bounds associated with complexity-theoretic reductions from hard problems [28], there appears to be a connection between expressiveness of query languages and the dimensionality of the data. The language of high-dimensional near-neighbour queries is expressive enough to ask hard questions.

We are interested in query languages that embed Boolean operators: where the ability to specify query sets Q_1 and Q_2 implies also being able to specify $Q_1 \cap Q_2$ (AND), $Q_1 \cup Q_2$ (OR), and $U \backslash Q_1$ (NOT). Including such operators makes the language sufficiently expressive that it may be hard to beat the naive solution; and performance suffers even further if a complicated query language makes testing an element against the query an expensive operation in itself.

The present work describes and experimentally tests some techniques for speeding up general database query with expressive query languages, when bit-vector filtering is already in use. The hope is to reduce the constant in the $\Omega(n)$ by not examining *every* element in the database individually; and to possibly get better than $\Omega(n)$ performance on easy queries while still supporting hard queries on the same data structure. Our approach builds upon a bit-vector filtering method described previously [23] and implemented in a mature free-software project. We have a specific application originating in computational linguistics, but the techniques studied here are intended for general use.

1.1 About the Application

The Tsukurimashou Project [22] develops parametric fonts for Japanese-language typesetting, and associated software tools. IDSgrep is one of these tools: a search utility for Han character dictionaries [23].

The term *Han script* refers generically to a set of tens or hundreds of thousands of characters used in varying forms to write text in East Asian languages. Use of Han script is current in Chinese and Japanese. Korean is now written primarily in an alphabetic script with limited use of Han characters. Vietnamese is written in Latin script today, but it was historically written with Han characters, and specifically Vietnamese forms for Han characters are standardized in Unicode. The present work focuses on *kanji*, the Japanese form. One feature of Han script is that characters can be analysed as hierarchical combinations of elements that may be shared with other characters, or may be characters in themselves, as shown in Fig. 1.

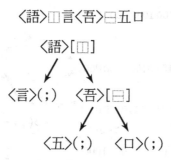

Fig. 1. A dictionary entry for the character 語 meaning "language" [23].

Font developers, language learners, and computational linguists each have reasons to query databases of these trees with Boolean criteria, as in "Find all characters that contain 言 on the left, and 口 anywhere, but not 上."

The details of the IDSgrep data model and query language are not new here and not relevant to the algorithmic considerations of the present paper, which treats the matching function as a black box. They are covered in earlier work [22, 23] in much greater detail than is possible in this space or appropriate to the present work. But to summarize for interested readers: the structure of each character is represented by an *EIDS tree* ("Extended Ideographic Description Sequence," an extension of Unicode's IDS concept [27]), like the one illustrated in Fig. 1. Each node of the tree is labelled with a *functor* describing the relation among its children, such as 目 for one above the other, and optionally a *head* such as 吾 naming the character represented by that subtree. Not all subtrees of characters are characters in themselves, so not all nodes have heads. Queries are also trees in the same format, entered using a simple prefix syntax.

The basic matching rule is that if a query and a dictionary entry both have heads, then they match if and only if the heads are identical; if they do not both have heads, then they match if their functors and all their children match, recursively. But there are also special matching operators invoked by special functors in the query. For instance, ? is a match-everything wildcard, allowing the query 目?心 to match any dictionary entry for a character with a top and bottom part in which the bottom matches 心. Special operators include match-everything; Boolean AND, OR, and NOT; match any subtree; an *associative* match which rearranges trees along the lines of the associative law in arithmetic; and a few others intended for special purposes. Evaluating the full matching function between one query and one dictionary entry is a relatively expensive operation (worst-case cubic in the size of the tree and the query, excluding certain matching operators that invoke third-party software with no time guarantees); and a naive implementation of search would do this evaluation on all of the $O(n)$ trees in the database.

To support filtering search, IDSgrep calculates a *bit vector* for each EIDS tree, as shown in Fig. 2. Bit vectors, like EIDS trees, are treated as opaque by the search algorithm, but internally each one is the concatenation of four 32-bit

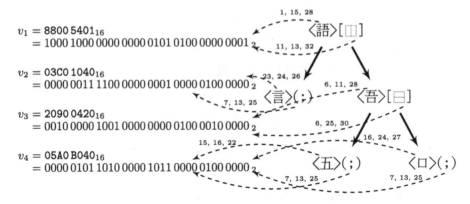

Fig. 2. Calculating a bit vector [23].

words which are Bloom filters of the tree's node labels. The first word, denoted v_1 in the figure, encodes the root of the tree. The functor ⊡ and arity (it is a binary node) are hashed to select bits 11, 19, and 32; the head 語 selects bits 1, 15, and 28. Similarly, the left child of the root determines v_2, the right determines v_3, and all other nodes select bits in v_4. Conditions on trees expressed in the query language, such as "must contain this label anywhere in the tree" imply conditions on the bit vectors, such as "at least three bits in this subset of the indices, must be set"; and the division into words allows computing the bit vector for a tree given its root labels and the vectors of the root's children.

1.2   Related Work

Bloom filtering [3] is well known. Guo et al. [9] describe the technique, also well known, of combining Bloom filters with Boolean AND and OR to perform the same operations on the sets the filters approximate. Our tree-matching problems are connected with *unification* in logic programming languages such as Prolog [4], and Aït-Kaci et al. [1] introduced bit vectors as a way to solve unification problems.

Advanced query languages for trees often take the form of modifications to the language of regular expressions. Polách describes such pattern matching in general abstract terms [18], and there is much work on regular expression-like tree matching specific to computational linguistics applications [7,14,15]. Some of our own work [24,25] applies bit vectors to unification in computational linguistics. Kaneta et al. [11] use them for another tree-matching problem. These references are described in more detail, with others related to *kanji* dictionaries and the linguistics application of IDSgrep, in our earlier IDSgrep paper [23].

The difficulty of high-dimensional queries has become known as the *curse of dimensionality*: exact query problems in high-dimensional spaces, across a wide range of different kinds of problems, have a strong tendency toward costs in time, space, or both that are exponential in the dimension. Approximation techniques have become standard in efforts to achieve practical results for high dimensions [10]. Theoretical work like that of Williams [28] links the difficulty of similarity search to the Strong Exponential Time Hypothesis, essentially saying that (for exact queries, in the worst case), we cannot do better than looking at $\Omega(n)$ points without proving unexpected deep results in complexity theory. Query languages that include Boolean logic have obvious direct application to satisfiability-type $\mathcal{NP}$-hard problems, and work like that of Frances and Litman [8] and our own [21,26] links similarity-based queries to $\mathcal{NP}$-hardness. Since advanced query languages seem doomed to hardness when the dimension is high enough, there is interest in at least *measuring* dimensionality of real data to detect when that phenomenon occurs [17,20].

The Binary Decision Diagram (BDD) is an interesting data structure in its own right, used here as a black box. Knuth describes its workings in detail [13]. IDSgrep's implementation uses the BuDDy library, due to Lind-Nielsen [16].

1.3   Notation

Although the software manipulates bit vectors as constant-sized objects using CPU bitwise instructions, we write them as if they were sets (implicitly, the sets of indices containing 1 bits) as a notational convenience. Thus, for bit vectors u and v, we write $u \cap v$ for the bitwise AND of u and v; $u \cup v$ for the bitwise OR; and $u \subseteq v$ for the statement that v contains a 1 bit at every index where u contains a 1 bit. We also write $|u|$ for the Hamming weight (or population count) of u, which is the number of 1 bits.

2   Bit-Vector Search with Enhancements

IDSgrep [23] performs a general database query (Problem 1) on opaque objects representing *kanji* dictionary entries and queries against them. Both happen to be EIDS trees as mentioned earlier, but the data structure is deliberately abstracted from the query algorithm. From the search algorithm's point of view, there is simply a database of arbitrary objects and a relatively expensive function $match(N, H)$ which returns a Boolean value true if and only if N (the *needle*, an EIDS tree representing the query) is considered to match H (the *haystack*, an object from the database). With only that abstract interface, nothing better than linear search would possible.

However, the underlying query language also provides bit vectors for the database objects, and filtering functions from the bit vectors to Boolean results, guaranteed to return true for all vectors of objects that match the query. A more efficient, but still linear, search can evaluate the filter functions first, and only invoke $match(N, H)$ when all filters return true. The bit vectors and filtering functions remain opaque, abstracted from the search algorithm.

The first layer of filtering uses a *lambda filter*, which is a pair (m, λ) consisting of a bit vector m and an integer λ, considered to match a dictionary entry's vector v if $|m \cap v| > \lambda$. Note that setting $\lambda = -1$ would match everything, giving a correct but unhelpful filter; the filter generator heuristically attempts to maximize λ and minimize m, while maintaining correctness.

The second layer of filtering uses a binary decision diagram directly encoding a monotonic function from bit vectors to Boolean truth values, again attempting to heuristically make that function return true as rarely as possible while still including all vectors of matching EIDS trees. This *BDD filter* is evaluated only if the lambda filter matches; then only if the BDD filter also matches does IDSgrep proceed to evaluating the exact EIDS matching function. As described in previous work, use of these filtering layers results in a significant improvement in query time on real-life data [23].

The new issue we address in the present work is that although the filtering can avoid many invocations of $match(N, H)$, the lambda filter at least is still tested for every entry in the database; and so there is a $\Omega(n)$ lower bound on all queries, though with an improved constant because of the filtering. How can the search algorithm on these opaque objects avoid looking at every entry?

2.1   Blocks with Bounds

Suppose the index file is divided into *blocks* and we record some information about each block summarizing logical statements that apply to all vectors in the block. If, based on that information, we can infer that *no* vector in a block could possibly match the query, then we can skip over examining the entire block. If the number of blocks is asymptotically smaller than the number of entries in the database, and we end up accepting (not skipping) less than a constant fraction of them, then the query time can break the $\Omega(n)$ barrier.

Bearing in mind the monotonic nature of the filtering layers, which implies that only 0 bits in a vector are really useful for excluding entries, we would like the summary to give maximal information about the 0 bits of the vectors in the block. We choose to store *containment* and *cardinality* bounds: for each block, a bit vector b and integer μ such that for every vector v in the block, $v \subseteq b$ and $|v| \leq \mu$. The optimal values are b equal to the OR of all vectors in the block and μ equal to the greatest Hamming weight. Note the similarity in structure between these bounds and the lambda filtering function.

Given bounds (b, μ) for a block and a lambda filter (m, λ), if $\min\{|b \cap m|, \mu\} \leq \lambda$ then it is impossible for any vector in the block to match the filter, and we can skip the entire block. Similarly, we can detect cases where an entire block fails to match a BDD filter, by a traversal of the BDD.

We report experimental results for a range of block sizes up to putting the entire database in a single block. Doing that is equivalent to not using blocks at all, except in the rare case of a match-nothing query that can be proven to match nothing from the whole-database containment and cardinality bounds.

2.2 Sorting

Since in our application the order of entries in the database is not significant, we can use that order to enhance search. In particular, it would cost very little to sort the index into lexicographically increasing order. If we are also splitting the index into blocks, we can sort within each block.

Given a lambda or BDD filter, we can easily compute the lexicographic range of vectors it could match, and then at query time, start with a binary search to find the first vector in the index or block that is within the range. Unfortunately, since both filters are monotonic, the all-ones vector will be matched by every filter that matches anything, and so only the lower end of the range usefully limits the search, and this improvement can only improve the constant in the search time. Nonetheless, it seems an inexpensive way to remove a few more filter checks from the search.

We compute the lexicographic bounds just once per query. In principle, a more detailed calculation could find lexicographic bounds on the vectors that match the query *and* that also obey the containment and cardinality bounds of a block, and this way we could usefully generate a lexicographic upper as well as a lower bound. But such a calculation would be more expensive in itself and would have to be repeated for every block instead of being done once per query. It seems unlikely to give much benefit in practice over the other ways we are already applying the containment and cardinality bounds. We leave testing that for future work.

Sorting the index data by bit vector implies that our eventual access to the dictionary file will be in randomized order, and random seeking could significantly increase the overall cost of the search when the dictionary file is stored on disk. To avoid this issue, when sorting the index we also generate a sorted dictionary with the entries arranged in the same order as the index; then accesses to dictionary entries during search will at least be sequential, if not consecutive.

2.3 Clustering

It ought to be the case that vectors in the same block are similar to each other, for some definition of similarity. Then queries are more likely to match either none or many of the vectors in the block, maximizing the chance that we can exclude the block on the basis of its summary information. If we must keep vectors in their original database order then we are stuck with making blocks be consecutive intervals of that order; but if we are sorting the index anyway, it makes sense to do clustering first and make the blocks correspond to clusters. Even the sorting itself, followed by making blocks out of consecutive entries, might be expected to provide some clustering benefit because similar vectors might tend to appear near each other in lexicographic order.

Noting the way the containment bound works as a bitwise OR of the vectors in the block, it seems the worst situation is when we include a vector in a large block that has a 1 at an index where all others in the block are 0. Then just because of including that vector, we have an additional 1 in the containment bound and fewer chances to skip all the other vectors in the block. We tested a variant of k-means clustering using a special similarity measurement that captures the idea of avoiding such situations.

Suppose, during clustering, we are considering moving a vector v into a cluster C_i, which is a multiset of vectors. If v is already assigned to C_i, let C_i' be the cluster with (one instance of) v removed; otherwise $C_i' = C_i$. Then we compute:

$$fit(v, C_i') = \frac{1}{|C_i'| + 1} \left[\min_{v_j=1} |\{w \in C_i' | w_j = 1\}| + \begin{cases} 1 & \text{if } v \text{ was assigned to } C_i, \\ 0 & \text{otherwise.} \end{cases} \right]$$

This says that how well a vector fits in a cluster is basically the fraction of other vectors in the cluster that share its least popular attribute; a vector will fit well in a cluster where all its 1 bits are already included in the cluster's bound as a result of many other vectors. The special handling for the case of v "already" assigned to C_i appears to be necessary for reliable termination; other variations we tried would loop indefinitely on some inputs. For similar reasons, we limited cluster size to at most twice the initial block size. Any cluster currently at the maximum size limit will not be considered as a possible destination for moving a vector during the optimization. That counteracted an observed tendency for the algorithm to put most of the database in a single huge cluster.

The clustering algorithm starts by assigning consecutive blocks of vectors from the index to be clusters (in input order if unsorted, or sorted order if we are sorting) and then iteratively attempts to move every vector to the cluster that maximizes its value of $fit(v, C_i')$, until no more such moves are possible. If using sorted indices, we sort within the clusters again after finding them.

3   Experimental Results

We extended the current version of IDSgrep to use the techniques described in the previous section, creating a special version for these experiments designated version 0.5.2, and we evaluated it using the same data, test queries, and hardware and operating system configurations used in our earlier work [23]. The test database contains 217,288 entries for decompositions of Han characters especially concentrating on Japanese *kanji*, gathered from the CJKVI [12], CHISE-IDS [6], KanjiVG [2], and Tsukurimashou [19] projects. There are 1,642 test queries, representing a spectrum of complexity and result set size from single exact-match character queries to more complicated Boolean operations. Speed testing was performed on a MacBook Pro equipped with a 2.3 GHz Intel iCore i7 CPU and 8 G of RAM, running Mac OS X 10.9.5. A package of our code and data is available to assist in reproducing the results, and includes the experimental results that were omitted here for space reasons.

The test query set from our earlier work [23] was designed to demonstrate the dictionary application, and includes many queries for which lambda and BDD filtering are ineffective. Those queries also tend to be relatively slow in other parts of the software; as a result, they dominate the total real running time for the test set taken as a whole when filtering is applied, and they make time measurements specific to filtering difficult. To better measure the filtering-specific techniques in the present paper, we separated out 512 "slow" queries, which are those with associative-match or match-anywhere operators at the root, or consisting solely of Boolean combinations of such queries. Such queries are easily recognized by testing that definition, and their important feature is that they generate filter functions which match nearly everything, so filtering search has little effect. The remaining 1,130 queries, where filtering search is expected to be of more use, are designated "fast." We tested each power of two block size from 4 to 262,144; that last, being larger than the database, effectively means no blocking at all.

Running the 1,130 queries in the fast query test set on the database of 217,288 entries means doing, or avoiding, a total of $1,130 \times 217,288 = 245,535,440$ tests of whether a query matches a database entry. Each test either results in the entry being discarded at some level of processing, or in a final match which returns a result. Figure 3 shows how frequently these outcomes occurred in our experiments, for selected choices of parameters. The categories shown in each stacked column represent increasingly expensive outcomes. A match may be discarded by skipping a block; by binary search (only for sorted indices); by the lambda filter; by the BDD filter; or it may be a BDD hit, which is true of 894,341 tests (0.36 %) for all parameter settings. Of those BDD hits, 29,606 will match in the final tree test, but that is too small a proportion (0.012 %) to usefully depict on the chart. Results shown in this figure are the same for all trials of each parameter set because the query algorithm is deterministic. A similar chart for the 512 slow queries is shown in Fig. 4. Here also, the final tree matches (137,415, or 0.12 % of $512 \times 217,288 = 111,251,456$ tests) form too small a proportion to be visible on the chart.

The running times for the fast and slow query test sets are shown in Figs. 5 and 6 respectively. Note these figures use logarithmic scales. The quantities plotted are sample means of user CPU seconds per loop as measured by IDSgrep's built-in statistics feature, with error bars representing intervals of ± 2 sample standard deviations, on 20 trials of each combination of parameter settings.

4   Discussion and Conclusions

The outcome counts of Fig. 3 show that the new approach of excluding blocks based on their containment and cardinality bounds allows the query to exclude a significant fraction of more expensive filter checks, for the queries in the "fast" set where filtering is effective. Without sorting or clustering, this effect is only significant at the smallest block sizes, but with sorting, clustering, or both, we can exclude many blocks even at block sizes up to thousands of vectors. For sorted indices, the lexicographic lower bound allows excluding a few more vectors, up to about 10 % on the largest block sizes.

However, these techniques have very little effect on the "slow" queries shown in Fig. 4. We are only excluding blocks at all at the smallest block sizes, we

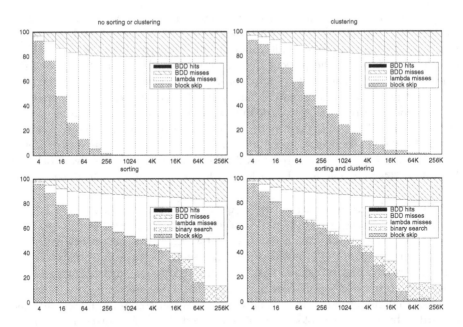

Fig. 3. Outcomes of testing dictionary entries against fast queries. Vertical axis: proportion in percent; horizontal: entries per block, 1K = 1,024.

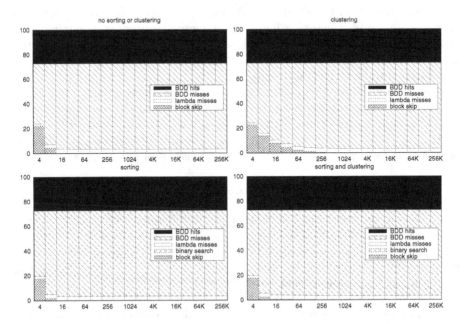

Fig. 4. Outcomes of testing dictionary entries against slow queries. Vertical axis: proportion in percent; horizontal: entries per block, 1K = 1,024.

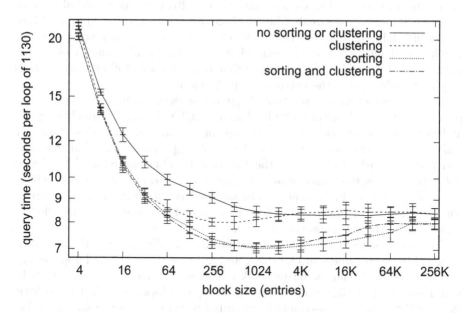

Fig. 5. Running times for fast queries.

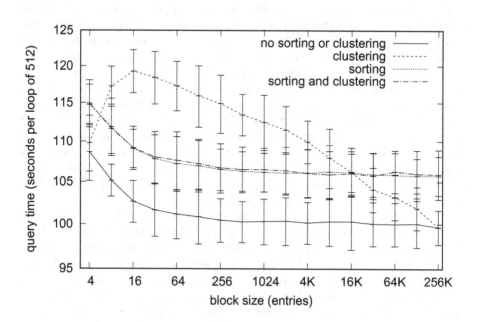

Fig. 6. Running times for slow queries.

are excluding very few blocks, and somewhat oddly, the clusters found without sorting first perform noticeably better than any other choice of options. These effects highlight the different nature of the "slow" queries: they are queries for which almost all vectors are lambda filter hits, and about a quarter of vectors are also BDD filter hits. Trimming the time consumption of filter *misses* makes very little difference to the bottom line performance.

To some extent we can say that IDSgrep is a victim of its own success. The plain $\Omega(n)$ search algorithm with lambda and BDD filtering is already so good, and in particular has such small constants for the filtering step, that it already shifts much of the running time away from filtering and into the final tree tests and ancillary tasks like parsing the file formats. For that reason, we should not expect to see the trends shown in Figs. 3 and 4 to be reflected strongly in the overall times of Figs. 5 and 6. The outcome counts measure only filtering, whereas the overall times also include parsing and tree matches, which do not vary between the experimental treatments.

The results suggest two directions for future work. First, more advanced variations of bit vector query are best targeted to larger data sets, and applications beyond *kanji* dictionaries; future work on bit vector queries in general might better use other applications for testing. Skipping blocks, even if it saves little time for IDSgrep, can reasonably be expected to help more when failing to skip blocks is more expensive, as in a very large database or one with an even more expensive query language than ours. It would be interesting to apply these techniques to advanced query languages on other, much larger, data collections—for instance, document databases with bits encoding keyword presence and Boolean queries over those, or image local-descriptor databases with very expensive similarity measurements. The smaller the result set in comparison to the overall database, the more we can realistically expect bit vector filtering to help.

Second, it may be appropriate to change the underlying bit vector and filter definitions in the specific application of *kanji* dictionaries. We are getting good filtering on some queries, but not on others, with 128-bit vectors; and the specific design and parameters of how EIDS trees generate bit vectors have not been changed or studied systematically since IDSgrep first introduced bit vector filtering. It would not cost much more time or space to switch to 256-bit vectors; could that bring more of the "slow" queries into the "fast" group? Can the definition of "slow" queries inform future bit vector designs that could exclude more entries and benefit more from the present work?

The difficulty appears to come from the associative and match-anywhere operators, which notably are not Boolean operators of the kind most likely to give hardness reductions. Maybe a redesigned bit vector definition specifically targeting those operators could move more queries into the "fast" class. In particular, the associative match operator currently generates a match-everything filter, but might be enhanced to generate a more restrictive filter. The match-anywhere operator in a query reduces to a Boolean OR of several other queries, naturally hitting many vectors, and an improved vector design might add some bits specifically to serve this operator better.

Requiring bit vector filters to be monotonic was an important technical aspect of the IDSgrep design. It made the difference between the feasibility and infeasibility of using BDDs. However, it limits the benefit of our new sorting technique, because monotonicity means we can only compute a useful lexicographic *lower* bound. The upper bound is always the all-ones vector. If we could make non-monotonic filters work with the rest of the system, it might help toward the goal of better than linear query time for easy queries. In particular, if a query matching only one entry could also match only one vector then we could expect the binary search to find that one vector after only $O(\log n)$ steps.

It is natural to extend the single-level blocking done here to a recursive tree-like structure, with a constant number (rather than constant size) of blocks each divided into smaller blocks, through as many levels as needed. A recursive data structure would lend itself to proving sublinear bounds for queries where that may be possible, while still degrading gracefully to linear time on harder queries where the lower bounds forbid anything better.

To conclude, we have proposed new techniques for filtering bit vector search, and tested them in the specific application of a *kanji* dictionary. The new techniques are experimentally shown to allow skipping a significant fraction of more expensive vector tests on some kinds of queries, but they cost enough to be of limited use for small databases or for queries where bit vectors are already failing. We have described technical issues relevant to our implementation. We have discussed these results and their implications both for bit vector query in general and the *kanji* dictionary application in particular.

References

1. Aït-Kaci, H., Boyer, R.S., Lincoln, P., Nasr, R.: Efficient implementation of lattice operations. ACM Trans. Program. Lang. Syst. **11**(1), 115–146 (1989)
2. Apel, U.: KanjiVG. http://kanjivg.tagaini.net/
3. Bloom, B.H.: Space/time trade-offs in hash coding with allowable errors. Commun. ACM **13**(7), 422–426 (1970)
4. Bramer, M.: Logic Programming with Prolog, 2nd edn. Springer, London (2013)
5. Chávez, E., Navarro, G., Baeza-Yates, R., Marroquín, J.L.: Searching in metric spaces. ACM Comput. Surv. **33**(3), 273–321 (2001)
6. CHISE project. http://www.chise.org/
7. Choi, Y.S.: Tree pattern expression for extracting information from syntactically parsed text corpora. Data Min. Knowl. Disc. **22**(1–2), 211–231 (2011)
8. Frances, M., Litman, A.: On covering problems of codes. Theor. Comput. Syst. **30**(2), 113–119 (1997)
9. Guo, D., Wu, J., Chen, H., Yuan, Y., Luo, X.: The dynamic bloom filters. IEEE Trans. Knowl. Data Eng. **22**(1), 120–133 (2010)
10. Indyk, P., Motwani, R.: Approximate nearest neighbors: towards removing the curse of dimensionality. In: Proceedings of the Thirtieth Annual ACM Symposium on Theory of Computing. pp. 604–613. ACM, New York (1998)
11. Kaneta, Y., Arimura, H., Raman, R.: Faster bit-parallel algorithms for unordered pseudo-tree matching and tree homeomorphism. J. Discrete Algorithms **14**, 119–135 (2012)

12. Kawabata, T.: IDS data for CJK unified Ideographs. https://github.com/cjkvi/cjkvi-ids
13. Knuth, D.E.: The Art of Computer Programming, Pre-fascicle 1B, vol. 4. Addison-Wesley, Reading (2009)
14. Lai, C., Bird, S.: Querying linguistic trees. J. Logic Lang. Inf. **19**(1), 53–73 (2010)
15. Levy, R., Andrew, G.: Tregex and Tsurgeon: tools for querying and manipulating tree data structures. In: Calzolari, N., Choukri, K., Gangemi, A., Maegaard, B., Mariani, J., Odijk, J., Tapias, D. (eds.) 5th International Conference on Language Resources and Evaluation (LREC 2006), Genoa, Italy, 22–28 May 2006
16. Lind-Nielsen, J.: BuDDy: a BDD package. http://buddy.sourceforge.net/manual/main.html
17. Ott, E.: Chaos in Dynamical Systems, 2nd edn. Cambridge University Press, Cambridge (2002)
18. Polách, R.: Tree pattern matching and tree expressions. Master's thesis, Czech Technical University in Prague (2011)
19. Skala, M.: Tsukurimashou font family and IDSgrep. http://tsukurimashou.osdn.jp/
20. Skala, M.: Measuring the difficulty of distance-based indexing. In: Consens, M., Navarro, G. (eds.) SPIRE 2005. LNCS, vol. 3772, pp. 103–114. Springer, Heidelberg (2005). doi:10.1007/11575832_12
21. Skala, M.: On the complexity of reverse similarity search. In: Chávez, E., Navarro, G. (eds.) First International Workshop on Similarity Search and Applications (SISAP 2008), Cancun, Mexico, 11–12 April 2008, pp. 149–156. IEEE (2008)
22. Skala, M.: Tsukurimashou: a Japanese-language font meta-family. TUGboat **34**(3), 269–278. In: Proceedings of the 34th Annual Meeting of the TEX Users Group (TUG 2013), Tokyo, Japan, 23–26 October 2013 (2014)
23. Skala, M.: A structural query system for Han characters. Int. J. Asian Lang. Process. **23**(2), 127–159 (2015)
24. Skala, M., Krakovna, V., Kramár, J., Penn, G.: A generalized-zero-preserving method for compact encoding of concept lattices. In: 48th Annual Meeting of the Association for Computational Linguistics (ACL 2010), Uppsala, Sweden, 11–16 July 2010, pp. 1512–1521. Association for Computational Linguistics (2010). http://www.aclweb.org/anthology/P10-1153
25. Skala, M., Penn, G.: Approximate bit vectors for fast unification. In: Kanazawa, M., Kornai, A., Kracht, M., Seki, H. (eds.) MOL 2011. LNCS (LNAI), vol. 6878, pp. 158–173. Springer, Heidelberg (2011). doi:10.1007/978-3-642-23211-4_10
26. Skala, M.A.: Aspects of metric spaces in computation. Ph.D. thesis, University of Waterloo (2008)
27. Unicode Consortium: Ideographic description characters. In: The Unicode Standard, Version 6.0.0, Section 12.2. The Unicode Consortium, Mountain View, USA (2011). http://www.unicode.org/versions/Unicode6.0.0/ch12.pdf
28. Williams, R.: A new algorithm for optimal 2-constraint satisfaction and its implications. Theor. Comput. Sci. **348**(2–3), 357–365 (2005)

Domain Graph for Sentence Similarity

Fumito Konaka[✉] and Takao Miura

Department of Advanced Sciences, HOSEI University,
3-7-2 KajinoCho, Koganei, Tokyo 184–8584, Japan
fumito.konaka.2t@stu.hosei.ac.jp, miurat@hosei.ac.jp

Abstract. In this work we propose a new method for word similarity. Assuming that each word corresponds to a unit of semantics, called *synset*, with categorical features, called *domain*, we construct a *domain graph* of a synset which is all the hypernyms which belong to the domain of the synset. Here we take an advantage of domain graphs to reflect semantic aspect of words. In experiments we show how well the domain graph approach goes well with word similarity. Then we extend the domain graph in sentence similarity independent of BOW. In addition we assess the execution time in terms of the task and show the significant improvements.

Keywords: Domain graph · Synsets · Similarity

1 Introduction

Nowadays we have a huge amount of digital information such as Web and Google Books with tags or some others. Typical examples are Social Network Service (SNS), twitters and BLOGs. Text in SNS are generally composed of short sentences with semantic ambiguity (synonymous/homonymous words and jargons), onomatopoeia (mimetic/words such as ding-dong, ba-dump, wwwww) and spelling inconsistency (or, *orthographic variants* of words) such as never/nevr, baby/babyyyyy. There can be no systematic formulation and we need computer-assisted approach to tackle with these problems.

A typical application is information retrieval and text mining by which we may get to the heart of interests in large datasets. In information retrieval, each document is described by a multiset over words, called *Bag-of-Words* (BOW). Here we construct a vector to each multiset where the column contains term frequency (TF) or the one multiplied by inverse document frequency (IDF). The approach is called *Vector Space Model* (VSM). BOW approach assumes that a multiset describes *stable* and *frequent* meaning. For example, a multiset {John, Dog, Bite} means "a Dog Bites John" but not "John Bites a Dog". All these mean, for example, that we can give document similarity and ranking using vector calculation.

However, VSM is not useful to short documents, since individual words may carry a variety of semantics and context by word-order. That's why VSM doesn't always go well with synonymous/homonymous situation and we hardly overcome WSD (Word Sense Disambiguity) issues.

L. Amsaleg et al. (Eds.): SISAP 2016, LNCS 9939, pp. 151–163, 2016.
DOI: 10.1007/978-3-319-46759-7_12

One of the difficulties is how to define similarity between sentences independent of VSM. We like to give sentence similarity not based on syntactic aspects but on semantic ones so that we examine more powerful retrieval on both long and short documents, including SNS texts.

This investigation contributes to the following points: First, we propose a new similarity between two words to reflect semantic aspects and to give indexing. Second, we improve query efficiency with the much simpler indices to words. Finally, we show the effectiveness of new similarity over SNS sentences.

The rest of the paper is organized as follows. In Sect. 2 we introduce several concepts and discuss why it is hard to achieve the definition of semantic similarity. In Sect. 3 we propose our approach and show some experimental results in Sect. 4 to see how effective our approach works. In Sect. 5 we conclude this investigation.

2 Word Similarity

To describe word similarity, there are two kinds of approaches, *knowledge-based* and *corpus-based*. Knowledge-based similarity means that, using semantic structures such as ontology, words are defined similar by evaluating the structure.

Usually *knowledge-base* consists of many *entry words*, each of which contains several units (*synsets*) of semantics, explanation sentences to each synset and relations (*links*) to other synsets. The links describe several semantic ties, called *ontology*, such as *hypernyms, synonyms, homonyms, antonyms* and so on. A synonym means several words share identical synset and a homonym means a single word carries multiple synsets. One of the typical examples is WordNet [14], an ontology dictionary containing 155,287 words which are divided into 117,659 groups (synsets), each of them corresponds to a synonymous group of words. Very often we see several links to other synsets of hypernyms (broader level) which have strong relationship of semantic similarity. For example, two words `Corgi` and `Bulldog` are similar because both are dogs where the synsets `corgi, bulldog` are defined in advance and they have links to a synset `dog`. In a same way, they are similar because both are mammals and because both are animals. However `Siamese` and `Bulldog` are not similar because both are not dog, but similar because both are mammals and because both are animals. We could even go so far as to say everything is similar because it is an object.

There have been several kinds of similarities proposed so far using WordNet, putting attention on the links [13] and some of them are available and open in WordNet::Similarity or NLTK[1]. Some of the similarity definitions are provided as *Path, Lch, WuPalmer, Res, Jcn* and *Lin* as follows.

$$Path = \max_{s_i, s_j \in w_1, w_2} -\log pathlen(s_i, s_j) \tag{1}$$

$$Lch = \max_{s_i, s_j} -\log \frac{pathlen(s_i, s_j)}{2 \times D} \tag{2}$$

---

[1] http://wn-similarity.sourceforge.net/, http://www.nltk.org/.

$$WuPalmer = \max_{s_i, s_j} \frac{2 \times depth(LCS(s_i, s_j))}{depth(s_i) + depth(s_j)} \tag{3}$$

$$Res = \max_{s_i, s_j} -\log P(LCS(s_i, s_j)) \tag{4}$$

$$Jcn = \max_{s_i, s_j} \frac{1}{2 \times \log P(LCS(s_i, s_j)) - (\log P(s_i) + \log P(s_j))} \tag{5}$$

$$Lin = \max_{s_i, s_j} \frac{2 \times \log P(LCS(s_i, s_j))}{\log P(s_i) + \log P(s_j)} \tag{6}$$

In the definitions, w_1, w_2 mean words and s_i, s_j synsets belonged to words.

While *Path*, *Lch* and *WuPalmer* are defined based on minimum path length, all of *Res, Jcn, Lin* are based on entropy. Both *WuPalmer* and *Jcn* assume synsets become similar when they locate at deep level.

There exist cyclic structures among *verb* relationship in WordNet 3.0 as in Fig. 1 [16]. Remember a *cycle* means a path (a sequence of arcs) such that there exist arcs $a_1, a_2, a_2, .., a_n$ and $a_1 = a_n$. We say a *loop* if $n = 1$. Then a graph is called *cyclic*. Otherwise *acyclic*. Also a *multiple path* means there are multiple distinct paths from a to b, or a node b has multiple parents[2]. Acyclic graphs may have multiple paths. Note that similarity based on minimum path length cannot be well-defined in a case of multiple paths as in the right of Fig. 1.

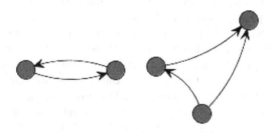

Fig. 1. Cycle and multiple path

As for corpus-based similarity, we take analytical information and apply characteristic features for similarity. One of the aproach take advantages of Latent Semantic Analysis (LSA) [5]. Here we build up a document matrix D over words and documents and decompose D into $D = U\Sigma V^T$ by Singular Value Decomposition (SVD). The technique is based on Principal Component Analysis and the *latent semantics* can be defined using co-occurrence of words and documents. Similarity corresponds to the one between two vectors of words over latent semantics.

3   Domain Graph and Similarity

Let us introduce a new similarity between words based on knowledge-base to capture their own sematic aspects. As we said previously, VSM means that

---

[2] Sometimes this is called a *ring*.

we interpret words and sentences in a "common" way, i.e., we have frequent interpretation to all the sentences and words even if we like to do that differently. The new similarity allows us to reflect role and relationship of the words.

Generally each word may correspond to a (non-empty) set of synsets with several features such as an ontological structure (considered as a directed graph) among synsets. To introduce similarity between two words, we discuss *Domain Graph* by which we take knowledge-base similarity into consideration, which mean we put our stress on relationship among words defined by knowledge-base. Generally word similarity can be defined with the one of synsets and ontology relationship among synsets: the stronger similarity means the closer relationship in a sense of path length or far apart distance. When two words are not similar, their synsets should be far apart with each other, the common synset has higher level in the ontology. Our discussion could have same motivation as *WuPalmar* and *Jcn*, but the similarity can be simple and efficient since we give the similarity in terms of graph structures.

Given a word, we assume there happen several synsets and each synset has *domain* feature as well as the explanation and links. A problem to decide which synset we think about, is called *Word Sense Disambiguation* (WSD) [15] and here we don't discuss WSD any more. Each synset belongs to several domains. For example, in WordNet, *Lexicographer File Names* (or *domains*) are defined as Table 1[3]. Note that every domain is complementary to ontology, i.e., a collection of short-cuts over paths, apart from levels,

The idea of *Domain Graph* comes from hypernym relationship consisting of nodes (synsets) and arcs (hypernym relationship) between synsets. We may consider domains as a new feature of a synset (a node). Given a word w with the synset s_w, a *Domain Graph* of w means all the hypernyms (ancestors) in such a way that every path belongs to one of the domains of s_w. In Domain Graph, it is assumed that every pair of synsets at shallow level may not be similar. This means a notion of domains allows us to ignore high levels of abstraction (such as `object`) and to overlap several parts in the ontology structure.

The graph can be described by *sub-graph* (nodes and arcs) in the directed graph. To examine how similar two words are, let us define similarity of graphs. Considered P, Q as two sets of nodes, the most common similarity is *Jaccard* coefficient, denoted by $Jacc(P, Q) = \frac{|P \cap Q|}{|P \cup Q|}$, where $|P|$ means the number of nodes in P. Note it may take time to obtain the co-efficient to large P, Q.

Let us define the *Domain Graph similarity* of two words w_1, w_2. Let s_1, s_2 be synsets corresponded to w_1, w_2 respectively. The DG similarity is defined to be the Jaccard similarity of $G(s_1)$ and $G(s_2)$ where $G(s)$ means all the nodes in the sub-graph of s in the domain graph of interests.

$$DGsimilarity(w_1, w_2) = Jacc(G(s_1), G(s_2))$$

---

[3] There are 45 *Lexicographer Files* based on syntactic category and logical groupings. They contain synsets during WordNet development. There is another approach *WordNet Domains* which is a lexical resource created in a semi-automatic way by augmenting WordNet with domain labels. To each synset, there exists at least one semantic domain label annotated by hands from 200 labels [1].

Minimum Hash (MinHash) function h provides us with efficient computation [3] for Jaccard coefficients. In fact, we can estimate the coefficient $Jacc(p, q)$ which is equal to the probability of min $h(p) = $ min $h(q)$. Given k MinHash functions and n function values to p, q that are matched, $Jacc(p, q)$ should be simply n/k, i.e., $\hat{J} = n/k$. Since we can obtain k hash values immediately, we can estimate Jaccard coefficients very quickly without any structural information such as indices.

Table 1. Excerpt from domains over synsets

ID	Domain	Description
00	adj.all	All adjective clusters
01	adj.pert	Relational adjectives (pertainyms)
02	adv.all	All adverbs
03	noun.Tops	Unique beginner for nouns
04	noun.act	Nouns denoting acts or actions
05	noun.animal	Nouns denoting animals
29	verb.body	Verbs of grooming, dressing and bodily care
30	verb.change	Verbs of size, temperature change, intensifying, etc.
31	verb.cognition	Verbs of thinking, judging, analyzing, doubting
44	adj.ppl	Participial adjectives

Let us discuss how to construct Domain Graph. Among others, we need WSD process to specify which synset we have to a word w. Figure 2 shows algorithms for "makeDomainGraph".

To select single synset to w, we do WSD process (*doWSD* in step 1) based on Lesk Algorithm [15] as shown in an algorithm for "scanDict" Here we examine how many relevant words we have with respect to a query, and choose the synset of the biggest ratio. In the algorithm for "makeDomainGraph", we select a synset s_w defined as below:

$$s_w = argmax_{s \in Synsets} \frac{|T \cap (gloss(s) \cup synonyms(s))|}{|gloss(s) \cup synonyms(s)|}$$

In the definition, given a word w in the algorithm for "makeDomainGraph", *Synsets* means all the synsets the word w has, T all the words appeared in a query, $gloss(s)$ all the words appear in the explanation (in WordNet) of a synset s and $synonyms(s)$ all the words containing s as its synset.

Let $D(s_w)$ be a domain (through WSD) which a synset s_w of a word w belongs to. Let c be a hypernym of s_w, then we follow the link to c as long as c belongs to $D(s_w)$. In short, a *Domain Graph* of w means all the hypernyms (ancestors) of the domain $D(s_w)$.

Let us show areas in Fig. 3. Let an area surrounded by solid lines be baseline synsets given by *Path* (formula 1), and an area by dotted lines be synsets in a

---

Algorithm 1 makeDomainGraph(Sentence T, Word $w \in T$)

Output: The Set of Subgraphs $DomainGraph(w)$

1: Synset $s_w \leftarrow doWSD(w,T)$, Domain $D(s_w)$

2: $DomainGraph(w) \leftarrow null$

3: $scanDict(s_w, D(s_w), DomainGraph(w))$

Algorithm 2 scanDict(Synset s, Domain D, $DomainGraph$)

Output: $DomainGraph$

1: $DomainGraph.add(subgraphs)$

2: The List of Hypernyms $HList \leftarrow getHypernyms(s)$

3: **for** each hypernym h in $HList$ **do**

4:     CurrentSynest $c \leftarrow h$, CurrentDomain $D(c)$

5:         **if** $D(c) = D$ **then**

6:             $scanDict(c, D, DomainGraph)$

7:         **end if**

8: **end for**

Fig. 2. Proposed algorithms

domain graph. We examine the similarity of synsets A and B in the left of Fig. 3, and the one of A and A' in the right of Fig. 3. Since a node A' has an arc to D but A doesn't, A' is more similar to B compared to A. In fact, in the baseline area, there are 2 arcs (AC and BC; A'C and BC) of the shortest path in Fig. 3 so that we have same similarity of AB and A'B. On the other hand, in the area by domain graph, we have different situation. We don't have same similarity AB and A'B because there are 3 arcs ACD and BD on left and 2 arcs A'D and BD on right.

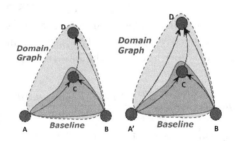

Fig. 3. Area by baseline and domain graphs

4   Experiments

In this section let us discuss experimental results to examine the proposed approach. First, we examine the effectiveness of domain graph by comparing several similarities among words through our approach with and without the domain graphs. Second we extend our approach to sentence similarity. We discuss Domain Graph Approach of words to sentences.

In these experiments, we assume WordNet 3.0 and its domains. We also assume $k = 10$ for a MinHash function *murmurhash3* (which is obtained by small experiments) through experimental Java libraries [8].

4.1 Similarity Among Words

Here we examine 4 corpus sets each of which has score values to each pair of words by hands: Li30 [11] RG65 [17] WS353 [7] and VP130 [19]. Once we obtain our similarity values, we compare them with the scores within by looking at Spearman order-correlations. In this case, we examine all the synsets of word pairs to obtain the maximum similarity, same as formulas (1)–(6).

We give similarity between two words with domain graph and without. As the baseline similarity values, we examine *Path*, *Lch*, *WuPalmer*, *Res*, *Jcn* and *Lin* (formulas (1)–(6)) in Natural Language Took Kit (NLTK). Also, as the ontology in WordNet, we apply WS4J[4] as baseline Paths.

We show the results in Table 2 which contains correlation values (ρ) and execution time (s). The tables shows that ρ results with domain graph are the best ones except VP130, slightly superior to the one without: +0.045(Li30), +0.004(RG65), +0.12(WS353) and +0.032(VP130). The half of the execution shows the best ones too.

Table 2. Word similarity and efficiency

		Corpus							
		Li30		RG65		WS353		VP130	
		ρ	s	ρ	s	ρ	s	ρ	s
Path		0.729	2.189	0.781	2.243	0.296	4.495	0.725	2.817
Lch		0.729	2.219	0.781	2.302	0.296	4.58	0.725	2.776
WuPalmer		0.705	2.186	0.755	2.3	0.329	4.699	0.728	2.839
Res		0.704	4.151	0.776	4.271	0.329	6.608	0.661	4.717
Jcn		0.742	4.24	0.775	4.331	0.280	6.981	0.695	4.878
Lin		0.761	4.168	0.784	4.369	0.296	7.01	0.689	4.859
Domain graph	(No Index)	0.776	1.108	0.798	1.345	0.406	6.343	0.693	3.863
	(Index)		0.127		0.208		0.778		0.721
No Domain		0.731	1.462	0.794	1.92	0.286	10.491	0.661	4.107

As seen easily, the domain graph approach in most of the corpus has better correlation values to others (ρ). Note we don't discuss WSD issue about s_w to construct domain graphs. There is no sharp distinction with and without

---

[4] https://code.google.com/archive/p/ws4j/.

Table 3. Sentence pairs in PIT2015

Similarity	Sentence 1	Sentence 2
0	What the hell is Brandon bass thinking	Brandon Bass is shutting Carmelo down
1	EJ Manuel is the 1st qb taken huh	1st QB off of the board
2	Aaron dobson is a smart pick	They pluck Aaron Dobson from the Herd
3	Please give the Avs a win	Come on avs be somebody
4	Barry Sanders really won the Madden cover	So Barry Sanders is on the cover
5	I liked that video body party	I like that body party joint from Ciara

domain graph in our approach, because the domain graphs contain few multiple paths (say, only 8 nodes have multiple parent nodes within whole relations in WordNer 3.0 [16]) so that the results by our approach become the best but no big difference. As for execution efficiency, our graph approach is superior to others in LI30 and RG65, and equal in VP130. Indexed one means all the hash values are prepared in advance and no CPU overhead arises.

4.2 Semantic Textual Similarity

Let us examine how well the proposed approach goes with short sentences. Here we examine PIT2015 corpus, i.e., we examine PIT-2015 Twitter Paraphrase Corpus [18] as a test corpus, which includes many short sentences extracted at more than 500 Twitter sites from April 24, 2013 to May 3, 2013. The corpus contain 17,790 pairs of sentences divided into 13,063 pairs for training and 4,727 pairs for development. And there are 972 pairs included for test. We examine these 13,063 pairs for training and the 972 pairs for test. Each pair has scored by Amazon Mechanical turk in terms of 0 (not similar) to 5 (most similar). Also morphological and proper noun information have been attached to each word. Table 3 shows some examples of the PIT corpus.

We conduct two experiments, calculating *sentence similarity* and *execution time*.

4.2.1 Sentence Similarity

We apply preprocessing (lowercase conversion and making original form by Tree-Tagger[5]) to the corpus and provide the feature information as well as the one by the corpus.

Here we add *character bigram, trigram* for characteristic words and *word unigram, bigram* for word sequences to every sentence in a form of Jaccard co-efficients as the feature values. Then we examine sentence similarity by Support Vector Regression (SVR) by using the features above. Given a sentence $n = 1, .., N$,

---

[5] http://www.cis.uni-muenchen.de/~schmid/tools/TreeTagger/.

let x_n be a feature vector of 5-dimension for the n-th sentence: x_n^1, x_n^2 for character bigram, character trigram respectively, x_n^3, x_n^4 for word unigram, word bigram respectively and x_n^5 for domain graph. Then $y(x_n)$ means the regression value to the 5 features of x_n through SVR. Using LIBSVM [2], we apply ε-SVR with default parameters by minimizing V targeted for better fitting as below:

$$V \;=\; C \sum_{n=1}^{N} (\xi_n + \hat{\xi}_n) + Z$$

$$\xi_n = \begin{cases} 0 \;\; if \;(t_n \le y(x_n) + \varepsilon) \\ \xi_n \;\; if \;(t_n > y(x_n) + \varepsilon) \end{cases} , \quad \hat{\xi}_n = \begin{cases} 0 \;\; if \;(t_n \ge y(x_n) + \varepsilon) \\ \hat{\xi}_n \;\; if \;(t_n < y(x_n) + \varepsilon) \end{cases}$$

In the definition of V, the first term shows a penalty to data beyond an allowable error ε for regression while the second term Z means its normalization. We put the similarity values to each pair by SVR and compare them with the one in the corpus and we evaluate the result by Pearson correlations.

As the baseline, we discuss ASOBEK [6] proposed by Eyecioglu which is based on SVR with character bigram (x_n^1) and word unigram (x_n^3), Logistic Regression (LR) based on word n-gram (x_n^3, x_n^4) by Das [4], and Weighted Textual Matrix Factorization (WTMF) by Guo [9]. We apply DomainGraph approach to all the features $(x_n^1, .., x_n^5)$, the one except character bigram/trigram $(x_n^3, .., x_n^5)$ and the one with only domain graph (x_n^5) feature.

Let us show the result in Table 4. Our approach with all the features scored the best because of domain graph feature (x_n^5). Compared to ASOBEK character 2-gram and (word 1-gram) and LR (word n-gram), this approach is at least 9.9 % better but comparable. So is true for our domain graph approach without character n-gram (14.9 % better). On the other hand, only one feature in word 1-gram and domain graph doesn't work well.

First of all, let us discuss why WTMF doesn't work well. As shown in Table 3, WTMF works poor (62 %), because the approach comes from word co-occurrence in documents and the situation can be hardly detected in short sentences. It seems hard to solve the problem by Matrix Factorization.

Table 4. Sentence similarity results (PIT2015)

	Model	Features	Correlation	Improvement
(1)	ASOBEK	x_n^1, x_n^3	0.504	0.90
(2)		x_n^3	0.071	0.13
(3)	LR	x_n^3, x_n^4	0.511	0.91
(4)	WTMF		0.35	0.62
(5)	Our approach	$x_n^1, ..., x_n^5$	0.561	1.00
(6)		x_n^3, x_n^4, x_n^5	0.488	0.87
(7)		x_n^5	0.071	0.13

In ASOBEK, two sentences in Table 5 (1) and (2) look similar to (2). Note (1′) and (2′) contain words/morphemes and look alike. The corpus gives the similarity 4 while the correlation is 0.727 by our approach. The major difference of two ASOBEK comes from character bigram (x_n^1). Looking into the detail, we have Jaccard coefficients of character bigram 0.5714, character trigram 0.5294, word unigram 0.4375, and word bigram 0.3125 and domain graph 0.7241. ASOBEK (1) takes x_n^1, x_n^3 into consideration and the x_n^1 is bigger than x_n^3, so ASOBEK (2) goes worse. On the other hand, our approach (5) is 14.9 % better than (6) since x_n^5 is dominant.

Table 5. Similar sentences

(1)	MHP wishes you a safe and happy Memorial Day weekend
(2)	We hope that everyone has a very safe and happy Memorial Day Weekend
(1′)	wish#verb, memorial#noun, day#noun, weekend#noun
(2′)	hope#verb, have#verb, memorial#noun, day#noun, weekend#noun

Our approach (7) works poor because every sentence contains many words and word unigram and bigram should be considered. It is said that *character n-gram* may work well for spelling inconsistency. Some examples in PIT2015 are the following: "The **ungeekedeliteschicago** Daily is out", "Lydia is a **GROOOOOOOOWN** woman", "I will **brin** them Taco Bell chipotle soo they let me stay."

Let us go into the detail of spelling inconsistency issue. We examine another corpus SemEval2012 MSRpar, MSRvid and SMTeuroparl for short sentences with training data, because they contain no spelling inconsistency. In Table 6 we show the comparison results with and without character n-gram features (x_n^1, x_n^2). As seen easily, there happen no difference and we can say *character n-gram* may not be useful for spelling inconsistency issue by our approach.

Table 6. Sentence similarity results (SemEval2012)

Model	Feature 1 x_n^3, x_n^4, x_n^5	Feature 2 $x_n^1, x_n^2, x_n^3, x_n^4, x_n^5$	Improvement
MSRpar	0.409	0.610	1.49
MSRvid	0.684	0.811	1.19
SMTeuroparl	0.501	0.552	1.10
PIT2015	0.488	0.561	1.15

4.2.2   Execution Time

We also assess the execution time of semantic textual similarity. In this experiment, we use 25,000 sentences randomly selected from the PIT corpus as a

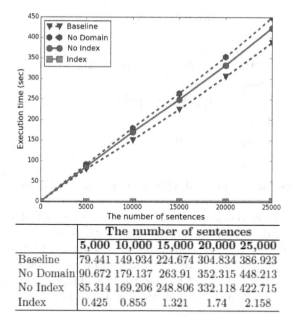

	The number of sentences				
	5,000	**10,000**	**15,000**	**20,000**	**25,000**
Baseline	79.441	149.934	224.674	304.834	386.923
No Domain	90.672	179.137	263.91	352.315	448.213
No Index	85.314	169.206	248.806	332.118	422.715
Index	0.425	0.855	1.321	1.74	2.158

Fig. 4. Execution times (sec)

dataset. The time is measured during calculating the similarity between all sentences and a query randomly selected from the dataset. Each system is compared with others in terms of the time which is measured ten times and averaged.

We consider four systems, *No Index* measuring the time from constructing the domain graph to calculating the similarity, *Index* measuring the time to calculate the similarity using the index, *No Domain* constructing the graph without domain likewise Sect. 4.1 and *Baseline*.

Measuring sentence similarity systems using WordNet tend to calculate the similarity in every conceivable combination and to use the maximum value [12]:

$$sim(S_1, S_2) = \sum_{w_i \in S_1} \max_{w_j \in S_2} sim_{WN}(w_i, w_j) \tag{7}$$

In Eq. 7, S_1, S_2 are sentences and im_{WN} is the similarity using WordNet. Thus we consider Eq. 7 is appropriate as *Baseline*. In this experiment sim_{WN} is *Path* implemented by WS4J[6].

We show the environment in Table 7 and the results in Fig. 4.

Figure 4 shows *No Index* execute 25,000 sentences with 0.94 times as much as *No Domain*. This improvement is caused by the neglect of structures at shallow level. And it is mentionable *No Domain* is inferior to *Baseline*. *Path* search the shortest path by recording the current shortest and updating it. In other word, *Baseline* can shift to the next step when *Path* cannot renew. However, Domain

---

[6] https://code.google.com/archive/p/ws4j/.

Table 7. Machine specs

Item	Spec
CPU	Intel(R) Xeon(R) X3430 2.40 GHz
Memory	16 GB
OS	Windows 7 64bit OS
Language	Java
Compiler	Eclipse Compiler for Java

Graph need more times compared with *Baseline* since it use all synsets meeting conditions.

Figure 4 also shows *Index* execute 25,000 sentences with 0.006 times as much as *No Index*. For calculating the sentence similarity, *Baseline* seeks for the maximum word similarity in all combinations. *Baseline* have to calculate all combinations with each sentence since the combinations depends on the input. By contrast, Domain Graph does not have to read the input again and calculate all combinations since the similarity is defined as Jaccard coefficients.

5 Conclusion

In this investigation, we have proposed a new similarity among words using domain graph. The similarity provides us with ontological aspects on similarity while avoiding trivial knowledge often appears at shallow level. Also we have discussed semantic properties of the similarity based on domain graph independent of BOW aspects. We have shown how to obtain features of domain graph by minimum hash techniques so that the approach can be useful for information retrieval.

We have shown the effectiveness of our approach by experiments. The experiments show that the results by our approach become the best (but no big difference because of WorNet ontology) while the execution efficiencies are comparable. By extending the approach for sentence similarity, we have also shown domain graph approach works best, say, at least improved 9.9 %, (because of domain graph feature) than other baseline.

All these show our approach is promising for query to short sentences.

Some problems remain unsolved. Often spelling inconsistency makes the similarity worse or incorrect, but no sharp solution is proposed until now. Character n-grams or any other techniques are not enough to improve queries, but domain graph with word normalization could help the situation better.

References

1. Bentivogli, L., Forner, P., Magnini, B., Pianta, E.: Revising WordNet domains hierarchy: semantics, coverage, and balancing. In: COLING 2004 Workshop on "Multilingual Linguistic Resources", pp. 101–108 (2004)

2. Chang, C.C., Lin, C.J.: LIBSVM: a library for support vector machines. ACM Trans. Intell. Syst. Technol. (TIST) **2**(3), 27 (2011)
3. Cohen, E., et al.: Finding interesting associations without support pruning. IEEE Trans. Knowl. Data Eng. **13**(1), 64–78 (2001)
4. Das, D., Smith, N.A.: Paraphrase identification as probabilistic quasi-synchronous recognition. In: Proceedings of the Joint Conference of the 47th Annual Meeting of the ACL and the 4th International Joint Conference on Natural Language Processing of the AFNLP, vol. 1, pp. 468–476. Association for Computational Linguistics (2009)
5. Deerwester, S., Dumais, S., et al.: Indexing by latent semantic analysis. J. Am. Soc. Inf. Sci. **41**(6), 391407 (1990)
6. Eyecioglu, A., Keller, B.: ASOBEK: Twitter paraphrase identification with simple overlap features and SVMs. In: Proceedings of SemEval (2015)
7. Finkeltsein, L., et al.: Placing search in context: the concept revisited. In: Proceedings of the 10th International Conference on World Wide Web. ACM, 2001. pp. 406–414
8. Finlayson, M.A.: Java libraries for accessing the Princeton WordNet: comparison and evaluation. In: Proceedings of the 7th Global Wordnet Conference, Tartu, Estonia (2014)
9. Guo, W., Diab, M.: Modeling sentences in the latent space. In: Proceedings of the 50th Annual Meeting of the Association for Computational Linguistics: Long Papers, vol. 1, pp. 864–872. Association for Computational Linguistics (2012)
10. Konaka, F., Miura, T.: Textual similarity for word sequences. In: Amato, G., Connor, R., Falchi, F., Gennaro, C. (eds.) SISAP 2015. LNCS, vol. 9371, pp. 244–249. Springer, Heidelberg (2015). doi:10.1007/978-3-319-25087-8_23
11. Li, Y., et al.: Sentence similarity based on semantic nets and corpus statistics. IEEE Trans. Knowl. Data Eng. **18**(8), 1138–1150 (2006)
12. Liu, H., Wang, P.: Assessing sentence similarity using wordnet based word similarity. J. Softw. **8**(6), 1451–1458 (2013)
13. Meng, L., Huang, R., Gu, J.: A review of semantic similarity measures in wordnet. Int. J. Hybrid Inf. Technol. **6**(1), 1–12 (2013)
14. Miller, G.A.: WordNet: a lexical database for English. Commun. ACM **38**(11), 39–41 (1995)
15. Navigli, R.: Word sense disambiguation: a survey. ACM Comput. Surv. (CSUR) **41**(2), 10 (2009)
16. Richens, T.: Anomalies in the WordNet verb hierarchy. In: Proceedings of the 22nd International Conference on Computational Linguistics, vol. 1, pp. 729–736. Association for Computational Linguistics (2008)
17. Rubenstein, H., Goodenough, J.B.: Contextual correlates of synonymy. Commun. ACM **8**(10), 627–633 (1965)
18. Xu, W., Callison-Burch, C., Dolan, W.B.: SemEval-2015 Task 1: paraphrase and semantic similarity in Twitter (PIT). In: Proceedings of the 9th International Workshop on Semantic Evaluation (SemEval) (2015)
19. Yang, D., Powers, W.M.W.: Verb similarity on the taxonomy of WordNet. Masaryk University (2006)

Context Semantic Analysis:
A Knowledge-Based Technique for Computing
Inter-document Similarity

Fabio Benedetti[✉], Domenico Beneventano, and Sonia Bergamaschi

Dipartimento di Ingegneria Enzo Ferrari,
Università di Modena e Reggio Emilia, Modena, Italy
{fabio.benedetti,domenico.beneventano,sonia.bergamaschi}@unimore.it

Abstract. We propose a novel knowledge-based technique for inter-document similarity, called *Context Semantic Analysis* (CSA). Several specialized approaches built on top of specific knowledge base (e.g. Wikipedia) exist in literature but CSA differs from them because it is designed to be portable to any RDF knowledge base. Our technique relies on a generic RDF knowledge base (e.g. DBpedia and Wikidata) to extract from it a vector able to represent the context of a document. We show how such a *Semantic Context Vector* can be effectively exploited to compute inter-document similarity. Experimental results show that our general technique outperforms baselines built on top of traditional methods, and achieves a performance similar to the ones of specialized methods.

Keywords: Knowledge graph · Knowledge base · Inter-document similarity · Similarity measures

1 Introduction

Recent years have seen growing number of knowledge bases that have been used in several domains and applications. Besides DBpedia [2], which is the heart of the Linked Open Data (LOD) cloud [5], other important examples includes: Wikidata [25], a collaborative knowledge base; YAGO [22], a huge semantic knowledge base, derived from Wikipedia, WordNet and GeoNames; Snomed CT [6], the best known ontology in the medical domain and AGROVOC [7], a multilingual agricultural thesaurus we used recently for annotating agricultural resources [4].

Recent research trends indicate that semantic information and knowledge-based approaches can be used effectively for improving existing techniques, as Natural Language Processing (NLP) and Information Retrieval (IR); on the other hand, much still remains to be done in order to effectively exploit these rich models in these fields [21]. For instance, in the context of inter-document similarity, which plays an important role in many NLP and IR tasks, the classic techniques rely solely on syntactic information and are usually based on Vector Space Models [23], where the documents, composed by words, are represented

L. Amsaleg et al. (Eds.): SISAP 2016, LNCS 9939, pp. 164–178, 2016.
DOI: 10.1007/978-3-319-46759-7_13

in a vector space having words as dimensions. Such techniques fail in detecting relationships among concepts like in these two sentences: "*The Rolling Stones with the participation of Roger Daltrey opened the concerts' season in Trafalgar Square*" and "*The bands headed by Mick Jagger with the leader of The Who played in London last week*". These two sentences contain highly related concepts which can be found by exploiting the knowledge and network structure encoded within knowledge bases such as DBpedia, even if they are not contained explicitly in the text.

In this paper, we present *Context Semantic Analysis* (CSA), a novel technique for estimating inter-document similarity, leveraging the information contained in a knowledge base. One of the main novelty of CSA w.r.t. other knowledge-based techniques for document similarity is its applicability to generic RDF knowledge bases, so that all datasets belonging to the LOD cloud [5] (more than one thousand) can be used.

CSA is based on the notion of *contextual graph* of a document, i.e. a subgraph of the knowledge base which contains the contextual information of the document. The contextual graph is then suitably weighted to capture the degree of associativity between its concepts, i.e., the degree of relevance of a property for the entities it connects. The vertices of such a weighted contextual graph are then ranked by using *PageRank* methods, so obtaining a *Semantic Context Vector*, which represents the *context* of the document. Finally, we estimate the similarity of two documents by comparing their Semantic Context Vectors with standard methods, such as the *cosine similarity*. By evaluating our method on a standard benchmark for document similarity (which consider correlations with human judges), we show how it outperforms almost all other methods and how it is portable to generic knowledge bases. Moreover we analyze and show its scalability in a clustering task with a larger corpus of documents.

The paper is structured as follows. Section 2 contains the related work, while Sect. 3 is devoted to some preliminaries useful for the rest of the paper. Then, CSA is described in Sect. 4 and Sect. 5 contains its evaluation. Finally, the last Section contains some conclusions.

2   Related Work

Text similarity has been one the main research topic of the last few years due to wide range of its applications in tasks such as information retrieval, text classification, document clustering, topic detection, etc. [11]. In this field a lot of techniques have been proposed but we can group them in two main categories, *content based* and *knowledge enriched* approaches, where the main difference is that the first group uses only textual information contained in documents while the second one enriches these documents by extracting information from other sources, usually knowledge bases.

The standard document representation technique is the *Vector Space Model* [23]. Each document is expressed as a weighted high-dimensional vector, the dimensions corresponding to individual features such as words. The result is

called the *bag-of-words* model and it is the first example of *content based* approach. The limitation of this model is that it does not address polysemy (the
same word can have multiple meanings) and synonymy (two words can represent the same concept). Another technique belonging the *content based* group
is Latent Semantic Analysis (LSA) [9], which assumes that there is a latent
semantic structure in the documents it analyzes. Its goal is to extract this latent
semantic structure by applying dimensionality reduction to the terms-document
matrix used for representing the corpus of documents.

Recently, a lot of effort has been employed in designing new techniques for
text similarity which use information contained in knowledge bases. A first example of this *knowledge enriched* approaches is Explicit Semantic Analysis (ESA)
[10], which indexes documents with Wikipedia concepts and it uses Wikipedia
hyperlink structure information for mapping any text as a weighted vector of
Wikipedia-based concepts. Another documents similarity technique that lever-
age the information contained in Wikipedia is WikiWalk [27], where the personalized PageRank on Wikipedia pages is used, with a personalization vector
based on the ESA weights on concepts detected in the documents, to produce
a vector used for estimating the similarity. A big drawback of this approach is
the computational cost, indeed, for each document we have to execute first ESA
and then compute the personalized PageRank on the whole Wikipedia. Another
remarkable approach is SSA, i.e. Salient Semantic Analysis [12]. This method
starts with Wikipedia for creating a corpus where concepts and saliency are
explicitly annotated, then, the authors use this corpus to build concept-based
word profiles, which are used to measure the semantic relatedness of words and
texts. These group of *knowledge enriched* approach are designed for using only
Wikipedia as source of knowledge and they are not portable to generic knowledge bases. Our method CSA differs from them because it aims to be a generic
approach that can use any knowledge bases expressed according to the Semantic Web standard, i.e. described in RDF, so that all datasets belonging to the
Linked Open Data cloud [5] (more than one thousand) can be used as source of
knowledge. To the best of our knowledge, the only approach portable to generic
knowledge bases is the one proposed in [21], where the authors represent documents belonging to a corpus as graphs extracted form a generic knowledge base.
It differs from CSA because it is based on a Graph Edit Distance (GED) graph
matching method to estimate similarity, while in our approach a document is
represented as a vector and the similarity can be estimated more effortlessly by
using cosine similarity.

3   Preliminaries

3.1   Inter-document Similarity

Vector Space Models are generally based on a co-occurrence matrix, a way of
representing how often words co-occur; in a *term-document matrix*, each row
represents a word and each column represents a document. Let C be a corpus
composed of n documents, where each document d_j is composed by a sequence

of terms. Let m be the number of terms in C; the *term-document matrix* T is a matrix $m \times n$ where its cell (i, j) contains the weight t_{ij} assigned to term i in the document j. A document d_j is then represented by the *vector* $\boldsymbol{d_j} = [t_{1j}, \ldots, t_{mj}]$. Different strategies of weighting exist (see, for example, [19]); where the weight t_{ij} is equal to the number of time the term i appears in the document j. the most famous weighting strategy is *td-idf* (*Term Frequency - Inverse Document Frequency*) [19].

The most common way of estimating the similarity of two documents is the *cosine similarity*, i.e., the cosine or angular distance between context vectors representing the two documents, because it has been shown to be effective in practice for many information retrieval applications [9].

3.2   Knowledge Base

We focus on RDF knowledge bases[1]; an RDF KB can be considered a set of facts (statements), where each fact is a triple of the form <*subject,predicate,object*>. A set of such triples is an *RDF graph* $KB = (V, E)$: a labeled, directed multi-graph, where subjects and objects are vertices and the predicates are labeled edges between them. According to [8], vertices are divided in 3 disjoint sets, URIs U, blank nodes B and literals L; literals cannot be the subjects of RDF triples.

The triples of an RDF KB can usually be divided into *A-Box* and *T-Box*; while the *A-Box* contains instance data (i.e. extensional knowledge), the *T-Box* contains the formal definition of the terminology (classes and properties) used in the *A-Box*; as an example, Fig. 1 shows an extract of DBpedia[2]. Our methods relies only on the extensional knowledge of a KB, i.e. only on the *A-Box*; for our experiments we choose two generic domain KBs: DBpedia [2] and Wikidata [25], due to their large coverage and variety of relationships at the extensional level.

3.3   PageRank

PageRank was first proposed to rank web pages [20], but the method is now used in several applications for finding vertices in a graph that are most relevant

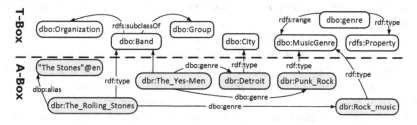

Fig. 1. Example of an RDF KB, with the *A-Box* and the *T-Box*.

---

[1] https://www.w3.org/TR/rdf-primer/.
[2] We abbreviate URI namespaces with common prefixes, see http://prefix.cc for details.

for a certain task. Let G be a graph with n vertices and d_i be the outdegree of the vertex i; the Standard PageRank algorithm computes the *PageRank vector* R defined by the equation:

$$R = cMR + (1 - c)v$$

where the *transition probability* matrix M is a $n \times n$ matrix given by $M_{ij} = 1/d_i$ if it exists an edge from i to j and 0 otherwise, c is the *damping factor*, a scalar value between 0 and 1 and the *personalization vector* v is a $n \times 1$ uniform vector in which each element is $1/n$. Standard PageRank uses just graph topology, but many graphs, as the ones in our case, come with weights on either nodes or edges, which can be used to *personalize* the PageRank algorithm. The *Personalized PageRank* [13] uses *node weights* to define a non-uniform vector v and thus biasing the computation of the *PageRank vector* R to be more influenced from heavier nodes. Another variant is the *Weighted PageRank* [26] which uses *edge weights* to define a custom transition probability matrix for influencing further the computation of the *PageRank vector* R.

4   Context Semantic Analysis

Given a corpus C of documents and an RDF knowledge graph KB, CSA is composed of the following three steps:

- **Contextual Graph Extraction:** a *Contextual Graph* $CG(d)$ containing the contextual information of a document d is extracted from the KB.
- **Semantic Context Vectors Generation:** the *Semantic Context Vector* $SCV(d)$ representing the context of the document d is generated analyzing its $CG(d)$.
- **Context Similarity Evaluation:** the *Context Similarity* is evaluated by comparing the context vectors of documents belonging to the corpus C.

4.1   Contextual Graph Extraction

Given a document d and a knowledge graph KB, the goal of this first step is to extract a subgraph from KB containing all the information about d. Our method relies only on the extensional knowledge of a KB, i.e. on its *A-Box*. More precisely, given a knowledge base KB, we consider the subgraph $KB_A = (V_A, E_A)$ where the triples are in the *A-Box* of the KB. We also exclude the triples containing literals, so, all the vertices V_A belongs to $(U \cup B)$ and every edge E_A corresponds to an *object property*. In Fig. 1 we have only 3 triples that belongs to KB_A: the ones containing the *dbo:genre* property.

The extraction of the Contextual Graph $CG(d)$ for a document d is a three-step process:

1. Starting Entities Identification: the entities of KG_A which are explicitly mentioned in the document d are identified: such set of entities is called *starting entities* of d, denoted by $SE(d)$. The problem to find the set $SE(d)$ is an instance of the well-known *Named Entity Recognition* problem [18]; it is out of scope of this work, we tested some of the already implemented techniques and on the basis of the obtained results, we empirically chosen DBpedia Spotlight [17] and TextRazor[3] to identify starting entities w.r.t. DBpedia and Wikidata, respectively.

2. Contextual Graph Construction: the Contextual Graph of the document d is defined as the subgraph of KG_A composed by all the triples that connect with a path of length l, at least 2 starting entities in $SE(d)$. More precisely, given a document d and a length $l > 0$, we define:

$$CG_l(d) = \{<s,p,o> \,|\, <s,p,o> \in KG_A \land <s,p,o> \in Path(s_1,s_2) \land$$
$$length(Path(s_1,s_2)) \leq l \land s_1, s_2 \in SE(d) \land s_1 \neq s_2\}$$

where $Path(s_1, s_2)$ is a path on KG_A from s_1 and s_2.

For example, let us consider the two sentences used in the introduction:
d_1: "***The Rolling Stones*** *with the participation of* ***Roger Daltrey*** *opened the concerts' season in* **Trafalgar Square**"
d_2: "*The bands headed by* **Mick Jagger** *with the leader of* ***The Who*** *played in* **London** *last week*".

It is easy to find as starting entities in DBpedia: $SE(d_1)$ {**The Rolling Stones, Roger Daltrey,Trafalgar Square**} and $SE(d_2)$ {**Mick Jagger, The Who,London**}. For example, by using $l = 2$ we obtain $CG_2(d_1)$ with 5 nodes and $CG_2(d_2)$ with 12 nodes; by using $l = 3$ we obtain $CG_3(d_1)$ with 141 nodes and $CG_3(d_2)$ with 66 nodes. The most significant portion of information shared between $CG_3(d_1)$ and $CG_3(d_2)$ is shown in Fig. 2.

3. Contextual Graph weighting: In the literature several graph weighting methods have been proposed to capture the degree of associativity between concepts in the graph, i.e., the degree of relevance of a property for the entities it connects [1,21]. The most common way of weighing a property p_i is to compute its *Information Content (IC)*, $IC(X = p_i) = -log(P(p_i))$, where $P(p_i)$ is the probability that a random variable X exhibits the outcome p_i; thus, $IC(p_i)$

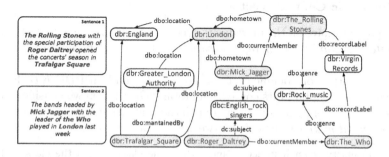

Fig. 2. Portion of DBpedia containing the most significant shared contextual information between the two sentences on the left

---

[3] https://www.textrazor.com/.

measures the specificity of the property p_i, regardless of the entities it actually connects. To take into account that the same property can connect more or less specific entities, $IC(obj_i|p_i)$ is computed in a similar way, where $P(obj_i|p_i)$ is the conditional probability that a node obj_i appears as object of the property p_i. This metric aims to provide an high weight to uncommon properties that points to uncommon object; the drawback is that it penalize infrequent object that occur with infrequent properties; for example, *dbo:Punk:Rock* is overall very infrequent, but it get an high probability when it occurs conditional on *dbo:genre*. The authors in [21] propose to mitigate this problem by computing the *Joint Information Content* $w_{jointIC} = IC(obj_i|p_i) + IC(p_i)$, and the *Combined Information Content* $w_{combIC} = IC(obj_i) + IC(p_i)$, making an independence assumption between property and object.

We introduce a new weighting function based on the fact that the importance of a property between two entities also depends on the classes to which such entities belong. For example, in Fig. 1, most people would agree that, for subjects which are instance of *dbo:Band*, the importance of *dbo:genre* increases when the object is an instance of *dbo:MusicGenre*. In fact, the 94 % of the *dbo:Band* instances are subject of a *dbo:genre* property that has as object, in 91 % of cases, an instance of *dbo:MusicGenre*, while only the 0.002 % of times, an instance of *dbo:City*. Taking in exam the triple $<s_i, p_i, o_i>$, we measure the correlation between a property p_i, the class of the subject s_i and the class of the object o_i by using the notion of *Total Correlation* [24], which is a method for weighting multi-way co-occurrences according to their importance:

$$TotalCorrelation(s_i, p_i, o_i) = -log(\frac{P(S_i, p_i, O_i)}{P(S_i)P(p_i)P(O_i)})$$

where S_i and O_i are the classes associated to the entities s_i and o_i, respectively[4].

Definitely, for contextual graphs we have three edge weights: *Total Correlation* (W_{TotCor}), Joint Information Content (W_{Joint}), and Combined Information Content (W_{Comb}).

4.2   Semantic Context Vectors

At this point we have all the ingredients necessary to define the notion of *Semantic Context Vector*, a vector representation of documents based on Contextual Graphs.

Given a corpus of documents $C = \{d_1, \ldots, d_n\}$ and an RDF KB, for each document $d \in C$ we build its contextual graph $CG_l(d)$; then we consider the set $E = \{e_1, \ldots, e_m\}$ of entities occurring in all the contextual graphs. Similar to the *term-document matrix* (see Sect. 3.1) we consider an *entity-document matrix M*,

---

[4] When an entity is an instance of more than one class we use the class with the minor number of instances because it better characterizes an entity; however if we filter the knowledge bases by excluding classes defined in external sources such as YAGO, GroNames, etc. only 6.4 % of entities in Dbpedia and 2.22 % in Wikidata are instances of more than one class.

a $m \times n$ matrix where the cell (i, j) contains the weight $s(e_i, d_j)$ of the entity $e_i \in E$ in the document $d_j \in C$. A document d_j is thus represented by the jth column of such matrix, called *Semantic Context Vector* of d_j and denoted by $SCV(d_j)$:

$$SCV(d_j) = (s(e_1, d_j), \ldots, s(e_m, d_j))$$

The weighting function $s(e_i, d_j)$ has to take into account for the importance of the entity e_i within $CG(d_j)$, by also considering the edge weights computed in the previous section. For this reason, we used the PageRank methods resumed in Sect. 3.3.

The *Semantic Context Vector* $SCV(d)$ of a document d is thus defined by 4 parameters:

1. *KB:* the RDF Knowledge Base; for example *KB=Dbpedia* and *KB=Wikidata*;
2. *CG-L:* the length for the Contextual Graph $CG_l(d)$; we used $l = 2$ and $l = 3$
3. *WeightMethod:* the edge weighting method for $CG_l(d)$: W_{Comb}, W_{Joint} and W_{TotCor}.

   Edge weights are used to set up the transition probability matrix M as a $k \times k$ matrix, where k is the number of nodes of $CG(d_j)$: $M_{pq} = \frac{w(p,q)}{\sum_{z=1}^{k} w(p,z)}$, where $w(p, q)$ returns the weight if an edge from p to q exists, otherwise it return 0.

   We denote with $W_{noweight}$ the case when edge weights are not used and the Standard PageRank algorithm is considered, where M is given by $M_{pq} = 1/d_p$ if it exists an edge from p to q and 0 otherwise (d_p be the outdegree of the vertex p).

4. *PageRankConfiguration:* the used *damping factor* and personalization vector. As *damping factor* we consider a range of values from 0.10 to 0.95 with a step of 0.05. As *personalization vector* we consider the following two cases:
   (a) *Standard PageRank:* in this case (denoted by r) there is no personalization vector, i.e., an uniform vector is considered;
   (b) *Personalized PageRank:* in this case (denoted by pr) the personalization vector $v = (v_1, \ldots, v_k)$ is setup to give an equal probability to starting entities: $v_i = 1/|SE(d)|$ if $e_i \in SE$ and 0 otherwise.
   With *r@50* and *pr@50* we denote Standard and Personalized PageRank, respectively, with a damping factor equal to .5; the same for other damping factor values.

As an example, for the documents d_1 and d_2 of Fig. 2, part of their *SCVs* are shown in Table 1; the *KB* is DBpedia and *CG-L* is equal to 3; both PageRank and Personalized PageRank are considered, with a damping factor equal to .75 (i.e. *r@75* and *pr@75*).

We can observe that PageRank tends to arrange weight in all the context graph's nodes, while with the Personalized PageRank all the weight is focused in the neighborhood of the starting entities.

4.3   CSA Similarity

The *CSA Similarity* between two documents d_1 and d_2 is computed as the *cosine similarity* between the Semantic Contextual Vectors $SCV(d_1)$ and $SCV(d_2)$; it is clear from the Semantic Context Vectors shown in Table 1 how the cosine similarity can detect some similarity between these two documents. In the next Evaluation Section we will analyze how the SCV's parameters affect the CSA similarity.

Table 1. Semantic Context Vectors of the two documents in Fig. 2

Entity	Document d_1		Document d_2	
	pr@75	r@75	pr@75	r@75
The Rolling Stones	**.187**	**.036**	.098	.082
Roger Daltrey	**.140**	**.018**	-	-
Trafalgar Square	**.155**	**.024**	-	-
London	.111	.048	**.225**	**.072**
Mick Jagger	.000	.024	**.155**	**.051**
The Who	.055	.028	**.175**	**.053**
England	.083	.050	.104	.090
Rock music	.072	.037	.098	.077

Linear combination of CSA with text similarity measures. The CSA similarity, sim_{CSA}, is only based on information extracted from a knowledge base; we used a linear combination of the CSA similarity with other similarity measures sim_{text} (such as LSA [15] and ESA [3]) to include in the final similarity measure also textual information:

$$sim_o = \alpha * sim_{CSA} + (1 - \alpha) * sim_{text}$$

where α is the weight parameter used for combining the two measures.

5   Evaluation

We evaluate the CSA performance in two different context: by considering its correlation with human judges, and, by analyzing its scalability in a clustering task.

5.1   Evaluation - Correlation with Human Judges

Experimental setup. The most common and effective way for evaluating techniques of inter-document similarity is to assess how the similarity measure produced emulates human judges. To this end, we use the dataset of documents LP50[5] [15], which contains 50 documents, selected from the Australian Broadcasting Corporation's news mail service, evaluated by 83 students of the University of Adelaide. The performance score is given by the *Pearson product-moment correlation* coefficient r [14] between the computed similarities and the ones assigned by human judges; the Pearson coefficient r measures the linear correlation between two variables.

Results and discussion. A summary of the results is shown in Fig. 3, which shows the Pearson coefficient r between the human gold standard and CSA

---

[5] https://webfiles.uci.edu/mdlee/LeePincombeWelsh.zip.

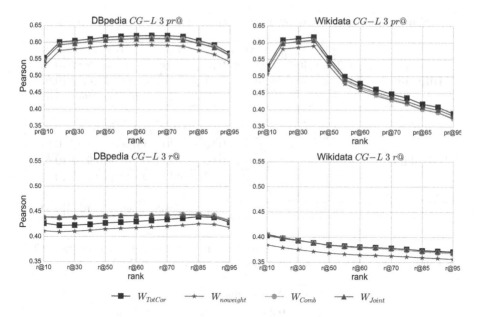

Fig. 3. Pearson correlation with human judgments (LP50 Dataset) of CSA, with different configurations.

by varying the parameters that define the Semantic Context Vectors, with the exception of *CG-L* that has been considered constant and equal to 3. One of the main result is that, for all the configurations, the Personalized PageRank (*pr*) outperforms the Standard PageRank (*r*); another interesting result is that, in almost all the configurations, the novel edge weighting function W_{TotCor} we proposed slightly outperforms the other ones, W_{Joint} and W_{Comb}. We can also appreciate different behaviors w.r.t the KB: DBpedia is more stable, while Wikidata exhibits a strong performance decay by increasing the damping factor, with the Personalized PageRank.

In particular, the CSA configuration with DBpedia, W_{TotCor}, Personalized PageRank with damping factor ranging from 0.30 to 0.85, is quite stable: it varies by only 2.5 % from the minimum (0.605 *pr*@30) to the maximum (0.62 *pr*@65); then such a CSA configuration is almost parameter free.

Table 2 shows the Pearson coefficient *r* for the best CSA configurations we found, by varying all the parameters.

In order to evaluate CSA we produced some baselines:

- Jaccard on *starting entities*: we used the *starting entities* collected for each document as descriptor of the document and we used the Jaccard similarity for estimating the similarity between documents, namely $sim(d_1, d2) = \frac{SE(d_1) \cap SE(d_2)}{SE(d_1) \cup SE(d_2)}$.
- Cosine (bag of words): we model the document corpus in a standard bag of words Vector Space Model and we compute the cosine similarity[6].

---

[6] Implemented as in [15] (only removing the stopwords).

Table 2. Results on the LP50 dataset (Pearson r correlation coefficient).

			$W_{noweight}$		W_{Comb}		W_{Joint}		W_{TotCor}		**Best**
DBpedia	CG-L	2	pr@40	0.57	pr@40	**0.59**	pr@60	0.58	pr@30	**0.59**	0.59
		3	pr@60	0.59	pr@65	0.61	pr@65	0.61	pr@65	**0.62**	0.62
								Jaccard on *starting entities*			0.49

			$W_{noweight}$		W_{Comb}		W_{Joint}		W_{TotCor}		
Wikidata	CG-L	2	pr@40	0.54	pr@40	0.56	pr@40	0.55	pr@40	**0.57**	0.57
		3	pr@40	0.59	pr@40	0.60	pr@40	0.60	pr@40	**0.61**	0.61
								Jaccard on *starting entities*			0.48

Cosine (bag of words)	0.41

Table 3. Best Pearson correlation obtained on the LP50 dataset by combining CSA ($l = 3$ and *Total Correlation* as weight function) with LSA and ESA

		Alpha value α					
		0.25		0.5		0.75	
DBpedia	CSA + LSA	pr@70	0.67	pr@70	0.67	pr@70	0.65
	CSA + ESA	pr@80	0.71	pr@65	**0.72**	pr@65	0.68
Wikidata	CSA + LSA	pr@40	0.67	pr@40	0.68	pr@40	0.65
	CSA + ESA	pr@40	**0.72**	pr@40	**0.72**	pr@40	0.67

CSA is able to outperform both baselines; we obtained a relative improvement of the 21 % (with either DBpedia and Wikidata) w.r.t. the Jaccard baseline[7]; this improvement is particularly significant because it is only due to information extracted from the *KBs* by CSA[8]. W.r.t. the Cosine baseline the margins are greater (34 % DBpedia and 33 % Wikidata); this result is not too surprising because this baseline utilize only the words contained in the text for estimating the similarity.

Table 3 shows the performance of the linear combination of CSA with the standard text similarity measures un-backgrounded LSA [15][9] and ESA reimplemented [3]. The best performance is obtained with $\alpha = 0.5$, and we can observe that the best configurations obtained in Table 2 for CSA (i.e. *pr@65* for DBpedia and *pr@40* for Wikidata) are also the best configurations of CSA combined with LSA and ESA.

Finally, in Table 4, CSA is compared with other literature techniques. The original performance of ESA reported in [10] on the LP50 dataset has been

---

[7] If not explicitly stated all the difference in performance are statistically significant at *p-value* < 0.05 using Fisher's Z-value transformation.

[8] The sets of starting entities are obtained by using NER APIs.

[9] With *td-idf* as weighting function.

Table 4. System comparison on the LP50 dataset

	Pearson coefficient r
CSA	0.62
CSA + LSA	0.65
CSA + ESA	0.72
Bag-of-Words [15]	0.41
Un-Backgrounded LSA [15]	0.52
Backgrounded LSA [15]	0.59
ESA original [10]	0.72
ESA reimplemented [3]	0.59
GED-based (Dbpedia) [21]	0.63
SSA [12]	0.68
WikiWalk + ESA [27]	0.77

criticized in [3] for being based on a cut-of value used to prune the vectors in order to produce better results on the LP50 dataset and, consequently, over-fit the approach to this particular dataset. In fact, a much lower performance has been obtained in [3, 12] by re-implementing ESA without adapting the cut-off value.

The main result of such comparison is that our CSA method is able to produce results comparable with well known techniques, like LSA and ESA, and it is able to achieve improvements when it is used in conjunction with them (for example, CSA + ESA obtains a correlation $r = 0.72$, so it attains a 16 % improvement). The Graph Edit Distance (GED) based approach of [21], which is the most similar to our, produces almost identical results but with GED the similarity measures are obtained in a much more computationally expensive way than in CSA (a deeper comparison is in the next Section). By taking in exam other *knowledge enriched* techniques built on top of a specific knowledge base (Wikipedia), CSA combined with ESA slightly outperforms SSA, but it does not reach the performance of WikiWalk + ESA.

5.2   Scalability Evaluation - Hierarchical Document Clustering

The goal of this evaluation is to estimate both the effectiveness and efficiency of CSA in a benchmark composed of a larger number of documents.

Experimental setup. We used a dataset (*re0*) of Reuters 21578[10], a collection of 1504 manually classified documents, which is commonly used for evaluating hierarchical clustering techniques. To build the clusters hierarchy we used

---

[10] Reuters collection is available at http://kdd.ics.uci.edu/databases/reuters21578/reuters21578.

Table 5. Results on the Reuters 21578 (*re0*) dataset (F-measure and execution time for building the cluster hierarchy)

	F-measure	Time
CSA	0.638	34 m
CSA + LSA	0.702	75 m
Jaccard on *starting entities*	0.415	22 m
LSA	0.611	42 m
GED-based similarity	NA	>100 h

a hierarchical clustering algorithm, based on a similarity measure and group-average-link [16]. In this test we used only DBpedia, since was before proved that it produce more stable results.

Performance is measured in terms of goodness of fit with existing categories by using *F measure*. As defined in [28], for an entire hierarchy of clusters the F measure of any class is the maximum value it attains at any node in the tree and an overall value for the F measure is computed by taking the weighted average of all values for the F measure as given by the following: $\sum_i \frac{n_i}{n} maxF(i, j)$, where the *max* is taken over all clusters at all levels, n is the number of documents and $F(i, j)$ is the F measure for the class i and the cluster j.

Results and discussion. First of all, for each document d we extracted its $CG_3(d)$ and we computed $SCV(d)$ for several configurations; then, we stored bot CGs and $SCVs$ on a file system[11]. The whole process took 32 hours, but we did not focus on improving the performance of this step, indeed, we can think of it as a preprocessing step. In Table 5 a summary of the results is shown; it includes the F measures and the average of the execution time obtained running 5 time the clustering algorithm. The configuration of CSA used for obtaining these results is GC-L=3, W_{TotCor} and pr@65, which proves to be the best configuration also in this test. We produced three different baselines: Jaccard on *starting entities*, LSA [19] and GED-based (DBpedia) [21]. We considered only the GED system since it is the most similar to our approach.

As a first observation, CSA outperforms all the considered baselines in terms of F-measure and the linear combination with LSA brings a 10 % improvement.

We were not able to successfully complete the test for GED due to its computational cost. Intuitively, to perform hierarchical clustering, we have to compute the inter-document similarity between all the documents of the corpus, i.e., 1501^2 measures of similarity for the *re0* dataset. While for CSA and LSA the cosine similarity is used, GED-similarity is based on a more expensive graph edit distance algorithm.

---

[11] We executed this experiment in a Ubuntu machine with 16 cores (Intel Xeon E312xx) and 98 Gb of RAM.

6   Conclusion and Future Work

In this paper, we proposed *Context Semantic Analysis* (CSA), a novel knowledge-based technique for estimating inter-document similarity. The technique is based on a Semantic Context Vector, which can be extracted from a Knowledge Base and stored as metadata of a document and used, when needed, for computing the Context Similarity with other documents. We showed the consistency of CSA respect to human judges and how it outperforms standard similarity methods. Moreover, we obtained comparable results w.r.t. other knowledge enriched approaches built on top of a specific KB (ESA, WikiWalk and SSA) with the advantage that our method is portable to any generic RDF KB (to the best of our knowledge CSA is the first system that shown its portability with two huge RDF KBs). Finally, we demonstrate its scalability and effectiveness performing hierarchical clustering with a larger corpus of documents.

To analyze the properties of CSA and to evaluate its performance we used two generic domain KBs, i.e. DBpedia and Wikidata; however, CSA is applicable to a generic RDF knowledge base. As a first future work, we are planning to test CSA with some domain specific KBs, such as the RDF version of AGROVOC[12] and Snomed CT. Then, we will analyze the time complexity needed to compute the Context Vector for any given document in order to judge the capability of CSA of dealing with web scale datasets in real/interactive time.

References

1. Anyanwu, K., Maduko, A., Sheth, A.: SemRank: ranking complex relationship search results on the semantic web. In Proceedings of the 14th International Conference on World Wide Web, pp. 117–127. ACM (2005)
2. Auer, S., Bizer, C., Kobilarov, G., Lehmann, J., Cyganiak, R., Ives, Z.: DBpedia: a nucleus for a web of open data. In: Aberer, K., et al. (eds.) ASWC/ISWC - 2007. LNCS, vol. 4825, pp. 722–735. Springer, Heidelberg (2007). doi:10.1007/978-3-540-76298-0_52
3. Bär, D., Zesch, T., Gurevych, I.: A reflective view on text similarity. In: RANLP, pp. 515–520 (2011)
4. Beneventano, D., Bergamaschi, S., Sorrentino, S., Vincini, M., Benedetti, F.: Semantic annotation of the cerealab database by the agrovoc linked dataset. Ecol. Inform. **26**, 119–126 (2015)
5. Bizer, C., Heath, T., Berners-Lee, T.: Linked data-the story so far. In: Sheth, A.P. (ed.) Semantic Services, Interoperability, Web Applications: Emerging Concepts, pp. 205–227. IGI Global, Hershey (2009)
6. Bos, L., Donnelly, K.: SNOMED-CT: the advanced terminology and coding system for eHealth. Stud. Health Technol. Inform. **121**, 279–290 (2006)
7. Caracciolo, C., Stellato, A., Morshed, A., Johannsen, G., Rajbhandari, S., Jaques, Y., Keizer, J.: The AGROVOC linked dataset. Semant. Web 4(3), 341–348 (2013)
8. Cyganiak, R., Wood, D., Lanthaler, M.: RDF 1.1 concepts, abstract syntax. W3C Recomm. **25**, 1–8 (2014)

---

[12] http://aims.fao.org/standards/agrovoc/linked-open-data.

9. Dumais, S.T.: Latent semantic analysis. Annu. Rev. Inf. Sci. Technol. **38**(1), 188–230 (2004)
10. Gabrilovich, E., Markovitch, S.: Computing semantic relatedness using wikipedia-based explicit semantic analysis. IJCAI **7**, 1606–1611 (2007)
11. Gomaa, W.H., Fahmy, A.A.: A survey of text similarity approaches. Int. J. Comput. Appl. **68**(13), 13–18 (2013)
12. Hassan, S., Mihalcea, R.: Semantic relatedness using salient semantic analysis. In: AAAI (2011)
13. Haveliwala, T.H.: Topic-sensitive pagerank. In: Proceedings of the 11th International Conference on World Wide Web, pp. 517–526. ACM (2002)
14. Lawrence, I., Lin, K.: A concordance correlation coefficient to evaluate reproducibility. Biometrics **45**, 255–268 (1989)
15. Lee, M., Pincombe, B., Welsh, M.: An empirical evaluation of models of text document similarity. In: Cognitive Science (2005)
16. Manning, C.D., Raghavan, P., Schütze, H., et al.: Introduction to Information Retrieval, vol. 1. Cambridge University Press, Cambridge (2008)
17. Mendes, P., Jakob, M., García-Silva, A., Bizer, C.: DBpedia spotlight shedding light on the web of documents. In: I-Semantics (2011)
18. Nadeau, D., Sekine, S.: A survey of named entity recognition and classification. Lingvisticae Investigationes **30**(1), 3–26 (2007)
19. Nakov, P., Popova, A., Mateev, P.: Weight functions impact on LSA performance. In: EuroConference RANLP, pp. 187–193 (2001)
20. Page, L., Brin, S., Motwani, R., Winograd, T.: The pagerank citation ranking: bringing order to the web (1999)
21. Schuhmacher, M., Ponzetto, S.P.: Knowledge-based graph document modeling. In: Proceedings of the 7th ACM International Conference on Web Search and Data Mining, pp. 543–552. ACM (2014)
22. Suchanek, F.M., Kasneci, G., Weikum, G.: YAGO: a core of semantic knowledge. In: Proceedings of the 16th International Conference on World Wide Web, pp. 697–706. ACM (2007)
23. Turney, P.D., Pantel, P., et al.: From frequency to meaning: vector space models of semantics. J. Artif. Intell. Res. **37**(1), 141–188 (2010)
24. Van de Cruys, T.: Two multivariate generalizations of pointwise mutual information. In Proceedings of the Workshop on Distributional Semantics and Compositionality, pp. 16–20. Association for Computational Linguistics (2011)
25. Vrandečić, D., Krötzsch, M.: Wikidata: a free collaborative knowledgebase. Commun. ACM **57**(10), 78–85 (2014)
26. Xing, W., Ghorbani, A.: Weighted pagerank algorithm. In: Second Annual Conference on Communication Networks and Services Research, 2004. Proceedings, pp. 305–314. IEEE (2004)
27. Yeh, E., Ramage, D., Manning, C.D., Agirre, E., Soroa, A.: WikiWalk: random walks on wikipedia for semantic relatedness. In Proceedings of the 2009 Workshop on Graph-Based Methods for Natural Language Processing, pp. 41–49. Association for Computational Linguistics (2009)
28. Zhao, Y., Karypis, G.: Evaluation of hierarchical clustering algorithms for document datasets. In Proceedings of the Eleventh International Conference on Information and Knowledge Management, pp. 515–524. ACM (2002)

Comparisons and Benchmarks

An Experimental Survey of MapReduce-Based Similarity Joins

Yasin N. Silva$^{(\boxtimes)}$, Jason Reed, Kyle Brown, Adelbert Wadsworth,
and Chuitian Rong

Arizona State University, Glendale, AZ, USA
{ysilva, jmreed3, kabrowl7, ajwadswo, crong5}@asu.edu

Abstract. In recent years, Big Data systems and their main data processing framework - MapReduce, have been introduced to efficiently process and analyze massive amounts of data. One of the key data processing and analysis operations is the Similarity Join (SJ), which finds similar pairs of objects between two datasets. The study of SJ techniques for Big Data systems has emerged as a key topic in the database community and several research teams have published techniques to solve the SJ problem on Big Data systems. However, many of these techniques were not experimentally compared against alternative approaches. This was the case in part because some of these techniques were developed in parallel while others were not implemented even as part of their original publications. Consequently, there is not a clear understanding of how these techniques compare to each other and which technique to use in specific scenarios. This paper addresses this problem by focusing on the study, classification and comparison of previously proposed MapReduce-based SJ algorithms. The contributions of this paper include the classification of SJs based on the supported data types and distance functions, and an extensive set of experimental results. Furthermore, the authors have made available their open-source implementation of many SJ algorithms to enable other researchers and practitioners to apply and extend these algorithms.

Keywords: Similarity joins · Big Data systems · Performance evaluation · MapReduce

1 Introduction

The processing and analysis of massive amounts of data is a crucial requirement in many commercial and scientific applications. Internet companies, for instance, collect large amounts of data such as content produced by web crawlers, service logs and click streams generated by web services. Analyzing these datasets may require processing tens or hundreds of terabytes of data. Big Data systems and MapReduce, their main data processing framework, constitute an answer to the requirements of processing

This work was supported by Arizona State University's SRCA and NCUIRE awards, the NSFC (No. 61402329), and the China Scholarship Council.

© Springer International Publishing AG 2016
L. Amsaleg et al. (Eds.): SISAP 2016, LNCS 9939, pp. 181–195, 2016.
DOI: 10.1007/978-3-319-46759-7_14

massive datasets in a highly scalable and distributed fashion. These systems are composed of large clusters of commodity machines and are often dynamically scalable, i.e., cluster nodes can easily be added or removed depending on the workload. Important examples of these Big Data systems are: Apache Hadoop [26]; Google's File System [10], MapReduce [9] and Bigtable [8]; and Microsoft's Dryad [11] and SCOPE/Cosmos [7].

The Similarity Join is one of the most useful operations for data processing and analysis. This operation retrieves all data pairs from two datasets (R and S) whose distances are smaller than or equal to a predefined threshold ε. Similarity Joins have been extensively used in domains like record linkage, data cleaning, sensor networks, marketing analysis, multimedia applications, recommendation systems, etc. A significant amount of work has been focused on the study of non-distributed implementations. Particularly, Similarity Joins have been studied as standalone operations [12–15], as operations that use standard database operators [16–18], and as physical database operators [1–5].

The study of Similarity Join techniques for Big Data systems has recently emerged as a key topic in the data management systems community. Several research teams have proposed and published different techniques to solve the Similarity Join problem on Big Data systems (e.g., [19–25]). Unfortunately, however, many of these techniques were not experimentally compared against alternative approaches. This was the case in part because some of these techniques were developed in parallel while others were not implemented even as part of their original publications. Consequently, while there are many techniques to solve the Similarity Join problem, there is not a clear understanding of: (1) how these techniques compare to each other, and (2) which technique to use in real-world scenarios with specific requirements for data types, distance functions, dataset sizes, etc. Furthermore, the need for comparative work in the area of data management was recently highlighted by the editors of a top journal in this area [6].

This paper addresses this problem by focusing on the study, classification and comparison of the Similarity Join techniques proposed for Big Data systems (using the MapReduce framework). The main contributions of this paper are:

- The classification of Similarity Join techniques based on the supported data types and distance functions.
- An extensive set of experimental results. These results include tests that compare the performance of alternative approaches (based on supported data type and distance function) under various dataset sizes and distance thresholds.
- The availability of the authors' open-source implementation of many Similarity Join algorithms [27]. Our goal is to enable other researchers and practitioners to apply and extend these algorithms.

The remaining part of this paper is organized as follows. Section 2 presents the description of all the algorithms considered in our study and a classification of the algorithms based on the supported data types and distance functions. Section 3 presents the experimental evaluation results and discussions (this section is divided into subsections that focus on specific data types and distance functions). Finally, Sect. 4 presents the conclusions.

2 MapReduce-Based Similarity Join Algorithms

2.1 Classification of the Algorithms

Table 1 presents the MapReduce-based Similarity Join algorithms considered in our study. For each algorithm, the table shows the supported data types and distance functions (DFs), and the data types that could be supported by extending the original algorithms. In order to systematically evaluate the different algorithms, we classify them based on the supported data types. The experimental section of this paper, compares all the algorithms that support a given data type and its associated distance functions.

Table 1. Similarity Join algorithms and supported distance functions and data types.

Algorithm	Supported Distance/ Similarity Functions	Supported Data Types			
		Text/String	Numeric	Vector	Set
Naïve Join	Any DF	•	•	*	•
Ball Hashing 1	Hamming Distance Edit Distance	•			
Ball Hashing 2	Hamming Distance Edit Distance	•			
Subsequence	Edit Distance	•			
Splitting	Hamming Distance Edit Distance	•			
Hamming Code	Hamming Distance	•			
Anchor Points	Hamming Distance Edit Distance	•	*	*	
MRThetaJoin	Any DF	•	•	•	•
MRSimJoin	Any metric DF	•	•	•	•
MRSetJoin	JS, TC, CC, Edit Distance*	*			•
Online Aggregation	JS, RS, DS, SC, VC				•
Lookup	JS, RS, DS, SC, VC				•
Sharding	JS, RS, DS, SC, VC				•
• Natively Supported					
* Can be extended to support this data type or distance function					
JS=Jaccard Similarity, TC=Tanimoto Coefficient, CC=Cosine Coefficient, RS=Ruzicka Similarity, DS=Dice Similarity, SC=Set Cosine Sim., VC=Vector Cosine Sim.					

2.2 Description of the Studied Similarity-Join Algorithms

Naïve Join. The Naïve Join algorithm [22] is compatible with all data types and distance functions, and works in a single MapReduce job. The algorithm uses a key space defined by a parameter J, which is proportional to the square root of the number of reducers (reduce tasks) to be used. During the Map phase, pairs of input data elements are assigned to a key pair with the form (i, j) where $0 \leq i \leq j \leq J$. For each input record X, the mapper (map task) outputs key-value pairs with the form $((i, j), X)$, such that any two records are mapped to at least one common key. The reducer receives all of the records for a given key and compares each pair of records outputting the pairs with distance smaller than or equal to ε (distance threshold). The algorithm proposed in [22] does not consider the case where two records are mapped to more than one common key. In this case, we solved the problem by outputting only when $i = j$.

Ball Hashing 1. The Ball Hashing 1 algorithm [22] takes a brute force approach to solving the Similarity Join problem. This algorithm assumes that the alphabet of the input value is finite and known. The Map phase takes in a given input record r and generates a ball of radius ε. In effect, for a given join attribute vr, it will generate a set Vr composed of every possible value within ε of vr. For each value Vr_i in Vr that is not equal to vr, the Map will emit the key-value pair $<Vr_i, r>$. The Map will additionally output the key-value pair $<vr, r>$. As vr is the join attribute in r, this ensures a collision in the Reduce phase with any matching pairs (links). Any Reduce group that contains such a record ($<vr, r>$) should be considered an active group and the record r should be considered native to that group. Any Reduce group that does not have a native record within it should be considered inactive and does not need to be processed further. In the active groups, the join matches are generated by combining the native members with each of the non-native members in the group. The original paper does not consider the possibility of multiple input records having the same join value. If this is the case, there is the additional need to join native members among each other as well as all native records against all non-native records. This algorithm supports string data with the Edit and Hamming distance functions.

Ball Hashing 2. Ball Hashing 2 [22] is an extension of the Ball Hashing 1 algorithm. The difference is that in the Map phase, it generates balls of size $\varepsilon/2$. Because of this, it is necessary to process every Reduce group. A brute force comparison is performed in each Reduce group to find any matches and eliminate the possibility of duplicate outputs. The algorithm supports string data with Edit and Hamming distance metrics.

Subsequence. Subsequence [22] is an algorithm proposed for string data and the Edit Distance. The Map phase generates all the $(b - \varepsilon/2)$-subsequences of each input string (b is the string length) and outputs pairs of the form $<subsequence, input\_string>$. The Reduce phase compares all the records sharing the same subsequence to identify the Similarity Join matches. The key idea behind this algorithm is that if two strings are within ε, they will share at least one identical subsequence.

Splitting. The Splitting algorithm [22] is composed of a single MapReduce job and is based on splitting strings into substrings. These substrings are then compared to other substrings generated from the input dataset. In order to be considered a Similarity Join match, a pair of strings only needs to share one common substring. In the Map task, each input string (with length b) is split into substrings of length $b/(\varepsilon + 1)$. The result will be composed of $b/(b/(\varepsilon + 1))$ substrings. Each substring will be outputted with a key consisting of its position (i) in the parent string, and the substring that was generated, s_i. The value that will be attached to the key is the parent string. Each reducer will compare (pair wise) all the substrings that have a matching key and output the pairs that are separated by a distance smaller than or equal to ε. To avoid the generation of duplicate pairs at multiple reducers, a match is generated only within the Reduce group associated with the position of the first common substring between two matching strings. This distance functions supported by this algorithm are Hamming and Edit Distance.

Anchor Points. This algorithm distributes the input data into groups where all the members of a group are within a certain distance of an anchor point [22]. The technique supports the Hamming and Edit Distance functions. In the case of Hamming Distance, the algorithm finds first a set of anchor points such that every input record is within ε from at least one anchor. This set is stored in a distributed cache and used at each mapper. For each input record s, the mapper outputs key-value pairs for every anchor point that is within 2ε of s. The mapper marks the closest anchor point to s as its home group. In the Reduce phase, the strings of a given home group will be compared to other strings from other groups that were sent to the same reducer. All strings in the home group will be compared as well. In the case of Edit Distance, the anchor points are a subset of the data such that every input record is within ε deletions from at least one anchor. This modified algorithm only works with fixed-length strings. This fact is not directly stated in the paper but was confirmed by the authors.

Hamming Code. The Hamming Code algorithm [22] is a SJ technique proposed for string data and the Hamming Distance. Since this algorithm only works when $\varepsilon = 1$ and the strings' length is one less than a power of 2, it is not included in our evaluation.

MRThetaJoin. MRThetaJoin [23] is a randomized Theta Join algorithm that supports arbitrary join predicates (including Similarity Join conditions). This approach uses a single MapReduce job and requires some basic statistics (input cardinality). The approach uses a model that partitions the input relations using a matrix that considers all the combinations of records that would be required to answer a cross product. The matrix cells are then assigned to reducers in a way that minimizes job completion time. A memory-aware variant is also proposed for the common scenario where partitions do not fit in memory. Since any Theta Join or Similarity Join is a subset of the cross-product, the matrix used in this approach can represent any join condition. Thus, this approach can be used to supports Similarity Joins with any distance function and data type. For the performance evaluation of Similarity Joins presented in this paper, we implemented an adaptation of the memory-aware 1-Bucket-Theta algorithm proposed in [25] that uses the single-node QuickJoin algorithm [15] in the reduce function.

MRSimJoin. The MRSimJoin algorithm [20, 21] iteratively partitions the data into smaller partitions, until each partition is small enough to be processed in a single node. The process is divided into a sequence of rounds, and each round corresponds to a MapReduce job. Partitioning is achieved by using a set of pivots, which are a subset of the records to be partitioned. There are two types of partitions, base partitions and window-pair partitions. Base partitions hold all of the records closest to a given pivot, rather than any other pivot. Window-pair partitions hold records within the boundary between two base partitions. If possible, e.g., Euclidean Distance, the window-pair partitions should only include the points within ε from the hyperplane separating adjacent base partitions. If this is not possible, a distance is computed to a generalized hyperplane boundary (lower bound of the distance). This algorithm can be used with any data type and metric. The experimental section in [20] shows that in most cases the number of pivots can be adjusted to ensure the algorithm runs in a single MapReduce job.

MRSetJoin. MapReduce Set-Similarity Join [19] is a Similarity Join algorithm that consists of three stages made up of various MapReduce jobs. In the first stage, data statistics are generated in order to select good signatures, or tokens, that will be used by later MapReduce jobs. In the second stage, each record has its record-ID and join-attribute value assigned to the previously generated tokens, the similarity between records associated with the same token is computed, and record-ID pairs of similar records are outputted. In the third stage, pairs of joined records are generated from the output of the second stage and the original input data. MRSetJoin supports set-based distance functions like Jaccard Distance and Cosine Coefficient. There are multiple options presented for each stage, however, the paper states that BTO-PK-BRJ is the most robust and reliable option. Thus, this option is used in this survey as the representative of this technique.

V-Smart-Online Aggregation. Online Aggregation [24] is a Similarity Join algorithm under the V-SMART-Join framework, which can be used for set and multiset data and set-based distance functions like Jaccard and Dice. In general, the V-SMART-Join framework consists of two phases, joining and similarity. Although the framework includes three different joining phase algorithms, Online Aggregation, Lookup, and Sharding, only one of the three was selected to participate in the survey. According to the experimental results in [24], Online Aggregation generally outperforms the Sharding and Lookup algorithms, and as such it was selected to represent this approach. The algorithm is based on the computation of $Uni(M_i)$ for each multiset M_i. $Uni(M_i)$ is the partial result of a unilateral function (e.g., $Uni(M_i) = |M_i|$). During the joining phase (one MapReduce job), the $Uni(M_i)$ of a given multiset M_i is joined to all the elements of M_i. The similarity phase, composed of two MapReduce jobs, builds an inverted index, computes the similarity between all candidate pairs, and outputs the Similarity Join matches.

3   Experimental Comparison

This section presents the experimental comparison of previously proposed MapReduce-based Similarity Join algorithms. One of the key tasks for this survey work was the implementation of the studied algorithms. While in some cases, the source code was provided by the original authors (MRSetJoin, MRSimJoin), in most cases, the source code was not available and consequently had to be implemented as part of our work (e.g., Ball Hashing 1, Ball Hashing 2, Naïve Join, Splitting, Online Aggregation, MRThetaJoin). We have made available the source code of all the evaluated algorithms in [27]. All the algorithms were implemented and evaluated using Hadoop (0.20.2), the most popular open-source MapReduce framework. The experiments were performed using a Hadoop cluster running on the Amazon Elastic Compute Cloud (EC2). Unless otherwise stated, we used a cluster of 10 nodes (1 master + 9 worker nodes) with the following specifications: 15 GB of memory, 4 virtual cores with 2 EC2 Compute Units each, 1,690 GB of local instance storage, 64-bit platform. The number of reducers was computed as: $0.95 \times \langle no.\ worker\ nodes \rangle \times \langle max\ reduce\ tasks\ per\ node \rangle = 25$. Table 2 shows configurations details for individual algorithms.

The experiments used a slightly modified version of the Harvard bibliographic dataset [28]. Specifically, we used a subset of the original dataset and augmented the record structure with a vector attribute to perform the tests with vector data. Each record contains the following attributes: unique ID, title, date issued, record change date, record creation date, Harvard record-ID, first author, all author names, and vector. The vector attribute is a 10D vector that was generated based on the characters of the title (multiplied against prime numbers). The vector components are in the range [0–999]. The minimum and maximum length (number of characters) of each attribute are as follows: unique ID (9, 9), title (6, 996), date issued (4, 4), record change date (15, 15), record creation date (6, 6), Harvard record-ID (10, 10), first author (6, 94), and all author names (6, 2462). The dataset for scale factor 1 (SF1) contains 200K records. The records of each dataset are equally divided between tables R and S.

The datasets for SF greater than 1 were generated in such a way that the number of matches of any Similarity Join operation in SFN is N times the number of matches in SF1. For vector data, the datasets for higher SF were obtained adding shifted copies of the SF1 dataset where the distance between copies were greater than the maximum value of ε. For string data, the datasets for higher SF were obtained adding a copy of the SF1 data where characters are shifted similarly to the process in [19].

We evaluate the performance of the algorithms by independently analyzing their execution time while increasing the dataset size (SF) and the distance threshold (ε). We did not include the execution time when an algorithms took a relatively long time (more than 3 h). We performed four sets of experiments for the following combinations of data types and distance functions: (1) vector data and Euclidean Distance, (2) variable-length string (text) data and Edit Distance, (3) fixed-length string data and Hamming Distance, and (4) set data and Jaccard Distance. Each algorithm was executed multiple times; we report the average execution times.

Table 2. Additional configuration details.

Algorithm	Configuration Details
Naïve Join	$J = \sqrt{\text{Number of Reduce Tasks}}$
MRThetaJoin	$K = ((|R|+|S|) \times b)/m$, where $|R|$ and $|S|$ are the cardinalities of R and S, b = size in bytes per record, m = memory threshold (64 MB).
MRSimJoin	Memory limit for in-memory SJ algorithm = 64 MB. Number of Pivots = $SF \times 100$.

3.1 Comparison of Algorithms for Vector Data – Euclidean Distance

This section compares the performance of the algorithms that support vector data, namely MRSimJoin and MRThetaJoin. We use the Euclidean Distance function and perform the distance computations over the 10D vector attribute of the Harvard dataset.

Increasing Scale Factor. Figures 1 and 2 compare the way MRSimJoin and MRThetaJoin scale when the data size increases (SF1–SF4). The experiments use 10D vectors. The experiments in Fig. 1 use a relatively small value of ε (5 % of the maximum possible distance) while the ones in Fig. 2 a relatively large value (15 %). Figure 1 shows that, for small values of ε (5 %), MRSimJoin performs significantly better than MRThetaJoin when the data size increases. Specifically, the execution time of MRThetaJoin grows from being 2 times the one of MRSimJoin for SF1 to 7 times for SF4. The execution time of MRThetaJoin is significantly higher than that of MRSimJoin because the total size of all the partitions of MRThetaJoin is significantly larger than that of MRSimJoin. Figure 2 shows that, for larger values of ε (15 %), MRSimJoin still performs better than MRThetaJoin in the case of larger datasets but is outperformed by MRThetaJoin for small datasets. Specifically, the execution time of MRThetaJoin is 0.7 times the one of MRSimJoin for SF1 and SF2; and 1.2 and 1.9 times for SF3 and SF4, respectively.

Increasing ε. Figures 3 and 4 show how the execution time of MRSimJoin and MRThetaJoin increase when ε increases (1 %–20 %). Figure 3 considers relatively smaller values of ε (1 %–5 %) while Fig. 4 considers larger values (5 %-20 %). The results in both figures show that the performance of MRSimJoin is better than the one of MRThetaJoin for all the evaluated values of ε. Specifically, in Fig. 3 the execution time of MRThetaJoin is between 7 ($\varepsilon = 5$ %) to 11 ($\varepsilon = 1$ %) times the one of MRSimJoin while in Fig. 4, the execution time of MRThetaJoin is between 1.6 ($\varepsilon = 20$ %) to 9.2 ($\varepsilon = 5$ %) times the one of MRSimJoin. We can observe that the performance of MRSimJoin tends to get closer to the one of MRThetaJoin for very large values of ε. In general, the execution time of both algorithms grows when ε grows. The increase in execution time is due to a higher number of distance computations in both algorithms and slightly larger sizes of window-pair partitions in the case of MRSimJoin.

From the results presented in this section, we can conclude that MRSimJoin is in general the best approach to perform Similarity Joins with vector data unless the dataset size is very small and the distance threshold is extremely large.

3.2 Comparison of Algorithms for Variable-Length String Data – Edit Distance

This section compares the performance of the Similarity Join algorithms using string data and the Edit Distance. The tests use the first author name (variable-length: 6–94, alphabet size: 27) as the join attribute. The evaluated algorithms are: MRSimJoin, Naïve Join, MRThetaJoin, and Ball Hashing 1. For this last algorithm, we were only able to obtain results for the test with $\varepsilon = 1$. Even using SF1, this algorithm took significantly longer than the other algorithms. Ball Hashing 2 and Anchor Points are not included since they do not support variable-length strings. Splitting and Subsequence were not included since the brief information included in [22] to support variable-length strings was not sufficient to implement this feature. Ball Hashing 2 and Splitting are evaluated in Sect. 3.2 with fixed-length strings. Regarding the Edit

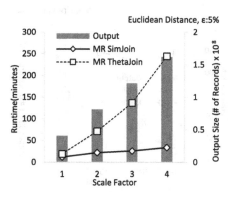

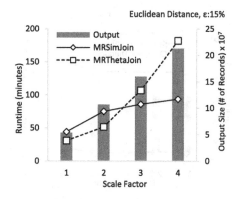

Fig. 1. Euclidean - Increasing SF (ε = 5 %)   **Fig. 2.** Euclidean - Increasing SF (ε = 15 %)

Distance metric, we consider the edit operations of insertion and deletion of a character. Both operations have a cost of 1. This is a common case of the Edit Distance and it is used in the specification of Naïve Join, Ball Hashing 1, and Ball Hashing 2. MRSimJoin and MRThetaJoin, which also support the Edit Distance with the character substitution operation, were adapted to support the metric with insertion and deletion. The maximum value of ε is 100.

Increasing Scale Factor. Figure 5 compares the performance of the algorithms when the dataset is incrementally scaled from SF1 to SF4. Naïve Join is the best performing algorithm for SF1 while MRSimJoin performs the best in all the other cases. For SF1, Naïve Join completed execution within 75 % of the execution time of MRThetaJoin, and 89 % of that of MRSimJoin. However, as the data size increased, MRSimJoin outperformed both Naïve Join and MRThetaJoin for SF2-SF4. For these values of SF, MRSimJoin's execution time is at most 74 % of that of MRThetaJoin, and at most 76 % of that of Naïve Join. Also, we observed that as the scale factor increased, the relative advantage of MRSimJoin improved too, and at SF4, MRSimJoin completed within 54 % of the execution time of MRThetaJoin and within 56 % of that of Naïve Join.

Increasing ε. Figure 6 compares the algorithms when the value of ε (distance threshold) increases from 1 to 4. For ε values of 1 and 2, MRSimJoin outperformed the other algorithms, completing always within 68 % of the execution time of MRThetaJoin and within 77 % of that of Naïve Join. The outlier on these tests was Ball Hashing 1. Specifically, its execution time was nine times the one of MRSimJoin for $\varepsilon = 1$. The Ball Hashing 1 tests using higher values of ε were cancelled after they took significantly longer than the other algorithms. For larger values of ε (3 and 4), Naïve Join outperformed the other algorithms. Specifically, Naïve Join completed within 78 % of MRSimJoin's execution time, and 89 % of MRThetaJoin's execution time for these larger values of ε.

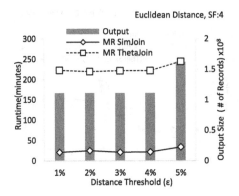

Fig. 3. Euclidean - Increasing ε (small)

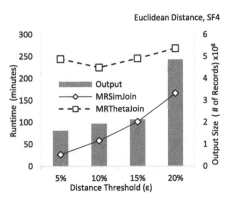

Fig. 4. Euclidean - Increasing ε (large)

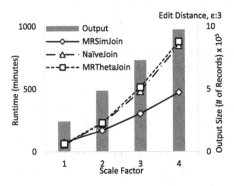

Fig. 5. Edit Dist. - Increasing SF

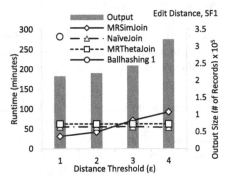

Fig. 6. Edit Dist. - Increasing ε

From these results, it can be concluded that MRSimJoin is, in general, the best approach to perform similarity joins with the Edit Distance (text data) when the dataset is large (greater than SF1 in our tests) or the distance threshold is relatively small (1 or 2 in our tests). For smaller datasets or larger distance thresholds, Naïve Join is the best approach among the evaluated algorithms.

3.3 Comparison of Algorithms for Fixed-Length Strings – Hamming Distance

The tests in this section perform Similarity Joins using Hamming Distance over the first 6 characters of the first author name (fixed-length: 6, alphabet: 27). The evaluated algorithms are: MRSimJoin, MRThetaJoin, Naïve Join, Splitting, Ball Hashing 1, and Ball Hashing 2. Anchor Points it is not included since the paper that introduced it showed that it is outperformed by other algorithms [22]. The maximum value of ε is 6.

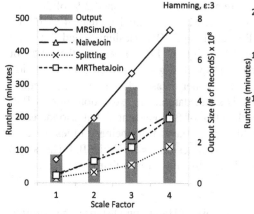

Fig. 7. Hamming Dist. - Increasing SF

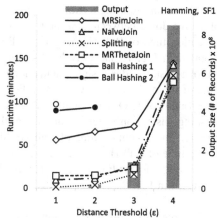

Fig. 8. Hamming Dist. - Increasing ε

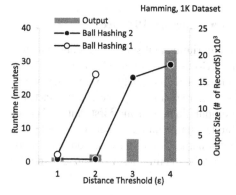

Fig. 9. Hamming Dist. - Increasing ε (1 k)

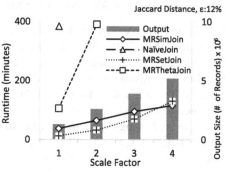

Fig. 10. Jaccard - Increasing SF

Increasing Scale Factor. The results of the experiments using increasing scale factors (SF1–SF4) are represented in Fig. 7. This figure shows that the Splitting algorithm outperforms all of the other algorithms for all the values of scale factor. Specifically, Splitting's execution time is at most 71 % of the one of MRThetaJoin, 60 % of Naïve Join, and 24 % of MRSimJoin. MRThetaJoin and Naïve Join have very similar results, with MRThetaJoin slightly outperforming Naïve Join for SF1, SF3 and SF4. The execution time of MRSimJoin is larger than the ones of the other algorithms compared in Fig. 7. Ball Hashing 1 and Ball Hashing 2 were excluded from the comparison as they did not complete within a reasonable amount of time.

Increasing ε. Figure 8 shows the results of comparing the algorithms with increasing values of the distance threshold. In these tests, the Splitting algorithm outperforms all other algorithms with the exception of $\varepsilon = 3$ where MRThetaJoin slightly outperforms it. Splitting's execution times are between 11 % ($\varepsilon = 1$) and 106 % ($\varepsilon = 3$) of those of

MRThetaJoin. Splitting's execution times are also between 15 % and 92 % of the ones of Naïve Join, between 3 % and 90 % of MRSimJoin, and less than 4 % of the execution time of Ball Hashing 1 and Ball Hashing 2. Ball Hashing 1 and 2 are not reported in Fig. 7 (and only have some data points in Fig. 8) because they did not return a result under a significantly long time (3 h). Figure 9 presents the execution time of these algorithms with a significantly smaller dataset (1K records) under multiple values of ε. Observe that even for this small dataset, the execution time of Ball Hashing 1 is not only significantly larger than that of Ball Hashing 2, but also increases rapidly. The execution time of Ball Hashing 1 increases from being 2 times the execution time of Ball Hashing 2 for $\varepsilon = 1$ to be 31 times for $\varepsilon = 2$. While Ball Hashing 2 clearly outperforms Ball Hashing 1, it is still significantly slower than other algorithms as shown in Fig. 8.

The results of this section show that in the case of Hamming Distance, the Splitting algorithm is the best choice for various values of dataset size and distance threshold. In most of the cases, Naïve Join and MRThetaJoin are the next best performing options.

3.4 Comparison of Algorithms for Set Data – Jaccard Distance

This section compares the performance of the algorithms that support set data, namely Naïve Join, MRThetaJoin, MRSimJoin, MRSetJoin, and Online Aggregation. The Lookup and Sharding algorithms were not included in our analysis since they were found to be generally outperformed by Online Aggregation [24]. We use the Jaccard Distance function and perform the distance computations over the First Author Name attribute. To this end, we first converted the author name into a proper set by deleting spaces and removing duplicates. For instance, the name "John Smith" is converted into the set {j, o, h, n, s, m, i, t}. The alphabet size and the maximum set size are 26. In this case, the range of ε is: 0 (0 %)–1 (100 %).

Increasing Scale Factor. Figure 10 compares the way the algorithms scale when the dataset size increases (SF1–SF4). Naïve Join is the slowest algorithm, having a SF1 runtime that is at least four times the ones of the other algorithms in this figure. It is also too slow to be executed with any of the higher scale factor values. MRThetaJoin was executed with SF1 and SF2 but its runtime was too long to be included for larger datasets. MRSimJoin and MRSetJoin have fairly similar execution times. MRSetJoin performs better with SF1–SF3 but its relative advantage decreases as the dataset size increases. Specifically, MRSetJoin's execution time is 33 %, 50 % and 73 % of those of MRSimJoin for SF1, SF2 and SF3, respectively. MRSimJoin outperforms MRSetJoin for SF4, where MRSimJoin's execution time is 89 % of the one of MRSetJoin. The results of the Online Aggregation algorithm were not included because these tests took too long and were cancelled or did not complete properly. We were able to successfully run this algorithm only with very small datasets ($\sim 1K$).

Increasing ε. Figure 11 shows how the execution time of the evaluated algorithms increases when ε increases (4 %–16 %). Naïve Join was the slowest algorithm and its runtime was at least 3.5 times of the ones of the other algorithms. MRSetJoin and

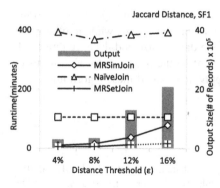

Fig. 11. Jaccard - Increasing ε (SF1)

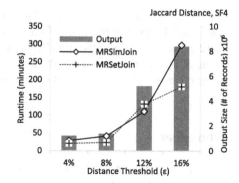

Fig. 12. Jaccard - Increasing ε (SF4)

MRSimJoin are the best performing algorithms. MRSetJoin's advantage over MRSimJoin tends to increase when ε increases. Specifically, MRSimJoin's execution time is 1.8 times the one of MRSetJoin for SF1 and 4.7 for SF4. Figure 12 provides additional details of the two best performing algorithms (MRSetJoin and MRSimJoin). This figure compares the algorithms' performance using SF4. Figure 12 shows that for a larger dataset (SF4), the relative advantage of MRSetJoin over MRSimJoin decreases. In this case, the execution time of MRSimJoin is between 0.8 and 1.8 of those of MRSetJoin.

The results presented in this section indicate that MRSetJoin is, in general, the best algorithm for set data and Jaccard Distance. MRSimJoin, which performed second in most tests, should be considered as an alternative particularly for very large datasets where it could, in fact, outperform MRSetJoin.

4   Conclusions

MapReduce is widely considered one of the key processing frameworks for Big Data and the Similarity Join is one of the key operations for analyzing large datasets in many application scenarios. While many MapReduce-based Similarity Join algorithms have been proposed, many of these techniques were not experimentally compared against alternative approaches and some of them were not even implemented as part of the original publications. This paper aims to shed light on how the proposed algorithms compare to each other qualitatively (supported data types and distance functions) and quantitatively (execution time trends). The paper compares the performance of the algorithms when the dataset size and the distance threshold increase. Furthermore, the paper evaluates the algorithms under different combinations of data type (vectors, same-length strings, variable-length strings, and sets) and distance functions (Euclidean Distance, Hamming Distance, Edit Distance, and Jaccard Distance). One of the key findings of our study is that the proposed algorithms vary significantly in terms of the supported distance functions, e.g., algorithms like MRSimJoin and MRThetaJoin support multiple metrics while Subsequence and Hamming Code support only one. There is also not a single algorithm that outperforms all the others for all the evaluated data types and distance functions. Instead, in some cases, an algorithm performs consistently better than the others for a given data type and metric, while in others, the identification of the best algorithm depends on the dataset size and distance threshold. The authors have made available the source code of all the implemented algorithms to enable other researchers and practitioners to apply and extend these algorithms.

References

1. Silva, Y.N., Aref, W.G., Ali, M.: The similarity join database operator. In: ICDE (2010)
2. Silva, Y.N., Pearson, S.: Exploiting database similarity joins for metric spaces. In: VLDB (2012)
3. Silva, Y.N., Aly, A.M., Aref, W.G., Larson, P.-A.: SimDB: a similarity-aware database system. In: SIGMOD (2010)
4. Silva, Y.N., Aref, W.G., Larson, P.-A., Pearson, S., Ali, M.: Similarity queries: their conceptual evaluation, transformations, and processing. VLDB J. **22**(3), 395–420 (2013)
5. Silva, Y.N., Aref, W.G.: Similarity-aware query processing and optimization. In: VLDB Ph.D. Workshop, France (2009)
6. Bernstein, P.A., Jensen, C.S., Tan, K.-L.: A call for surveys. SIGMOD Rec. **41**(2), 47 (2012)
7. Chaiken, R., Jenkins, B., Larson, P.-A., Ramsey, B., Shakib, D., Weaver, S., Zhou, J.: Scope: easy and efficient parallel processing of massive data sets. In: VLDB (2008)
8. Chang, F., Dean, J., Ghemawat, S., Hsieh, W.C., Wallach, D.A., Burrows, M., Chandra, T., Fikes, A., Gruber, R.E.: Bigtable: a distributed storage system for structured data. ACM Trans. Comput. Syst. **26**(2), 1–26 (2008)
9. Dean, J., Ghemawat, S.: MapReduce: simplified data processing on large clusters. In: OSDI (2004)
10. Ghemawat, S., Gobioff, H., Leung, S.-T.: The Google file system. In: SOSP (2003)
11. Isard, M., Budiu, M., Yu, Y., Birrell, A., Fetterly, D.: Dryad: distributed data-parallel programs from sequential building blocks. In: EuroSys (2007)
12. Dohnal, V., Gennaro, C., Zezula, P.: Similarity join in metric spaces using eD-index. In: Mařík, V., Štěpánková, O., Retschitzegger, W. (eds.) DEXA 2003. LNCS, vol. 2736, pp. 484–493. Springer, Heidelberg (2003). doi:10.1007/978-3-540-45227-0_48
13. Böhm, C., Braunmüller, B., Krebs, F., Kriegel, H.-P.: Epsilon grid order: an algorithm for the similarity join on massive high-dimensional data. In: SIGMOD (2001)
14. Dittrich, J.-P., Seeger, B.: GESS: a scalable similarity join algorithm for mining large data sets in high dimensional spaces. In: SIGKDD (2001)
15. Jacox, E.H., Samet, H.: Metric space similarity joins. ACM Trans. Database Syst. **33**, 7:1–7:38 (2008)
16. Chaudhuri, S., Ganti, V., Kaushik, R.: A primitive operator for similarity joins in data cleaning. In: ICDE (2006)
17. Chaudhuri, S., Ganti, V., Kaushik, R.: Data debugger: an operator-centric approach for data quality solutions. IEEE Data Eng. Bull. **29**(2), 60–66 (2006)
18. Gravano, L., Ipeirotis, P.G., Jagadish, H.V., Koudas, N., Muthukrishnan, S., Srivastava, D.: Approximate string joins in a database (almost) for free. In: VLDB (2001)
19. Vernica, R., Carey, M.J., Li, C.: Efficient parallel set-similarity joins using MapReduce. In: SIGMOD 2010 (2010)
20. Silva, Y.N., Reed, J.M., Tsosie, L.M.: MapReduce-based similarity join for metric spaces. In: VLDB/Cloud-I (2012)

21. Silva, Y.N., Reed, J.M.: Exploiting MapReduce-based similarity joins. In: SIGMOD (2012)
22. Afrati, F.N., Sarma, A.D., Menestrina, D., Parameswaran, A., Ullman, J.D.: Fuzzy joins using MapReduce. In: ICDE (2012)
23. Okcan, A., Riedewald, M.: Processing theta-joins using MapReduce. In: SIGMOD (2011)
24. Metwally, A., Faloutsos, C.: V-SMART-join: a scalable MapReduce framework for all-pair similarity joins of multisets and vectors. In: VLDB (2012)
25. Xiao, C., Wang, W., Lin, X., Yu, J.X.: Efficient similarity joins for near duplicate detection. In: WWW (2008)
26. Apache Hadoop. http://hadoop.apache.org/
27. SimCloud Project: MapReduce-based similarity join survey. http://www.public.asu.edu/~ynsilva/SimCloud/SJSurvey
28. Harvard Library: Harvard bibliographic dataset. http://library.harvard.edu/open-metadata

YFCC100M-HNfc6: A Large-Scale Deep Features Benchmark for Similarity Search

Giuseppe Amato, Fabrizio Falchi$^{(\boxtimes)}$, Claudio Gennaro, and Fausto Rabitti

ISTI-CNR, via G. Moruzzi 1, 56124 Pisa, Italy
{giuseppe.amato,fabrizio.falchi,claudio.gennaro,
fausto.rabitti}@isti.cnr.it

Abstract. In this paper, we present YFCC100M-HNfc6, a benchmark consisting of 97M deep features extracted from the Yahoo Creative Commons 100M (YFCC100M) dataset. Three type of features were extracted using a state-of-the-art Convolutional Neural Network trained on the ImageNet and Places datasets. Together with the features, we made publicly available a set of 1,000 queries and k-NN results obtained by sequential scan. We first report detailed statistical information on both the features and search results. Then, we show an example of performance evaluation, performed using this benchmark, on the MI-File approximate similarity access method.

Keywords: Similarity search · Deep features · Content-based image retrieval · Convolutional neural networks · YFCC100M

1 Introduction

The ability to efficiently search for similarity in large databases of images is a critical aspect for a number of content-based retrieval applications like web search engines, e-commerce, museum collections, medical image processing, etc. To address this problem, several approaches based on index methods have been proposed in the literature, such as approximate access methods based permutation-based indices [4,10,12,16,22]. However, an important issue in comparing performance of different access methods is the availability of realistic benchmarks, of very large size.

In this paper, we present YFCC100M-HNfc6, a similarity-search benchmark consisting of three types of features extracted from 97M images, and pre-computed similarity search results for k-NN searches ($k = 10,001$) on 1000 queries. To make scalability assessment of access methods, precomputed results for the 1000 queries were generated for increasing sizes of the dataset at intermediate steps of 1M. The features of the benchmark were extracted from the YFCC100M [21] dataset using state of the art Deep Convolutional Neural Networks.

Deep learning methods are "representation-learning methods with multiple levels of representation, obtained by composing simple but non-linear modules that each transform the representation at one level (starting with the raw input) into a representation at a higher, slightly more abstract level" [15]. Starting from

© Springer International Publishing AG 2016
L. Amsaleg et al. (Eds.): SISAP 2016, LNCS 9939, pp. 196–209, 2016.
DOI: 10.1007/978-3-319-46759-7_15

2012 [14], Deep Convolutional Neural Networks (DCCNs) have attracted enormous interest within the Computer Vision community because of the state-of-the-art results achieved in image classification tasks. The relevance of the internal representation learned by the neural network during training have been proved by recent works that the activation produced by an image within the intermediate layers can be used as a high-level descriptor of the image visual content [6,8,17,19].

The importance of having a very large dataset of publicly available features has been proven by the Content-based Photo Image Retrieval (CoPhIR) [7][1] we released on 2009, which has been used by many scientists working in the field of very large scale similarity search algorithms. The CoPhIR dataset consists of MPEG-7 features extracted from about 107M Flickr! images. Given the impressive improvement recently achieved in many Content-Based Image Retrieval applications by using Deep Features, we decided to create a new benchmark based on features obtained as activations of DCNNs. To accomplish to this task we had the opportunity to have access to an already publicly available image dataset, i.e., the YFCC100M [21], from which we extracted the deep features, and we collaborated with the team of the Multimedia Commons Initiative [2] that made our deep features also available through their website.

The availability of extracted features from large datasets contributes to the research in the field in three ways. First, it allows a fair comparison between similarity access methods. In fact, details in the feature extractions could result in slightly different features obtained by various research group making impossible comparing the performance measures obtained. Second, the extraction of some features, as the deep features, is computational demanding. Extracting features from a collection of 100M documents can take months on a standard PC or special hardware, as clusters of GPUs, are required to do it in days. Third, when the images of the dataset are public available online (as for CoPhIR and YFCC100M), having the features allows researchers to index without storing them locally but just pointing to them whenever results have to be shown.

The rest of the paper is organized as follows. In Sect. 2, we briefly describe related work. Detailed information about the dataset and the features extracted are given in Sect. 3. In Sect. 4, we give statistical information about both features and search results. Section 5 provides some general metrics for measuring the quality of approximate results and a case study of their application. Section 6 concludes the paper.

2   Related Work

There are other very large datasets that can be used as a benchmark for assessing the performance of similarity search access methods. Among the most relevant we mention the CoPhIR and the TEXMEX dataset.

The CoPhIR dataset [7] consists of 107 millions MPEG-7 features extracted from images. Not all the images in CoPhIR have a creative common license.

---

[1] http://cophir.isti.cnr.it/.

Moreover, Deep Features have been proved to outperform previous approach as MPEG-7 in many tasks [17].

The ANN_SIFT1B dataset from the TEXMEX corpus[2] consists of one billion SIFT local features of 128 dimensions extracted from about 1 million images. Any image has about 1,000 features and each of them would be a query for the system. Moreover, a ground truth with image as queries was note defined. It is worth to mention that, in terms of images, our proposed dataset is 2 order of magnitude bigger and the proposed deep features have larger dimensionality.

Deep Convolutional Neural Networks (DCNNs) have recently become state-of-the-art approach for many computer vision task such as image classification [14,20], image retrieval [6,11,14,17,20] and object recognition [11]. The use of the activation of intermediate layers as a high-level descriptor of the image visual content has been also proved to be effective by many recent works [6,8,17,19]. Rectified Linear Unit (ReLU) is part of almost all of the DCNN models and is typically applied also for extracting deep features from images [8,11]. However, there are works in which the ReLU was omitted [6,17,19]. The L2 Normalization of the feature in order to and compare using the Euclidean distance is a standard de-facto for deep features [8,19]. It is worth to mention, that the resulting ranking of similarity search is equivalent to the cosine similarity. Principal Component Analysis has been successfully used in [6,18].

3   The YFCC100M-HNfc6 Dataset

The Yahoo Flickr Creative Commons 100M (YFCC100M) [21] consists of approximately 99.2 million photos and 0.8 million videos, all uploaded to Flickr between 2004 and 2014 and published under a Creative Commons commercial or non commercial license. Metadata for the YFCC100M dataset are publicly available through Yahoo! Webscope[3]. YFCC100M images can be obtained directly from Flickr! using the information reported in the metadata or through the Multimedia Commons Initiative website[4] using the hash of the original image also reported in the metadata.

The YFCC100M-HNfc6 was obtained by extracting, from the YFCC100M images, the HNfc6 Deep Features, described in Sect. 3.1, and by pre-computing the k-NN results for 1,000 queries. We selected the first 1,000 images of our ordering (which is random) as similarity search queries. We performed k-NN search with $k = 10,001$ sequentially scanning the data. We report results at intermediate steps of 1 million objects in order to allow scalability measures of access methods. The query itself is between the results generated at any intermediate step. The results of the k-NN queries can be used as ground-truth results for ranges up to 0.629 for the Euclidean distance,and up to 606 for the Hamming. We did not to consider images smaller than 4 KB; therefore, we extracted features from 96,976,180 of the 99,206,564 images in YFCC100M [21]. YFCC100M-HNFc6 features are available

---

[2] http://corpus-texmex.irisa.fr/.

[3] https://webscope.sandbox.yahoo.com/catalog.php?datatype=i&did=67.

[4] https://multimediacommons.wordpress.com.

in 97 zipped text files in which the objects have been randomly ordered. Thus, any subset of the full dataset contains random objects from the full dataset. The features, the k-NN results, demos of on-line systems indexing the features, and other information is available on the deep features website [1]. As a result of different processing of the neuron activation, we give three distinct features: *ReLU-L2Norm*, which is the reference feature for the dataset; *Binary*, which is intended for high efficiency and *Raw*, which allows other researchers to test other type of processing of the neurons activations. Information about the extracted features is given in following section.

3.1   The HNfc6 Deep Features

Deep features have been extracted with a trained model publicly available for the popular Caffe framework [13]. Many deep neural network models and in particular trained models are available for this framework at[5]. Among them, we chose the HybridNet for several reasons: first, its architecture is the same of the famous AlexNet [14]; second, the HybridNet has been trained not only on the ImageNet subset used for ILSVRC competitions (as many others), but also on the Places Database [23]; last, but not least, experiments conducted on various datasets demonstrate the good transferability of the learning [5,8,23]. We decided to use the activation of the first fully connected layer (i.e., fc6) given the results reported on [6,8,11]. It is worth to mention that the activation of the second fully connected layer (i.e., fc7) can be obtained from the fc6 activation with a simple matrix operation using the pre-trained weights [1].

   As a result of different processing of the neuron activation, we give three distinct features described in the following.

ReLU-L2Norm. The reference features of our YFCC100M-HNfc6 data are 4,096 dimensional L2 Normalized vectors corresponding to the activation of the neurons of the HybridNet *fc6* layer after the ReLU. This activation function, which is part of the HybridNet Convolutional Neural Network, simply sets to zero all the elements of the vectors that are negative. The distance to be used to compare is the Euclidean (aka L2 distance). The ranking of the results for a query using a combination of the L2 normalization and Euclidean distance is the same to the ones using the Cosine similarity of the original feature vectors.

Binary. We provide a binary version of the deep features consisting of 4,096 bits (i.e., 512 bytes). We simply encoded the positive values of the activations as 1 s, while zeros and negative values as 0 s. We evaluated k-NN results also for these features using the Hamming distance.

Raw. The activations of the fc6 neurons without the ReLU are given for anyone that would like to try approaches different from the previous ones and can be

---

[5] https://github.com/BVLC/caffe/wiki/Model-Zoo.

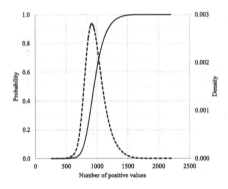 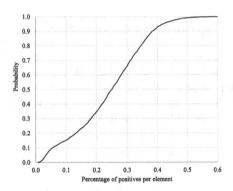

Fig. 1. Cumulative distribution function and probability density of n. positive elements per image.

Fig. 2. Cumulative distribution function of percentage of positives per feature element

useful to researchers willing to test new processing of neuron activations and experiments. We also give a k-NN results for these features comparing them with the Euclidean distance after the L2 Normalization. Please note that the values given on the deep features website [1] are not L2 Normalized.

Please note that the fc7 layer activation can be obtained from the raw fc6 data quite easily and without the use of any external library. On the deep features website, we give all the necessary information and a code example for anyone interested.

4 Statistical Analysis

In this Section, we report detailed statical information for the proposed benchmark. As reported in the previous Section, YFCC100M-HNfc6 consists of three version of deep features, which are the results of different processing of the exact values of the neuron activation.

We first show some statistics about the sign of the activation of the neurons that constitute the deep features. Each neuron of a hidden layer, as the *fc6* we considered, basically reports whether a specific high level feature, learned during the training phase, is found in the input image. The sign of the activation reports if the specific feature was found or not, while the absolute value is a measure of the confidence on this information. These statistic also illustrate the sparsity of both the *ReLU-L2Norm* and *Binary* features. Please note that the sparsity of the features is relevant for any access methods that would consider compression or techniques that require sparse information as inverted files.

In Fig. 1, we report the cumulative distribution function and probability density function for the number of positive values in features. This statistic is the very same for the *ReLU-L2Norm*, *Raw*, and *Binary*. In particular, the amount of positives has the following statistics: $min = 221$, $mode = 919$, $median = 972$, $mean = 972$ and $max = 2,201$. The results show that on average, each image has about 25 % of positive elements. Thus, the *ReLU-L2Norm* vectors are quite sparse.

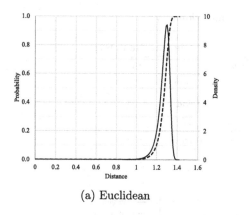
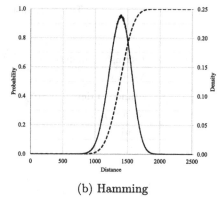

(a) Euclidean   (b) Hamming

Fig. 3. Cumulative distribution function (probability density as dotted line) for both Euclidean and Hamming distances

We now consider each vector component (i.e., each neuron in the *fc6* layer) individually. The goal of this analysis is evaluating the sparsity and estimating the usefulness of each particular element of the feature vector. As an example, an element whose values is always zero would be useless. Moreover, the overall sparsity of the data (shown in Fig. 1) could be the results of very different distribution of the values across the vector elements. In Fig. 2, we report the cumulative distribution related to the percentage of positives for an element all over the dataset. The graph shows that there is a 10 % of the vector components that have positive values in less than 5 % of the images, while there is 10 % of elements that are positive in more than 40 % of the images (that is, there is 90 % of elements that are positive in less than 40 % of the images). Overall, the results show that the sparsity is not equally spread across the feature element. Moreover, there are neurons which are almost never activated by the input images. The opposite, neurons activated by almost any image, do not exist.

We now focus on the distribution of distances. In Fig. 3, we report the cumulative and probability density functions for both the Euclidean distance applied to the *ReLU-L2Norm* features (a) and the Hamming distance applied to the features of *Binary*. In Table 1, we report the metric space intrinsic dimensionality [9] defined as $\mu^2/(2\sigma^2)$, and other information also related to these distributions.

The *ReLU-L2Norm* features in conjunction with the Euclidean distance appear to be very hard to index. The course of dimensionality is revealed in the graph and confirmed by the high *intrinsic dimensionality*. On the contrary, the *Binary* features combined with the Hamming distance reveal an intrinsic dimensionality of only 35 and the distribution is very similar to a Gaussian.

In Fig. 4, we analyse the amount of intersection between the results of the 1,000 *k*-NN queries we performed by sequentially scanning the dataset for the aim of creating the ground-truths that we made public available on our website. We compare the results obtained with the *ReLU-L2Norm* and *Binary* features

Table 1. Intrinsic Dimensionality

	Euclidean	Hamming
Intrinsic dimensionality ($\mu^2/(2\sigma^2)$)	276	35
Variance	0.0029	27057
Standard deviation (σ)	0.054	164.5
Mean (μ)	1.27	1383
Mode	1.28	1388

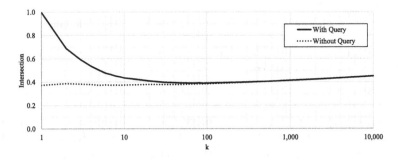

Fig. 4. Intersection between Euclidean and Hamming k-NN results varying k used for the k-NN search (with and without the query in the dataset).

reporting the average intersection varying k. We also considered the cases in which the query is between the results or is removed. When the query is in the result, at least the query itself is in the intersection. So, for instance, we have 100 % intersection for $k = 1$. The most interesting curve is the one in which we do not consider the query itself in the results. We obtained an intersection of about 0.4 for k between 1 and 1,000. It is also worth to note that the intersection increase with k.

In Fig. 5, we consider the distance of the results in our k-NN queries over the *ReLU-L2Norm* feature varying k reporting *min, mean, max, 10-th* and *90-th* percentiles. The graph shows that 90 % of the first 10 results for each query have a distance below 1.0. Unfortunately this distance is the same we obtain as mean for the 10,000 result. Moreover, for 10 % of the queries we have a first result at a distance greater than 0.9, which is the mean value we obtained for the result at position 1,000. In other words, while deep features have been proven to be effective in ranking, the value of the distance itself does not appear to be meaningful. This is also an effect of the high intrinsic dimensionality of the space and the curse of dimensionality.

4.1   Online CBIR Systems Using YFCC100M-HNfc6

We have also created and put on-line two different CBIR systems that use the YFCC100M-HNfc6 benchmark. One system is based on the the Metric Inverted

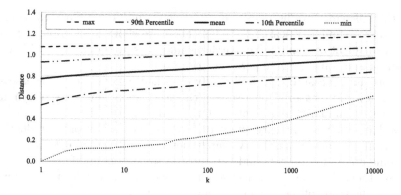

Fig. 5. *Mean, min, max,* 10-th and 90-th the *k*-th results distances.

File (MI-File) technique [4]. MI-File uses an inverted file to store relationships between permutations, and some approximations and optimizations to improve both efficiency and effectiveness. The basic idea is that entries (the lexicon) of the inverted file are the set of permutants (or pivots) P. The posting list associated with an entry $p_i \in P$ is a list of pairs $(o, \Pi_o^{-1}(i))$, $o \in C$, i.e. a list where each object o of the dataset C is associated with the position of the pivot p_i in Π_o.

The second system is based on the LuQ approach [3]. LuQ represents each DNCC feature as a text document and uses a NoSQL database (Apache Lucene) for efficiently indexing and searching purposes. It exploits the quantization of the vector components of the DCNN features, in which each real-valued vector component x_i is transformed in a natural numbers n_i given by $\lfloor Qx_i \rfloor$; where $\lfloor \rfloor$ denotes the floor function and Q is a multiplication factor > 1 that works as a *quantization factor*. n_i are then used as term frequencies for the "term-components" of the text document representing the feature vectors.

All the 97M features vectors of YFCC100M-HNfc6 were indexed using MI-File and LuQ approaches. The corresponding on-line demos are available at http://mifile.deepfeatures.org and http://melisandre.deepfeatures.org.

The whole Lucene 5.5 archive of LuQ approach is also available for download from the deep features website [1]. The advantage of this representation is that can be directly queried with Lucene by simply extracting the term vectors from the archive.

5   Performance Evaluation of Similarity Search Techniques

Performance assessment and comparison of similarity search techniques requires, in addition to a common dataset, also a common methodology to the execution of the experiments. Experiments must be executed using objective measures and should be reproducible so that other researcher can validate them and compare against other techniques.

In the following, we discuss some useful performance measures that can be used to assess similarity search methods. Then, we show an example of performance assessment of a similarity search technique using the YFCC100M-HNfc6 benchmark and some of these performance measures.

Performance measures can be broadly classified into measures for assessing the efficiency and measures for assessing the accuracy of similarity search algorithms. Exact similarity search algorithms, that is algorithms that retrieve all objects that satisfy a similarity query, require just assessment of their efficiency. However, many popular similarity search methods are approximate, i.e., might loose some qualifying objects and can retrieve some non-qualifying objects. In this case, therefore, the assessment of the quality of the results is relevant, to determine the trade-off between efficiency and accuracy.

A very obvious measure to assess efficiency of similarity search algorithms is measuring the *average query processing time* computed on a reasonable number of different queries. Unfortunately, this measures is not easily reproducible and objective. Query processing time depends on several aspects that are not easily controllable. Different hardware architectures, software installed, operating system, running environments, programming languages, can significantly affect the results.

A more objective measure, when data to be searched cannot be stored in main memory, is the *number of disk block reads*. All hard disks read data in blocks, for instance of size 4 K bytes. Reading more disk blocks means more time spent transferring data from disk to memory. This measure can give significant information both for traditional hard disks and for solid state disks. In case of traditional hard disks, it can be useful to distinguish between *consecutively stored disk block reads* and *randomly stored disk block reads*. Consecutively stored disk block reads do not require disk seeks, so they are order of magnitude faster than randomly stored disk block reads. However, this differentiation is not relevant in case of solid state disks, where there is practically no disk seek cost.

Another useful objective measure is the *number of object reads*. Similarity search algorithms generally access data from the disk either for accessing data structures of the index or for retrieving objects to be checked against the query conditions. The number of object reads gives an idea of the improvement of performance with respect to an exhaustive sequential scan of the entire dataset. A similar performance measure is the *number of distance computations*. In many realistic cases, the distance computation has an high cost as well. Counting the number of computed distances is useful as well, especially in those cases where part or all objects are stored in main memory.

Approximate similarity search methods offer improvement of efficient of some orders of magnitude with respect to exact similarity search algorithms, at the expense of some degradation of the accuracy. To assess the quality of approximate similarity search methods, a ground-truth must be available for the dataset, built using exact similarity search queries. In YFCC100M-HNfc6 the ground-truth is composed of 1,000 different queries for which the 10,001 nearest neighbours were retrieved. Ground-truth was generated using the entire dataset, and also smaller portions with size multiple of one million, to allow researchers to both test on a portion of the dataset and study scalability. A discussion on measures for assessing the quality of approximate similarity search can be found in [22]. Here we report some of those measures.

Two popular measures to assess quality of search results are the *Precision* and *Recall* measures. These can also be effectively used in our case. Let Q be a similarity query, for instance a range search or a k-NN search. Let ER_Q be the sorted exact similarity search result set, and AR_Q be the sorted approximate similarity search result set. Let $|\cdot|$ denote the size of a set. The precision P is the ratio between the number of correct results in the approximate result set, by the total size of the approximate result set:

$$P = \frac{|AR_Q \cap ER_Q|}{|AR_Q|}.$$

The recall R is the ratio between the number of correct results in the approximate result set and the number of correct results that should have been retrieved:

$$R = \frac{|AR_Q \cap ER_Q|}{|ER_Q|}.$$

An additional useful way to assess the quality of approximate results is to evaluate the discrepancy between the sorted approximate result set and the exact result set. This can be measured in terms of the difference in position of the objects between these sets. Let X be the entire dataset, X_Q the entire dataset sorted according to the distance from query object in query Q, $o = AR_Q[i]$ is the i-th object in the sorted approximate result set AR_Q, $X_Q(o)$ is the position of object o in sorted sets X_Q. The *Error on the Position, EP*, can be defined as a normalized version of the Induced Sperman Footrule distance as follows:

$$EP = \frac{\sum_{i=1}^{|AR_Q|} |X_Q(AR_Q[i]) - i|}{|AR_Q| \cdot |X|}.$$

The error on position measures the difference in positions between the approximate result and the exact result, averaged for all retrieved objects, and normalized dividing by the size of the dataset. Suppose the error on position is $EP = 10^{-5}$, on a dataset of 1 millions objects. This means that on average, the difference in position of retrieved objects, between the approximate and exact results, is $10^{-5} \cdot 1,000,000 = 10$.

Clearly all above measures should be averaged on several queries. As we said before, YFCC100M-HNfc6 offers k-NN results for 1,000 queries, so these measures can be averaged on those queries.

5.1   Performance Evaluation Example

In the following, we show some an example of performance assessments obtained using the full binary YFCC100M-HNfc6 benchmark and some of the performance measures mentioned in the previous section. The method that we test is the MI-File approximate similarity search index [4]. MI-File is a permutation based method that uses inverted files to perform fast approximate execution of k-NN queries. MI-File offers the following parameters to trade efficiency with accuracy:

- Amplification factor amp: when searching for the k-NN the MI-File retrieves a candidate set of $k' = amp \cdot k$ objects, reorders it according to the original distance function, and returns the top-k objects. The larger amp, the higher the search cost, and the higher the accuracy.
- data object permutation length k_i: the permutation representing a data object is obtained using the k_i closest reference objects out of the total set of reference objects. The value of k_i determines the number of posting lists containing a reference to the object being inserted.
- Query permutation length k_s: the permutation representing the query is obtained using the k_s closest reference objects out of the total set of reference objects. The value of k_s determines the number of posting lists accessed during a query execution.
- Maximum position difference mpd: posting lists are scanned considering entries referring objects whose reference objects position difference in their permutation, with respect to the query permutation, is at most mpd. The higher mpd, the more entries are retrieved from the posting lists.

Please see [4] for further details on the MI-File and its parameters usage.

In our experiments, we indexed the entire binary YFCC100M-HNfc6 dataset, using $k_i = 100$. The total number of reference objects for building permutations is 20,000. The queries were executed with amp ranging from 1 to 70. The values used for k_s ranged from 1 to 50 and $mpd = k_s$. We executed 100-NN queries using the 1,000 queries of the ground truth, and performance measures were obtained as average of the measures computed for all query. Results are shown in Fig. 6.

The two upper figures show the relationships between the number of disk blocks accessed and the quality of results. Disk block size is 4 K bytes. Every plot corresponds to different setting for amp, and the amount of disk blocks accessed was tuned by setting the k_s parameter. MI-File reaches a recall of 75 % with a number of disk block accesses around 30,000. For the same disk access cost, we have an error on position of $2 \cdot 10^{-6}$. This means that the on average the difference in position of retrieved objects, between the approximate and exact results, on a dataset of 100M objects, is around 200 positions. However, given that the recall is 0.75 %, so 75 out of 100 objects are correctly retrieved and their difference in position can be at most 25, most of the position error is due to the few objects (25 %) that were erroneously retrieved in place of the exact results.

The two bottom graphs show the relationships between the number of database objects accessed and the quality of the results. Here, the index access cost is not taken into account. In this case, 7,000 objects out of 100M total objects have to be accessed to have a recall of almost 75 % and a position error of $2 \cdot 10^{-6}$.

An objective estimation of the average query processing time can be obtained from the number of disk blocks accessed and the Input/Output Operations Per

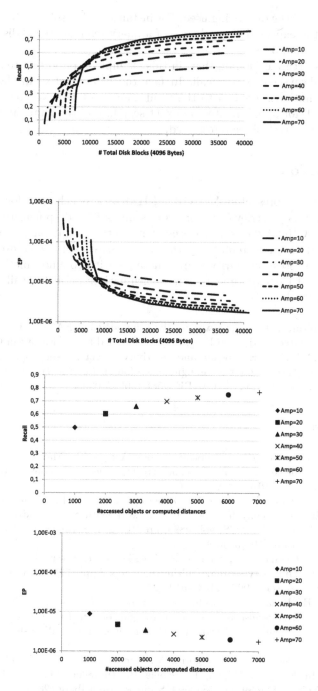

Fig. 6. Experiments executed indexing the YFCC100M-HNfc6 *binary* features with the MI-File considering k-NN search with $k = 100$

Second (IOPS) of the disk being used. For instance with a solid state disk having 8,600 IOPS, the expected elapsed time of a query reading 30,000 disk blocks of 4 K each, is around 3.5 s.

It is also worth mentioning that, in this example of performance evaluation, precision is always equal to recall. In fact the denominator, in the precision and recall definitions, is equal to the total number of retrieved objects, which is $k = 100$ both for approximate and exact search, and the numerator is always the number of correct objects retrieved.

6   Conclusion

In this paper, we presented YFCC100M-HNfc6: a benchmark for evaluating content-based image retrieval systems consisting of 97M deep features extracted from the YFCC100M dataset. Together with detailed statical information for the proposed dataset, we reported performance assessment and comparison that can be used to assess similarity search methods with common methodology to the execution of the experiments. The benchmark is publicly available on the Deep Features website [1].

Acknowledgments. This work was partially founded by: EAGLE, Europeana network of Ancient Greek and Latin Epigraphy, co-founded by the European Commision, CIP-ICT-PSP.2012.2.1 - Europeana and creativity, Grant Agreement n. 325122; and Smart News, Social sensing for breakingnews, co-founded by the Tuscnay region under the FAR-FAS 2014 program, CUP CIPE D58C15000270008.

References

1. Deep features. http://www.deepfeatures.org. Accessed 23 May 2016
2. The multimedia commons initiative. https://multimediacommons.wordpress.com/. Accessed 23 May 2016
3. Amato, G., Debole, F., Falchi, F., Gennaro, C., Rabitti, F.: Large scale indexing and searching deep convolutional neural network features. In: Madria, S., Hara, T. (eds.) DaWaK 2016. LNCS, vol. 9829, pp. 213–224. Springer, Heidelberg (2016). doi:10.1007/978-3-319-43946-4_14
4. Amato, G., Gennaro, C., Savino, P.: MI-File: using inverted files for scalable approximate similarity search. Multimedia Tools Appl. **71**(3), 1333–1362 (2014). http://dx.doi.org/10.1007/s11042-012-1271-1
5. Azizpour, H., Razavian, A., Sullivan, J., Maki, A., Carlsson, S.: From generic to specific deep representations for visual recognition. In: Proceedings of the IEEE Conference on Computer Vision and Pattern Recognition Workshops, pp. 36–45 (2015)
6. Babenko, A., Slesarev, A., Chigorin, A., Lempitsky, V.: Neural codes for image retrieval. In: Fleet, D., Pajdla, T., Schiele, B., Tuytelaars, T. (eds.) ECCV 2014, Part I. LNCS, vol. 8689, pp. 584–599. Springer, Heidelberg (2014)
7. Bolettieri, P., Esuli, A., Falchi, F., Lucchese, C., Perego, R., Piccioli, T., Rabitti, F.: CoPhIR: a test collection for content-based image retrieval. CoRR abs/0905.4627v2 (2009). http://cophir.isti.cnr.it

8. Chandrasekhar, V., Lin, J., Morère, O., Goh, H., Veillard, A.: A practical guide to cnns and fisher vectors for image instance retrieval. arXiv preprint arXiv:1508.02496 (2015)

9. Chávez, E., Navarro, G., Baeza-Yates, R., Marroquín, J.L.: Searching in metric spaces. ACM Comput. Surv. (CSUR) **33**(3), 273–321 (2001)

10. Chavez, G., Figueroa, K., Navarro, G.: Effective proximity retrieval by ordering permutations. IEEE Trans. Pattern Anal. Mach. Intell. **30**(9), 1647–1658 (2008)

11. Donahue, J., Jia, Y., Vinyals, O., Hoffman, J., Zhang, N., Tzeng, E., Darrell, T.: Decaf: a deep convolutional activation feature for generic visual recognition. arXiv preprint arXiv:1310.1531 (2013)

12. Gennaro, C., Amato, G., Bolettieri, P., Savino, P.: An approach to content-based image retrieval based on the lucene search engine library. In: Lalmas, M., Jose, J., Rauber, A., Sebastiani, F., Frommholz, I. (eds.) ECDL 2010. LNCS, vol. 6273, pp. 55–66. Springer, Heidelberg (2010). doi:10.1007/978-3-642-15464-5_8

13. Jia, Y., Shelhamer, E., Donahue, J., Karayev, S., Long, J., Girshick, R., Guadarrama, S., Darrell, T.: Caffe: convolutional architecture for fast feature embedding. arXiv preprint arXiv:1408.5093 (2014)

14. Krizhevsky, A., Sutskever, I., Hinton, G.E.: Imagenet classification with deep convolutional neural networks. In: Advances in neural information processing systems, pp. 1097–1105 (2012)

15. LeCun, Y., Bengio, Y., Hinton, G.: Deep learning. Nature **521**(7553), 436–444 (2015)

16. Mohamed, H., Marchand-Maillet, S.: Quantized ranking for permutation-based indexing. Inf. Syst. **52**, 163–175 (2015)

17. Razavian, A.S., Azizpour, H., Sullivan, J., Carlsson, S.: CNN features off-the-shelf: an astounding baseline for recognition. In: 2014 IEEE Conference on Computer Vision and Pattern Recognition Workshops (CVPRW), pp. 512–519. IEEE (2014)

18. Razavian, A.S., Sullivan, J., Maki, A., Carlsson, S.: A baseline for visual instance retrieval with deep convolutional networks. arXiv preprint arXiv:1412.6574 (2014)

19. Sermanet, P., Eigen, D., Zhang, X., Mathieu, M., Fergus, R., LeCun, Y.: Overfeat: integrated recognition, localization and detection using convolutional networks. arXiv preprint arXiv:1312.6229 (2013)

20. Simonyan, K., Zisserman, A.: Very deep convolutional networks for large-scale image recognition. arXiv preprint arXiv:1409.1556 (2014)

21. Thomee, B., Elizalde, B., Shamma, D.A., Ni, K., Friedland, G., Poland, D., Borth, D., Li, L.J.: YFCC100M: the new data in multimedia research. Commun. ACM **59**(2), 64–73 (2016)

22. Zezula, P., Amato, G., Dohnal, V., Batko, M.: Similarity Search: The Metric Space Approach, Advances in Database Systems, vol. 32. Springer, Heidelberg (2006)

23. Zhou, B., Lapedriza, A., Xiao, J., Torralba, A., Oliva, A.: Learning deep features for scene recognition using places database. In: Advances in neural information processing systems, pp. 487–495 (2014)

A Tale of Four Metrics

Richard Connor[(✉)]

Department of Computer and Information Sciences, University of Strathclyde,
Glasgow G1 1XH, UK
richard.connor@strath.ac.uk

Abstract. There are many contexts where the definition of similarity in multivariate space requires to be based on the correlation, rather than absolute value, of the variables. Examples include classic IR measurements such as TDF/IF and BM25, client similarity measures based on collaborative filtering, feature analysis of chemical molecules, and biodiversity contexts.

In such cases, it is almost standard for Cosine similarity to be used. More recently, Jensen-Shannon divergence has appeared in a proper metric form, and a related metric Structural Entropic Distance (SED) has been investigated. A fourth metric, based on a little-known divergence function named as Triangular Divergence, is also assessed here.

For these metrics, we study their properties in the context of similarity and metric search. We compare and contrast their semantics and performance. Our conclusion is that, despite Cosine Distance being an almost automatic choice in this context, Triangular Distance is most likely to be the best choice in terms of a compromise between semantics and performance.

1 Introduction

Comparing the similarity of two vectors of numbers is a very common occupation across many branches of computer and indeed other sciences. However the choice of similarity measure is often not taken very seriously. In situations where the correlation among dimensions is required, Cosine similarity is most commonly used, and is often adopted without further thought. There are however literally hundreds of other choices which have been used in various circumstances.

"Correlation" similarity is a requirement when the data values represent a number of independent dimensions, each of which represents a probability, or a number whose ratio with other numbers is more important that its absolute value. The most commonly known domain is information retrieval where document search is usually performed based on the relative frequency of terms within documents and the collections; it is important in many other domains as well.

Large-scale search by similarity requires some strong properties to benefit from available techniques. It is important for a similarity function to be coercible into a proper distance function for techniques such as exact metric indexing, and indeed for most forms of approximate metric indexing. It would be nice if the metric used had a well-defined semantics with respect to the value space under

© Springer International Publishing AG 2016
L. Amsaleg et al. (Eds.): SISAP 2016, LNCS 9939, pp. 210–217, 2016.
DOI: 10.1007/978-3-319-46759-7_16

consideration, a property often entirely overlooked. Hilbert embeddability gives further important guarantees for metric search [1,2]. Finally, efficient evaluation is also of clear importance when large data sets are being considered.

We are not aware of any metric which fits all of these requirements. Table 1 gives an overview of these properties for the four metrics we have considered; none has all the desirable properties. However, the little-known Triangular Distance meets all requirements other than semantic; furthermore, we show that it is an excellent estimator for Jensen-Shannon, and hence Structural Entropic, distances, and thus should inherit their semantic basis. Our conclusion is that this relatively unknown function is probably the best starting place for data sets without a known ground truth.

Table 1. Outline properties of the four metrics

Metric	Abbreviation used in text	Proper metric[a]	Hilbert embeddable	Semantic basis	Efficient evaluation
Cosine	Cos	Yes	No	No	Yes
Jensen-Shannon	Jsd	Yes	Yes	Yes	No
Structural Entropic	Sed	Yes	No	Yes	No
Triangular	Tri	Yes	Yes	No	Yes

[a]In fact all of these metrics require some massaging over their "natural" form to achieve this, see Sect. 2.

2 Definitions

The common names used for the four functions are not uniquely defined and are used differently according to context. In particular, similarity coefficients with these names have been used as well as divergence and distance (implying semi-metric and metric properties, respectively) functions. If all that is required is a ranking function over the input, these details are unimportant as the semantics will be indistinguishable. To allow a proper comparison, in this paper we will compare a form of each function that is a proper metric, and whose range is bounded in $[0, 1]$.

For all functions we consider the domain of values $V \in (\mathbb{R}^+)^n$ for some fixed n. Values v within this space have the properties $v_i \geq 0$ for all i, and $\sum_i v_i = 1$, and as such can represent probability distributions over n distinct events[1]. All of the metrics bar Cosine distance are defined over probability distributions in any case, and the normalisation of a vector space to achieve this property does not affect the outcome of Cosine distance. It should be noted that Cosine distance

---

[1] Some functions are formally undefined in the presence of zero values, requiring either $0 \log 0$ or $0/0$. In each case, there is in fact a good mathematical argument for treating these terms as 0 rather than *undefined*.

alone is well-defined over negative domains; we disregard this property in our context although note that this makes the function more generally applicable.

Terse definitions follow for the four functions under consideration. In most cases these are not the most familiar forms of each function, but in each case carry the same rank ordering as more familiar forms.

Preliminaries

$$h(x) = -x \log_2 x$$
$$\mathcal{K}(v, w) = 1 - \tfrac{1}{2} \sum_i (h(v_i) + h(w_i) - h(v_i + w_i))$$

Cosine Distance

$$D_{cos}(v, w) = \left(\frac{2}{\pi}\right) \cos^{-1}\left(\frac{v \cdot w}{|v||w|}\right) \tag{1}$$

Jensen-Shannon Distance

$$D_{jsd}(v, w) = \sqrt{\mathcal{K}(v, w)} \tag{2}$$

Structural Entropic Distance

$$D_{sed}(v, w) = (2^{\mathcal{K}(v,w)} - 1)^p \tag{3}$$

for any constant p with a value of $\leq 0.486^2$.

Triangular Distance

$$D_{tri}(v, w) = \sqrt{\tfrac{1}{2} \sum_i \frac{(v_i - w_i)^2}{v_i + w_i}} \tag{4}$$

3   Metric Properties

To be a proper metric, a function requires to be a semi-metric with the added property of triangle inequality. The semi-metric properties for all functions are evident directly from their definitions above.

Cosine Distance as defined here over $(\mathbb{R}^+)^n$ is simply the angle between two vectors defined in n-dimensional Cartesian space, adjusted into $[0, 1]$. Any three vectors in any dimensionality may be transformed into three-dimensional space whilst preserving the angles between them, from which it is immediately evident that, for any three such vectors in any dimensionality, triangle inequality holds.

Jensen-Shannon Distance as defined here has been shown to be a proper metric in [6, 10].

Structural Entropic Distance has been shown to be a proper metric in [4].

Triangular Distance is very little studied, the term was probably first coined in [13]. The fact that the form given here is a proper metric is alluded to in [12]; while the authors of this clearly knew it, there is no published proof we can find before one given in [1].

---

[2] See [4] for an explanation of this constant.

4   Semantics

All functions have the same endpoints: a distance of 0 occurs if and only if the vectors are identical, and a distance of 1 occurs if and only if no dimension has a non-zero value in both vectors, i.e. there is no correlation among the dimensions. Beyond this observation, we give a brief overview of the semantics of the functions.

Cos While Cosine distance has a very clear meaning in vector space, it is essentially meaningless when used for multivariate analysis, other than when approaching the common endpoints at 0 and 1.

The origins of the use of Cosine distance in Information Retrieval is hard to trace. Singhal [11] states: "Typically, the angle between two vectors is used as a measure of divergence between the vectors, and cosine of the angle is used as the numeric similarity (since cosine has the nice property that it is 1.0 for identical vectors and 0.0 for orthogonal vectors)." It would appear that cosine similarity has been in use in this field for 50 years without much more attention than this being paid to its semantics. This lack of meaning is directly related to the "long document" problem in information retrieval. Mathematically, this corresponds to the following observation.

Consider two probability distributions P_1 and P_2 whose distance $d(P_1, P_2)$ is measured by some metric d. Consider an event e such that $P_1(e) = 0$ and $P_2(e) \neq 0$. It is then realised that e can be refined into two separate events e_1 and e_2, i.e. $P_i(e) = P_i(e_1) + P_i(e_2)$. It is clearly desirable that $d(P_1, P_2)$ would not be affected by this refinement. However, when applied to Cosine distance this refinement decreases the measured distance, as the magnitude of the vectors is deceased, but the "dot product" is not affected. By simple observation of the other three metrics, distance is indeed preserved in this situation.

This observation is expanded in [3]. The problem is addressed in Information Retrieval by including parameters such as document length as a proxy for the number of terms contained in e.g. the Okapi BM formulae [5,8] shows how an equivalent parameterless construction based on Jensen-Shannon distance semantically outperforms BM25 and other cosine-based models.

Jsd was originally derived from Kullback-Leibler divergence, which has a well-defined semantics. However its massaging into the proper metric form of Jensen-Shannon distance is arbitrary [9] and the original semantics is lost.

Topsøe has found a meaningful, if somewhat obscure, semantics given in [7], based on the comparative entropy of information streams.

Sed was originally designed as an information distance in its own right, and as such is the only one of these functions with a semantic derivation [4]. The initial formulation is

$$D_{sed}(v, w) = \frac{C(v + w)}{\sqrt{C(v) \cdot C(w)}} - 1$$

where C is a measure of information content[3]. Comparing this with the initial formulation of Jensen-Shannon distance:

$$D_{jsd}(v, w) = K(v, m) + K(w, m), \quad m = (v + w)/2$$

where K is the Kullback-Leibler divergence, it is initially quite surprising that both can be written as monotonic transforms over the same kernel function as shown in Eqs. 2 and 3. However as this preserves rank order, Jensen-Shannon distance can safely be deemed to inherit the same semantic basis.

Tri Triangular distance has no obvious semantics over probability distributions but Topsøe has shown surprisingly tight upper and lower bounds for Jensen-Shannon distance expressed in terms of Triangular Discrimination [12]. This is where our interest in this function derives from. When the bounds given are manipulated for the proper metric forms of both functions, it turns out that

$$D_{tri}(v, w) \leq D_{jsd}(v, w) \leq \sqrt{2\ln(2)} \cdot D_{tri}(v, w)$$

It seems that in evenly distributed high-dimensional spaces Triangular Distance should be an excellent estimator for (the much more expensive) Jensen-Shannon Distance.

5   Runtime Evaluation Costs

Various optimisations are available for all of these metrics under various circumstances; here we simply give the brute-force cost of evaluating the different metrics over two different data sets, which appears to be usefully indicative of cost.

Table 2. Raw execution times

Data set		Cos	Jsd	Sed	Tri
Colors	"Cold" times	540	2681	1962	245
Colors	"Hot" times	532	2671	1883	245
10dim	"Cold" times	610	814	463	51
10dim	"Hot" times	608	794	434	47

Table 2 shows outline execution times in milliseconds for ten million distance evaluations from the SISAP *colors*[4] data and randomly generated 10-dimensional space. There are a few points to note:

---

[3] In fact Shannon's entropy raised to the power of the logarithm base, see [4] for details.

[4] 118 dimensions.

1. Results are indicative only; accurate performance depends on particular implementation details. However the code used[5] has been refined over some time, is believed to be the most generally efficient form of each function coded at a high level, and shares object representations over all metrics.
2. The distinction between *jsd* and *sed* is artificial; as noted, they may be calculated as functions of each other via the kernel definition. Here they are evaluated according to their original definitions as given in Sect. 4; there is a significant advantage in evaluating the complexity of each vector and the vector mean, rather than just the pairwise entropy differences.
3. The differences between "hot and "cold" evaluation times are explained by parts of the evaluation that can be memo-ised within the data structure used to represent the point; for example magnitude (*cos*) and complexity (*sed*).
4. *colors* is a set of relatively sparse vectors, containing a significant number of zero values, whereas *10dim* contains no zero values. The effect on the performance is very clear; however it should be noted that specific optimisations are available for all metrics over sparse spaces.

The tests were run on a laptop with a 2.9 GHz Intel Core i7 processor; only the relative figures and the effect of sparsity should be taken seriously. The only general outcome is that (a) there can be a very major difference in performance, and (b) Triangular Distance is always by far the cheapest of those tested.

6   Query Evaluation Costs

To give an idea of how well the metrics work in practice, a nearest-neighbour search was conducted using a balanced vantage point tree. Nearest-neighbour was chosen here to avoid difficulties with comparing different query thresholds. For the 10dim data set, a tree containing one million values was queried by 1,000 different queries, and for the SISAP *colors* data, 10 % of the data was selected randomly for use as queries. Total elapsed time for these queries is given in milliseconds.

While cosine distance performs best over the long vectors of the SISAP *colors* set, it is impossible to tell the value of this without having a ground truth for similarity; Cosine distance is well known to give artificially low values for high-dimensional data, and the presence of many small distances will clearly reduce the cost of any nearest-neighbour search.

The relative times of Triangular distance with both SED and Jensen-Shannon, for a function that should be expected to return very similar results, are notable (Table 3).

7   Triangular Distance as an Approximation of JSD/SED

It is apparent that Triangular distance is so far, at least in terms of efficiency, the most desirable of the metrics; however it has no clear semantics. However our primary interest in the metric was the observation explained above, that it should

---

[5] Aailable at https://bitbucket.org/richardconnor/metric-space-framework.

Table 3. Metric Index build and query times

Data set	Size		Cos	Jsd	Sed	Tri
Colors	101414	Build time	1209	4710	3693	862
Colors	(10.1k queries)	Query time	125355	1134417	673602	281702
10dim	1M	Build time	13512	17643	12766	3252
10dim	(1k queries)	Query time	21038	35842	25591	13266

SISAP colors Jsd vs Tri

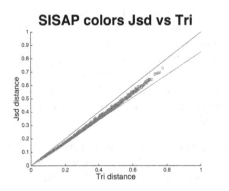

SISAP colors Jsd vs Cos

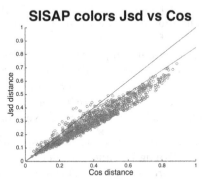

Fig. 1. Scatter plot showing the tight correlation between Triangular and Jensen-Shannon distances over randomly selected values from the SISAP *colors* data. Lines indicate the theoretical bounds for Jsd/Tri.

provide an excellent estimator for Jensen-Shannon and Structural Entropic distance, which we now test.

The left plot in Fig. 1 shows, for the SISAP *colors* data set, scatter plots of 1,125 randomly selected pairs of points measured against each other with the two metrics. For comparison, the right plot shows the same data with Cosine distance applied instead of Triangular. The measured correlation of Triangular distance is 0.9993, against 0.965 for Cosine distance. When a linear function intersecting the origin is applied, R^2 values are 0.9991 and 0.929 respectively. This test has been extensively re-performed with other real-world and generated data sets, and these correlation figures are found to be typical.

It is therefore evident that Triangular Distance is an excellent estimator for Jensen-Shannon Distance, and in turn is likely to inherit the semantic ordering assured by Structural Entropic Distance.

8   Conclusions

The simple conclusion of this work is that, when correlation is the desired basis of comparison, no ground truth is available, and runtime costs are an issue, then Triangular Distance should probably be the metric of choice. This little-known metric deserves to be used as default, rather than the almost-ubiquitous Cosine distance: it is more efficient, and semantically superior.

References

1. Connor, R., Cardillo, F.A., Vadicamo, L., Rabitti, F.: Hilbert Exclusion: Improved Metric Search Through Finite Isometric Embeddings. ArXiv e-prints, accepted for publication ACM TOIS, April 2016
2. Connor, R., Cardillo, F.A., Vadicamo, L., Rabitti, F.: Supermetric Search with the Four-Point Property. Accepted for publication SISAP, Tokyo, Japan, October 2016
3. Connor, R., Moss, R.: A multivariate correlation distance for vector spaces. In: Navarro, G., Pestov, V. (eds.) SISAP 2012. LNCS, vol. 7404, pp. 209–225. Springer, Heidelberg (2012)
4. Connor, R., Simeoni, F., Iakovos, M., Moss, R.: A bounded distance metric for comparing tree structure. Inf. Syst. **36**(4), 748–764 (2011)
5. Connor, R., Moss, R., Harvey, M.: A new probabilistic ranking model. In: Proceedings of the 2013 Conference on the Theory of Information Retrieval, ICTIR 2013, p. 23: 109–23: 112, NY, USA (2013). http://doi.acm.org/10.1145/2499178. 2499185
6. Endres, D., Schindelin, J.: A new metric for probability distributions. IEEE Trans. Inf. Theor. **49**(7), 1858–1860 (2003)
7. Fuglede, B., Topsoe, F.: Jensen-Shannon divergence and Hilbert space embedding. In: Proceedings of International Symposium on Information Theory, ISIT 2004, p. 31 (2004)
8. Jones, K.S., Walker, S., Robertson, S.E.: A probabilistic model of information retrieval: development and comparative experiments: part 2. Inf. Process. Manag. **36**(6), 809–840 (2000)
9. Lin, J.: Divergence measures based on the Shannon entropy. IEEE Trans. Inf. Theor. **37**(1), 145–151 (1991)
10. Österreicher, F., Vajda, I.: A new class of metric divergences on probability spaces and and its statistical applications. Ann. Inst. Stat. Math. **55**, 639–653 (2003)
11. Singhal, A.: Modern information retrieval: a brief overview. IEEE Data Eng. Bull. **24**(4), 35–43 (2001)
12. Topsoe, F.: Some inequalities for information divergence and related measures of discrimination. IEEE Trans. Inf. Theor. **46**(4), 1602–1609 (2000)
13. Topsøe, F.: Jenson-Shannon divergence and norm-based measures of discrimination and variation. Preprint math.ku.dk (2003)

Hashing Techniques

Fast Approximate Furthest Neighbors
with Data-Dependent Candidate Selection

Ryan R. Curtin[(✉)] and Andrew B. Gardner

Center for Advanced Machine Learning, Symantec Corporation,
Atlanta, GA 30338, USA
ryan@ratml.org, andrew_gardner@symantec.com

Abstract. We present a novel strategy for approximate furthest neighbor search that selects a candidate set using the data distribution. This strategy leads to an algorithm, which we call `DrusillaSelect`, that is able to outperform existing approximate furthest neighbor strategies. Our strategy is motivated by an empirical study of the behavior of the furthest neighbor search problem, which lends intuition for where our algorithm is most useful. We also present a variant of the algorithm that gives an absolute approximation guarantee; under some assumptions, the guaranteed approximation can be achieved in provably less time than brute-force search. Performance studies indicate that `DrusillaSelect` can achieve comparable levels of approximation to other algorithms while giving up to an order of magnitude speedup. An implementation is available in the **mlpack** machine learning library (found at http://www.mlpack.org).

1 Introduction

We concern ourselves with the problem of *furthest neighbor search*, which is the logical opposite of the well-known problem of nearest neighbor search. Instead of finding the nearest neighbor of a query point, our goal is to find the furthest neighbor. This problem has applications in recommender systems, where furthest neighbors can increase the diversity of recommendations [1,2]. Furthest neighbor search is also a component in some nonlinear dimensionality reduction algorithms [3], complete linkage clustering [4,5] and other clustering applications [6]. Thus, being able to quickly return furthest neighbors is a significant practical concern for many applications.

However, it is in general not feasible to return exact furthest neighbors from large sets of points. Although this is possible with Voronoi diagrams in 2 or 3 dimensions [7], and with single-tree or dual-tree algorithms in higher dimensions [8], these algorithms tend to have long running times in practice. Therefore, approximate algorithms are often considered acceptable in most applications.

For approximate neighbor search algorithms, hashing strategies are a popular option [9–11]. Typically hashing has been applied to the problem of nearest neighbor search, but recently there has been interest in applying hashing techniques to furthest neighbor search [12,13]. In general, these techniques are

© Springer International Publishing AG 2016
L. Amsaleg et al. (Eds.): SISAP 2016, LNCS 9939, pp. 221–235, 2016.
DOI: 10.1007/978-3-319-46759-7_17

based on random projections, where random unit vectors are chosen as projection bases. This allows probabilistic error guarantees, but the entirely random approach does not use the structure of the dataset.

In this paper, we first consider the structure of the furthest neighbors problem and then conclude that a data-dependent approach can be used to select a small set of candidate points that work for all query points. This allows us to develop:

- DrusillaSelect, an algorithm that selects candidate points based on the data distribution and outperforms other approximate furthest neighbors approaches in practice.
- A modified version of DrusillaSelect which satisfies rigorous approximation guarantees, and under some assumptions will provably outperform the brute-force approach at search time. However, it is not likely to be useful in practice.

Our empirical results in Sect. 7 show that the DrusillaSelect algorithm demonstrably outperforms existing solutions for approximate k-furthest-neighbor search.

2   Notation and Formal Problem Description

The problem of furthest neighbor search is easily formalized. Given a set of *reference points* $S_r \in \mathcal{R}^{n \times d}$, a set of *query points* $S_q \in \mathcal{R}^{m \times d}$, and a distance metric $d(\cdot, \cdot)$, the problem is to find, for each query point $p_q \in S_q$,

$$\text{argmax}_{p_r \in S_r} \, d(p_q, p_r). \tag{1}$$

A trivial way to solve this algorithm is by brute-force: for each query point, loop over all reference points and find the furthest one. But this algorithm takes $O(nm)$ time, and does not scale well to large S_r or S_q. In this paper, we will consider the ϵ-approximate form of the furthest neighbor search problem.

Given a set of *reference points* $S_r \in \mathcal{R}^{n \times d}$, a set of *query points* $S_q \in \mathcal{R}^{m \times d}$, an approximation parameter $\epsilon \geq 0$, and a distance metric $d(\cdot, \cdot)$, the ϵ-approximate furthest neighbor problem is to find a furthest neighbor candidate $\hat{p}_{fn}$ for each query point $p_q \in S_q$ such that

$$\frac{d(p_q, p_{fn})}{d(p_q, \hat{p}_{fn})} < 1 + \epsilon \tag{2}$$

where p_{fn} is the true furthest neighbor of p_q in S_r. When $\epsilon = 0$, this reduces to the exact furthest neighbor search problem. This form of approximation is also known as relative-value approximation.

3   Related Work

There have been a number of improvements over the naive brute-force search algorithm suggested above. Exact techniques based on Voronoi diagrams can solve the furthest neighbor problem. In 1981, Toussaint and Bhattacharya proposed building a furthest-point Voronoi diagram to solve the furthest neighbors problem in $O(m \log n)$ time [14]. But in high dimensions, Voronoi diagrams are not useful because of their exponential memory dependence on the dimension.

Another approach to exact furthest neighbor search uses space trees [8]. A tree is built on the reference points S_r, and nodes that cannot contain the furthest neighbor of a given query point are pruned. This is essentially equivalent to many algorithms for nearest neighbor search, such as the algorithm for nearest neighbor search with cover trees [15], but with inequalities reversed (i.e., prune nearby nodes, not faraway nodes). This can be done in a dual-tree setting, by also building a tree on the query points S_q. Dual-tree nearest neighbor search has been proven to scale linearly in the size of the reference set under some conditions [16], but no similar bound has been shown for dual-tree furthest neighbor search. It would be reasonable to expect similar empirical scaling. Unfortunately, tree-based approaches tend to perform poorly in high dimensions, and the tree construction time can cause the algorithm to be undesirably slow.

Further runtime acceleration can be achieved if approximation is allowed. It is easy to modify the single-tree and dual-tree algorithms to support this, in the manner suggested by Curtin for nearest neighbor search [17]. Although this is shown to accelerate nearest neighbor search runtime by a significant amount (depending on the allowed approximation), the setup time of building the trees can still dominate. A similar approach to this strategy is the fair split tree, designed by Bespamyatnikh [18]. But this approach suffers from the same issues.

The fastest known algorithms for approximate furthest neighbor search are hashing algorithms. Indyk [13] proposed a hashing algorithm based on random projections that is able to solve a slightly different problem: this algorithm is able to determine (approximately) whether or not there exists a point in S_r farther away than a given distance. This can be reduced to the approximate furthest neighbor problem we are interested in, but this is complex to implement.

Pagh et al. [12] refine this approach to directly solve the approximate furthest neighbor problem; this improves on the runtime of Indyk's algorithm and is easy to implement. This algorithm, called QDAFN ('query-dependent approximate furthest neighbor'), has a guaranteed success probability. A user must specify the number of projections and the number of points stored for each projection; usually, this number is generally low. But in very high-dimensional settings, the random projections can fail to capture important outlying points. This motivates us to investigate the point distribution as a path towards a better algorithm.

4   Furthest Neighbor Point Distribution

The furthest neighbor problem is quite different from the nearest neighbor problem, which has received significantly more attention [8,9,17,19–22]. This difference is perhaps somewhat counterintuitive, given that the furthest neighbor problem is simply an argmax over S_r, not an argmin. But this change causes the problem to have surprisingly different structure with respect to the results.

As a first observation of the differences between the two problems, consider that for any set S_r, the furthest neighbor of every point can be made to be a single point simply by adding a single point sufficiently far from every other point in S_r. There is no analog to this in the nearest neighbor search problem. Indeed, it is often true that for a furthest neighbor query with many query points, the results may contain the same reference point. This is easily demonstrated.

Define the **rank** of a reference point p_r for some query point p_q as the position of p_r in the ordered list of distances from p_q. That is, if the rank of p_r for some query point p_q is k, then p_r is the k-furthest neighbor from p_q.

We can obtain insight into the behavior of furthest neighbor queries by observing the average rank of points on some example datasets from the UCI dataset repository [23]. Figure 1 contains scatterplots displaying the average rank of a reference point versus the mean-centered norm of the reference point for the all-furthest-neighbors problem (that is, each point in the reference set is used as a query point).

Figure 1 shows that there is a clear and unmistakable correlation between the norm of a point and its average rank for the all-k-furthest-neighbor problem. For the `ozone` dataset, we can see that there are only a few points with high norm, and all of these have much lower average rank than the rest of the points.

This correlation is related to the phenomenon of *hubness* in the nearest neighbor search literature [24]; specifically, points with low average rank may be seen to be related to *anti-hubs* and distance-based outliers. In higher dimensions, more anti-hubs may be expected [25]—thus we may conclude that high-norm

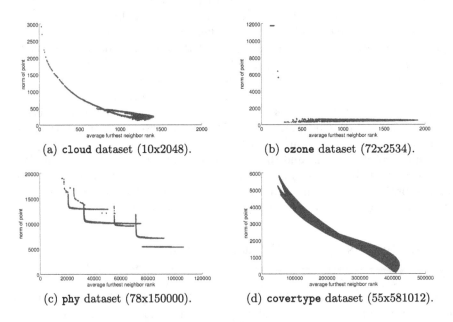

(a) **cloud** dataset (10x2048).

(b) **ozone** dataset (72x2534).

(c) **phy** dataset (78x150000).

(d) **covertype** dataset (55x581012).

Fig. 1. Average rank vs. norm for a handful of datasets. Observe that a large norm is correlated with a low rank.

points (which have low average rank and are related to anti-hubs) are increasingly important in high-dimensional settings. Therefore, an effective furthest neighbors algorithm for high-dimensional data should take this structure into account: *high-norm points are more important than low-norm points.*

5   The Algorithm: DrusillaSelect

Our collective observations motivate an algorithm for approximate furthest neighbor search, which we introduce as DrusillaSelect in Algorithm 1. The algorithm constructs a small collection of points by repeatedly choosing projection bases from the data points with largest norm.[1] Then, the other points in the dataset are projected onto the basis and are selected if they are good candidates.

Algorithm 1. DrusillaSelect: fast approximate k-furthest neighbor search.

1: **Input:** reference set S_r, query set S_q, number of neighbors k, number of projections l, set size m

2: **Output:** array of furthest neighbors $N[]$

3: {Pre-processing: mean-center data.}
4: $m \leftarrow \frac{1}{n} \sum_{p_r \in S_r} p_r$
5: $S_r \leftarrow S_r - m; S_q \leftarrow S_q - m$

6: {Pre-processing: build DrusillaSelect sets.}
7: **for all** $p_r \in S_r$ **do** $n[p_r] \leftarrow \|p_r\|$ {Initialize norms of points.}
8: **for all** $i \in \{0, 1, \ldots, l\}$ **do**
9: $\quad p_i \leftarrow \mathrm{argmax}_{p_r \in S_r} n[p_r]$ {Take next point with largest norm.}
10: $\quad v_i \leftarrow p_i / \|p_i\|$

11: $\quad$ {Calculate distortions and offsets.}
12: $\quad$ **for all** $p_r \in S_r$ such that $n[p_r] \neq 0$ **do**
13: $\quad\quad O[p_r] \leftarrow p_r^T v_i$
14: $\quad\quad D[p_r] \leftarrow \|p_r - O[p_r]v_i\|$
15: $\quad\quad s[p_r] \leftarrow |O[p_r]| - D[p_r]$

16: $\quad$ {Collect points that are well-represented by p_i.}
17: $\quad R_i \leftarrow$ points corresponding to largest m elements of $s[\cdot]$
18: $\quad$ **for all** $p_r \in R_i$ **do** $n[p_r] = 0$ {Mark point as used.}
19: $\quad$ **for all** $p_r \in S_r$ such that $\mathrm{atan}(D[p_r]/O[p_r]) \geq \pi/8$ **do**
20: $\quad\quad n[p_r] = 0$ {Mark point as used.}

21: {Search for furthest neighbors.}
22: **for all** $p_q \in S_q$ **do**
23: $\quad$ **for all** $R_i \in R$ **do**
24: $\quad\quad$ **for all** $p_r \in R_i$ **do**
25: $\quad\quad\quad$ **if** $d(p_q, p_r) > N_k[p_q]$ **then**
26: $\quad\quad\quad\quad$ update results $N[p_q]$ for p_q with p_r

---

[1] This is where the algorithm gets its name; the first author's cat displays the same behavior when selecting a food bowl to eat from.

After this collection is built, each query point is simply compared with all points in the collection in order to determine a good furthest neighbor candidate.

DrusillaSelect depends on two parameters: l, the number of projections, and m, the number of points taken for each projection. Empirically we observe that values in the range of $l \in [2,15]$ and $m \in [1,5]$ produce acceptably good approximations for most datasets, with approximation levels between $\epsilon = 0.01$ and $\epsilon = 1.1$.

The primary intuition of the algorithm is that we want to collect points in the sets R_i that are likely to be furthest neighbors of any query point. We know from our earlier experiments that points with high mean-centered norms are likely to be good furthest neighbor candidates. Thus, we start by selecting the highest-norm mean-centered point p_i as the primary point of the set R_i, and collect m points that are not too distorted by a projection onto the unit vector v_i which points in the direction of p_i. Any points that are not too distorted by this projection but not collected are ignored for future projections (line 18). In addition, points that lie within a cone pointing in the direction of v_i are also ignored (line 20). The value of $\pi/8$ was chosen for its decent empirical performance, but it would be reasonable to select different values.

The words "not too distorted" deserve some elaboration: we wish to find high-norm points that are well-represented by p_i, but we do not wish to find high-norm points that are *not* well-represented by p_i. Ideally, those points will be selected as the primary point of another set R_j. Therefore, for each point p_j, we calculate the offset $O[p_j]$; this is the norm of the projection of p_j onto v_i. Similarly, we calculate the distortion $D[p_j]$. Figure 2 displays a simple example of offset and distortion.

Our goal is to balance two objectives in selecting points for R_i:

- Select high-norm points.
- Select points that are well-represented by v_i.

The solution we have used here is to construct a score $s[p_j]$ which is just the distortion subtracted from the offset (see line 15). Figure 3 displays an example

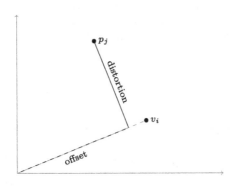

Fig. 2. Distortion and offset for p_j with base vector v_i.

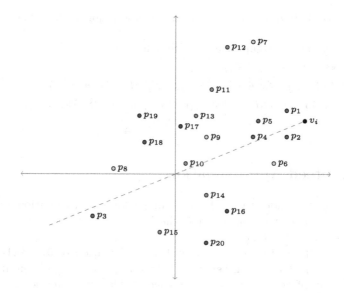

Fig. 3. Example scores for a set of points; red: highest scores, blue: lowest scores. (Color figure online)

v_i with 20 points; each point is indexed by its position in the ordered score set $s[\cdot]$. In the context of DrusillaSelect, if we took $m = 6$ (so, 6 points were selected for each v_i), then v_i and the five red points p_1 through p_5 would be selected to make up the set R_i. Then, p_7 would be chosen as v_{i+1} because it is the point with largest norm that has not been selected (line 9).

Once we have constructed the sets R_i, then our actual search is a simple brute-force search over every point contained in each set R_i. Because the total number of points in R is only lm, brute-force scan is sufficient.

DrusillaSelect has a somewhat similar structure to the QDAFN algorithm [12]; except for three important differences: *(i)* the vectors v_i are drawn using properties of the reference set, *(ii)* there is no priority queue structure when scanning the sets, and *(iii)* the projection bases chosen cannot be too similar. Although DrusillaSelect can involve more setup time, our empirical simulations show it is able to provide better results with fewer sets and points in each sets, resulting in better overall performance for a given level of approximation.

Table 1 gives a comparison of the runtimes of different approximate furthest neighbor algorithms. Note that DrusillaSelect and QDAFN have the same asymptotic setup time for the same l and m; but in practice, the overhead of DrusillaSelect setup time is higher than QDAFN for equivalent l and m. But again it must be noted that to provide the same results accuracy, l and m may generally be set smaller with DrusillaSelect than QDAFN.

Table 1. Runtimes of approximate furthest neighbor algorithms.

Algorithm	Setup time	Search time
`DrusillaHash`	$O(ld\|S_r\|\log\|S_r\|)$	$O(\|S_q\|dlm)$
QDAFN [12]	$O(ld\|S_r\|\log\|S_r\|)$	$O(\|S_q\|d(l\log l + m\log l))$
Indyk [13]	$O(ld\|S_r\|\log\|S_r\|)$	$O(l\|S_q\|(d + \log\|S_r\|)\log d\log\log d)$
Brute-force	None	$O(\|S_q\|\|S_r\|)$

6   Guaranteed Approximation

Next, we wish to consider the problem of an absolute approximation guarantee: in what situations can we ensure that the furthest neighbor returned is an ϵ-approximate furthest neighbor?

It turns out that this is possible with a modification of `DrusillaSelect`, given in Algorithm 2 as `GuaranteedDrusillaSelect`. This algorithm, instead of taking a number of projections l, takes an acceptable approximation level ϵ. The algorithm uses a utility quantity, $\delta = \epsilon/(6 + 3\epsilon)$.

The algorithm is roughly the same as `DrusillaSelect`, except for that more sets are added until all points with norm greater than $\delta\max_{p_r \in S_r}\|p_r\|$ are contained in some set R_i, and an extra point called the *shrug point* is held. The shrug point is set to be any point within the small zero-centered ball of radius $\delta\max_{p_r \in S_r}\|p_r\|$. This is needed to catch situations where p_q is close to every point in some R_i, and serves to provide a "good enough" result to satisfy the approximation guarantee.

Because `GuaranteedDrusillaSelect` collects potentially huge numbers of sets that may contain most of the points in S_r, the algorithm is primarily of theoretical interest. Although the algorithm will outperform brute-force search as long as the sets do not contain nearly all of the points in S_r, it is not likely to be practical for large S_r.

Now we may present our theoretical result. First, we need a utility lemma.

Lemma 1. *Given a mean-centered set S_r and a query point p_q with true furthest neighbor p_{fn}, if $\|p_q\| \leq \frac{1}{3}\max_{p_r \in S_r}\|p_r\|$, then $\|p_{fn}\| \geq \frac{1}{3}\max_{p_r \in S_r}\|p_r\|$.*

Proof. This is a simple proof by contradiction: suppose $\|p_{fn}\| < \frac{1}{3}\max_{p_r \in S_r}\|p_r\|$. Then, the maximum possible distance between p_q and p_{fn} is bounded above as $d(p_q, p_{fn}) < \frac{2}{3}\max_{p_r \in S_r}\|p_r\|$. But the minimum possible distance between p_q and the largest point in S_r is bounded below as

$$d(p_q, \underset{p_r \in S_r}{\mathrm{argmax}}\|p_r\|) \geq \max_{p_r \in S_r}\|p_r\| - \frac{1}{3}\max_{p_r \in S_r}\|p_r\| = \frac{2}{3}\max_{p_r \in S_r}\|p_r\|. \tag{3}$$

This means that the largest point in S_r is a further neighbor than p_{fn}, which is a contradiction. □

We may now prove the main result.

Algorithm 2. `GuaranteedDrusillaSelect`: guaranteed approximate k-furthest neighbor search.

1: **Input:** reference set S_r, query set S_q, number of neighbors k, acceptable approximation level ϵ, set size m
2: **Output:** array of furthest neighbors $N[]$

3: {Pre-processing: mean-center data.}
4: $m \leftarrow \frac{1}{n} \sum_{p_r \in S_r} p_r$; $S_r \leftarrow S_r - m$; $S_q \leftarrow S_q - m$

5: {Pre-processing: build GuaranteedDrusillaSelect sets.}
6: **for all** $p_r \in S_r$ **do** $n[p_r] \leftarrow \|p_r\|$ {Initialize norms of points.}
7: $\delta \leftarrow \frac{\epsilon}{6+3\epsilon}$
8: **while** $\max_{p_r \in S_r} n[p_r] > \delta \max_{p_r \in S_r} \|p_r\|$ **do**
9:     $p_i \leftarrow \text{argmax}_{p_r \in S_r} n[p_r]$ {Take next point with largest norm.}
10:    $v_i \leftarrow p_i / \|p_i\|$

11:    {Calculate distortions and offsets.}
12:    **for all** $p_r \in S_r$ such that $n[p_r] \neq 0$ **do**
13:        $O[p_r] \leftarrow p_r^T v_i$
14:        $D[p_r] \leftarrow \|p_r - O[p_r]v_i\|$
15:        $s[p_r] \leftarrow |O[p_r]| - D[p_r]$

16:    {Collect points that are well-represented by p_i.}
17:    $R_i \leftarrow$ points corresponding to largest m elements of $s[\cdot]$
18:    **for all** $p_r \in R_i$ **do** $n[p_r] = 0$ {Mark point as used.}

19: {Set shrug point (if we can).}
20: $p_{sh} \leftarrow \emptyset$
21: **if** there is any point such that $n[p_r] \neq 0$ **then**
22:    $p_{sh} \leftarrow$ some point such that $n[p_r] \neq 0$

23: {Search for furthest neighbors.}
24: **for all** $p_q \in S_q$ **do**
25:    **for all** $R_i \in R$ **do**
26:        **for all** $p_r \in R_i$ **do**
27:            **if** $d(p_q, p_r) > N_k[p_q]$ **then**
28:                update results $N[p_q]$ for p_q with p_r
29:        **if** $p_{sh} \neq \emptyset$ and $d(p_q, p_{sh}) > N_k[p_q]$ **then**
30:            update results $N[p_q]$ for p_q with p_{sh}

Theorem 1. *Given a set S_r and an approximation parameter $\epsilon < 1$ and any set size $m > 0$,* `GuaranteedDrusillaSelect` *will return, for each query point p_q, a furthest neighbor $\hat{p}_{fn}$ such that*

$$\frac{d(p_q, p_{fn})}{d(p_q, \hat{p}_{fn})} < 1 + \epsilon \qquad (4)$$

where p_{fn} is the true furthest neighbor of p_q in S_r. That is, $\hat{p}_{fn}$ is an ϵ-approximate furthest neighbor of p_q.

Proof. We know from Lemma 1 that if the norm of p_q is less than or equal to $1/3$ of the maximum norm of any point in S_r, then the true furthest neighbor must have norm greater than or equal to $1/3$ of the maximum norm of any point in S_r. Since δ is always less than $1/3$ in Algorithm 2, we know that any such point will be contained in some set R_i, and thus the algorithm will return the exact furthest neighbor in this case.

The only other case to consider, then, is when the norm of the query point is large: $\|p_q\| > \frac{1}{3} \max_{p_r \in S_r} \|p_r\|$. But we already know due to the way the algorithm works, that if $\|p_{fn}\| \geq \delta \max_{p_r \in S_r} \|p_r\|$, then p_{fn} will be contained in some set R_i and the algorithm will return p_{fn}, satisfying the approximation guarantee.

But what about when $\|p_{fn}\|$ is smaller? We must consider the case where $\|p_{fn}\| < \delta \max_{p_r \in S_r} \|p_r\|$. Here we may place an upper bound on the distance between the query point and its furthest neighbor:

$$d(p_q, p_{fn}) \leq \|p_q\| + \|p_{fn}\| < \|p_q\| + \delta \max_{p_r \in S_r} \|p_r\|. \tag{5}$$

We may also place a lower bound on the distance between the query point and its returned furthest neighbor using the shrug point p_{sh}. The distance between p_q and p_{sh} is easily lower bounded: $d(p_q, p_{sh}) \geq \|p_q\| - \delta \max_{p_r \in S_r} \|p_r\| > 0$. This is also a lower bound on $d(p_q, \hat{p}_{fn})$. We may combine these bounds:

$$\frac{d(p_q, p_{fn})}{d(p_q, \hat{p}_{fn})} < \frac{\|p_q\| + \delta \max_{p_r \in S_r} \|p_r\|}{\|p_q\| - \delta \max_{p_r \in S_r} \|p_r\|}. \tag{6}$$

Now, define the convenience quantity α as

$$\alpha = \frac{\max_{p_r \in S_r} \|p_r\|}{\|p_q\|}. \tag{7}$$

Because of our assumptions on p_q, we know that $\alpha < 3$. Using these inequalities, we may further simplify Eq. 6.

$$\frac{d(p_q, p_{fn})}{d(p_q, \hat{p}_{fn})} < \frac{1 + \delta\alpha}{1 - \delta\alpha} \tag{8}$$

$$= 1 + \frac{2\delta\alpha}{1 - \delta\alpha} \tag{9}$$

$$< 1 + \frac{6\delta}{1 - 3\delta} \tag{10}$$

and because $\delta = \frac{\epsilon}{6+3\epsilon}$, Eq. 10 simplifies to the result,

$$\frac{d(p_q, p_{fn})}{d(p_q, \hat{p}_{fn})} < 1 + \epsilon \tag{11}$$

and therefore the theorem holds.   □

Note that the theorem holds if we set δ to the simpler quantity of $\epsilon/9$; but the quantity $(\epsilon/(6+3\epsilon))$ provides a tighter bound.

Although `GuaranteedDrusillaSelect` does not guarantee better search time than brute force under all conditions, it does in most conditions. As one example, consider a large dataset where the norms of points in the centered dataset are uniformly distributed. Some of these points will have norm less than $(\epsilon/15)\max_{p_r \in S_r} \|p_r\|$. These points (except the shrug point p_{sh}) will not be considered by the `GuaranteedDrusillaSelect` algorithm, and this means that the `GuaranteedDrusillaSelect` algorithm will inspect fewer points at search time than the brute-force algorithm.

Next, consider the extreme case, where there exists one outlier p_o with extremely large norm, such that the next largest point has norm smaller than $(\epsilon/(6+3\epsilon))\|p_o\|$. Here, `GuaranteedDrusillaSelect` with $m = 1$ will only need to inspect two points: the extreme outlier, and the shrug point p_{sh}.

On the other hand, there do exist cases where `GuaranteedDrusillaSelect` gives no improvement over brute-force search, and every point must be inspected. If the dataset is such that all points have norm greater than $(\epsilon/(6+3\epsilon))\max_{p_r \in S_r} \|p_r\|$, then the sets R_i will contain every single point in the dataset.

These theoretical results show that it is possible to give a guaranteed ϵ-approximate furthest neighbor in less time than brute-force search, if the distribution of norms of S_r are not worst-case. But due to the algorithm's storage requirement, it is not likely to perform well in practice and so we do not investigate its empirical performance.

7   Experiments

Next, we investigate the empirical performance of the `DrusillaSelect` algorithm, comparing with brute-force search, QDAFN [12], and dual-tree exact furthest neighbor search as described by Curtin et al. [8]. Note that both brute-force search and the dual-tree algorithm return exact furthest neighbors; QDAFN and `DrusillaSelect` return approximations. Each implementation is either from **mlpack** [26] or is built using **mlpack**. We test the algorithms on a variety of datasets from the UCI dataset repository and **randu**, which is uniformly randomly distributed. These datasets and their properties are given in Table 2.

First, we compare runtimes across all four algorithms. The approximate algorithms are tuned to return, on average across the query set, $\epsilon = 0.05$-approximate furthest neighbors (using the parameters from Table 2). Table 3 shows the average runtimes of each of the four algorithms on each dataset across ten trials with the dataset randomly split into 30 % query set, 70 % reference set. I/O times are not included; the runtime only includes the time for the search itself, including preprocessing time (building hash tables, sets, or trees).

The `DrusillaSelect` algorithm provides average $\epsilon = 0.05$-approximate furthest neighbors up to an order of magnitude faster than any other competing algorithm, and it also needs to inspect fewer points to return an accurate approximate furthest neighbor (with the exception of the **pokerhand** dataset). In many

Table 2. Datasets and parameters.

Dataset	n	d	QDAFN params		DrusillaSelect params	
			l	m	l	m
Cloud	2048	10	30	60	2	1
Isolet	7797	617	40	40	2	1
Gisette	12500	5000	40	40	2	2
Corel	37749	32	5	5	2	1
p53	48192	5409	25	25	3	2
Randu	100000	10	15	15	5	2
Miniboone	130064	50	125	200	2	1
Phy	150000	78	12	12	4	1
Covertype	581012	55	15	20	6	2
Pokerhand	1000000	10	15	50	50	8
Susy	5000000	18	18	18	2	2
Higgs	11000000	28	32	32	2	2

Table 3. Runtimes for $\epsilon = 0.05$-approximate furthest neighbor search

Dataset	Brute-force	Dual-tree	QDAFN	DrusillaSelect
Cloud	0.039 s	0.040 s	0.011 s	0.001 s
Isolet	6.754 s	7.706 s	0.165 s	0.041 s
Gisette	141.923 s	141.963 s	1.875 s	0.549 s
Corel	10.292 s	1.030 s	0.021 s	0.021 s
p53	2258.331 s	270.341 s	3.475 s	2.734 s
Randu	42.392 s	28.004 s	0.316 s	0.0619 s
Miniboone	187.262 s	4.105 s	2.165 s	0.104 s
Phy	370.061 s	58.720s	0.203 s	0.189 s
Covertype	4077.922 s	144.993 s	1.244 s	0.203s
Randu	–	16.715 s	0.069 s	0.043 s
Pokerhand	–	852.001 s	11.749 s	8.035 s
Susy	–	88.295 s	21.678 s	2.4467 s
Higgs	–	425.053 s	56.094 s	12.694 s

cases, DrusillaSelect only needs to inspect fewer than 10 points to find good furthest neighbor approximations, whereas QDAFN must inspect 50 or more.

Our datasets have two extreme examples: the miniboone dataset, where the data lies on a low-dimensional manifold, and the randu dataset.

For the miniboone dataset, DrusillaSelect is able to easily recover only four points that provide average 1.05-approximate furthest neighbors. But

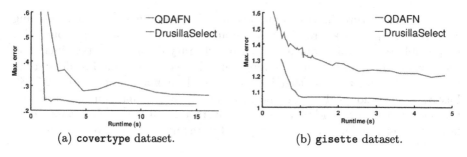

(a) **covertype** dataset.   (b) **gisette** dataset.

Fig. 4. Maximum error for QDAFN and `DrusillaSelect` as a function of runtime.

because QDAFN chooses random projection bases, it takes very many to have a high probability of recovering good furthest neighbors. In our experiments, we were not able to achieve good approximation reliably until using as many as 125 projection bases. This effect was also observed with the **covertype** dataset.

`DrusillaSelect` also outperforms other approaches on the **randu** dataset, despite there being no structure for `DrusillaSelect` to exploit. But the algorithm is still able to outperform others; this is because the algorithm specifically ensures that projection bases are not too similar (see lines 18–20).

Another important property of `DrusillaSelect` is that it gives a small maximum error compared to QDAFN. Figure 4 shows the maximum error of each approach as the number of points scanned increase on the **covertype** dataset. For QDAFN, we have swept with $l = m$ from $l = 20$ to $l = 250$, and for `DrusillaSelect`, we have set $m = l/3$ and swept l from 6 to 60.

Our experimental results have shown that `DrusillaSelect` gives excellent approximation while only needing to scan few points. Whereas QDAFN seems to perform poorly in high-dimensional settings where the data lie on a low-dimensional manifold (because projection bases are random), `DrusillaSelect` effectively captures the low-dimensional structure with few projection bases.

8 Conclusion

We have proposed an algorithm, `DrusillaSelect`, that builds a candidate set for approximate furthest neighbor search by using the properties of the dataset. This algorithm design is motivated by our empirical analysis of the structure of the approximate furthest neighbor search problem, and the algorithm performs quite compellingly in practice. It scales better with dataset size than other techniques.

We have also proposed a variant, `GuaranteedDrusillaSelect`, which is able to give an absolute approximation guarantee. Under some assumptions, this algorithm will provably outperform the brute-force approach at search time. This is a benefit that no other furthest neighbor search scheme is able to provide. However, this variant is not likely to be useful in practice due to the large number of points it must search to satisfy the guarantee.

Interesting future directions for this line of research may include combining a random projection approach with the approach outlined here. It would also be possible to generalize our approach to arbitrary distance metrics, including those where the points lie in an unrepresentable space. This could be done using techniques similar to some that have been used for max-kernel search [27,28]. Lastly, we have focused on high-norm points as 'important'; but a study connecting hubness (or anti-hubness) to the average furthest-neighbor rank would be enlightening and may potentially guide future improvements to this approach.

References

1. Said, A., Kille, B., Jain, B.J., Albayrak, S.: Increasing diversity through furthest neighbor-based recommendation. In: Proceedings of the Fifth International Conference on Web Search and Data Mining (WSDM 2012), p. 12 (2012)
2. Said, A., Fields, B., Jain, B.J., Albayrak, S.: User-centric evaluation of a k-furthest neighbor collaborative filtering recommender algorithm. In: Proceedings of the 2013 Conference on Computer Supported Cooperative Work, pp. 1399–1408. ACM (2013)
3. Vasiloglou, N., Gray, A.G., Anderson, D.V.: Scalable semidefinite manifold learning. In: Proceedings of the 2008 IEEE Workshop on Machine Learning for Signal Processing, 2008 (MLSP. 2008), pp. 368–373. IEEE (2008)
4. Defays, D.: An efficient algorithm for a complete link method. Comput. J. **20**(4), 364–366 (1977)
5. Schloss, P.D., Westcott, S.L., Ryabin, T., Hall, J.R., Hartmann, M., Hollister, E.B., Lesniewski, R.A., Oakley, B.B., Parks, D.H., Robinson, C.J., Sahl, J.W., Stres, B., Thallinger, G.G., Van Horn, D.J., Weber, C.F.: Introducing mothur: open-source, platform-independent, community-supported software for describing and comparing microbial communities. Appl. Environ. Microbiol. **75**(23), 7537–7541 (2009)
6. Veenman, C.J., Reinders, M.J.T., Backer, E.: A maximum variance cluster algorithm. IEEE Trans. Pattern Anal. Mach. Intell. **24**(9), 1273–1280 (2002)
7. Cheong, O., Shin, C.-S., Vigneron, A.: Computing farthest neighbors on a convex polytope. Theoret. Comput. Sci. **296**(1), 47–58 (2003)
8. Curtin, R.R., March, W.B., Ram, P., Anderson, D.V., Gray, A.G., Isbell Jr., C.L.: Tree-independent dual-tree algorithms. In: Proceedings of the 30th International Conference on Machine Learning (ICML 2013) (2013)
9. Datar, M., Immorlica, N., Indyk, P., Mirrokni, V.S.: Locality-sensitive hashing scheme based on p-stable distributions. In: Proceedings of the Twentieth Annual Symposium on Computational Geometry (SoCG 2004), pp. 253–262. ACM (2004)
10. Indyk, P., Motwani, R.: Approximate nearest neighbors: towards removing the curse of dimensionality. In: Proceedings of the Thirtieth Annual ACM Symposium on Theory of Computing (STOC 1998), pp. 604–613. ACM (1998)
11. Andoni, A., Indyk, P.: Near-optimal hashing algorithms for approximate nearest neighbor in high dimensions. In: 47th Annual IEEE Symposium on Foundations of Computer Science (FOCS 2006), pp. 459–468. IEEE (2006)
12. Pagh, R., Silvestri, F., Sivertsen, J., Skala, M.: Approximate furthest neighbor in high dimensions. In: Amato, G. (ed.) SISAP 2015. LNCS, vol. 9371, pp. 3–14. Springer, Heidelberg (2015). doi:10.1007/978-3-319-25087-8_1

13. Indyk, P.: Better algorithms for high-dimensional proximity problems via asymmetric embeddings. In: Proceedings of the Fourteenth Annual ACM-SIAM Symposium on Discrete Algorithms (SODA 2003), pp. 539–545. Society for Industrial and Applied Mathematics (2003)

14. Toussaint, G.T., Bhattacharya, B.K.: On geometric algorithms that use the furthest-point voronoi diagram. School of Computer Science, McGill University, Technical report No. 81.3 (1981)

15. Beygelzimer, A., Kakade, S., Langford, J.: Cover trees for nearest neighbor. In: Proceedings of the 23rd International Conference on Machine Learning (ICML 2006), pp. 97–104. ACM (2006)

16. Curtin, R.R., Lee, D., March, W.B., Ram, P.: Plug-and-play dual-tree algorithm runtime analysis. J. Mach. Learn. Res. **16**, 3269–3297 (2015)

17. Curtin, R.R.: Faster dual-tree traversal for nearest neighbor search. In: Amato, G. (ed.) SISAP 2015. LNCS, vol. 9371, pp. 77–89. Springer, Heidelberg (2015). doi:10.1007/978-3-319-25087-8_7

18. Bespamyatnikh, S.: Dynamic algorithms for approximate neighbor searching. In: Proceedings of the 8th Canadian Conference on Computational Geometry (CCCG 1996), pp. 252–257 (1996)

19. Bentley, J.L.: Multidimensional binary search trees used for associative searching. Commun. ACM **18**(9), 509–517 (1975)

20. Arya, S., Mount, D.M., Netanyahu, N.S., Silverman, R., Wu, A.Y.: An optimal algorithm for approximate nearest neighbor searching in fixed dimensions. J. ACM (JACM) **45**(6), 891–923 (1998)

21. Gionis, A., Indyk, P., Motwani, R., et al.: Similarity search in high dimensions via hashing. In: Proceedings of the Twenty-Fifth International Conference on Very Large Data Bases (VLDB 1999), vol. 99, pp. 518–529 (1999)

22. Gray, A.G., Moore, A.W.: N-Body problems in statistical learning. In: Advances in Neural Information Processing Systems 14 (NIPS 2001), vol. 4, pp. 521–527 (2001)

23. Lichman, M.: UCI machine learning repository, University of California Irvine, School of Information and Computer Sciences (2013). http://archive.ics.uci.edu/ml

24. Radovanoić, M., Nanopoulos, A., Ivanović, C.: Hubs in space: popular nearest neighbors in high-dimensional data. J. Mach. Learn. Res. **11**(Sep), 2487–2531 (2010)

25. Tomasev, N., Radovanović, M., Mladenic, D., Ivanović, M.: The role of hubness in clustering high-dimensional data. IEEE Trans. Knowl. Data Eng. **26**(3), 739–751 (2014)

26. Curtin, R.R., Cline, J.R., Slagle, N.P., March, W.B., Ram, P., Mehta, N.A., Gray, A.G.: MLPACK: a scalable C++ machine learning library. J. Mach. Learn. Res. **14**(1), 801–805 (2013)

27. Curtin, R.R., Ram, P., Gray, A.G.: Fast exact max-kernel search. In: Proceedings of the 2013 SIAM International Conference on Data Mining (SDM 2013), pp. 1–9. SIAM (2013)

28. Curtin, R.R., Ram, P.: Dual-tree fast exact max-kernel search. Stat. Anal. Data Min. **7**(4), 229–253 (2014)

NearBucket-LSH: Efficient Similarity Search in P2P Networks

Naama Kraus[1]([✉]), David Carmel[2], Idit Keidar[1,2], and Meni Orenbach[1]

[1] Viterbi EE Technion, Haifa, Israel
naamakraus@gmail.com
[2] Yahoo Research, Haifa, Israel

Abstract. We present NearBucket-LSH, an effective algorithm for similarity search in large-scale distributed online social networks organized as peer-to-peer overlays. As communication is a dominant consideration in distributed systems, we focus on minimizing the network cost while guaranteeing good search quality. Our algorithm is based on Locality Sensitive Hashing (LSH), which limits the search to collections of objects, called buckets, that have a high probability to be similar to the query. More specifically, NearBucket-LSH employs an LSH extension that searches in near buckets, and improves search quality but also significantly increases the network cost. We decrease the network cost by considering the internals of both LSH and the P2P overlay, and harnessing their properties to our needs. We show that our NearBucket-LSH increases search quality for a given network cost compared to previous art. In many cases, the search quality increases by more than 50 %.

1   Introduction

User *similarity search* in *Online Social Networks (OSNs)* is the task of effectively finding OSN users that are similar to a given user based on common *interests*. It is used for many applications including recommending new friends [23,28], as well as for recommending content based on preferences of similar users [2]. In this work, we consider *Peer-to-Peer (P2P)* OSNs (e.g., [6,9,21,24]), which offer increased scalability and avoid control by a single authority.

A similarity search algorithm in P2P OSNs faces several challenges: The algorithm should be decentralized in order to fit the P2P architecture. As network cost is a dominant consideration in P2P networks, the algorithm should be network-efficient, while preserving a good search quality. Furthermore, the similarity search should cope with the dynamic nature of OSNs: users join or leave, and users dynamically modify their interests. We present a similarity search algorithm in P2P OSNs that meets these requirements.

We base our algorithm on *Locality Sensitive Hashing (LSH)* [14,16], which is a widespread randomized method for efficient similarity search in high-dimensional spaces. LSH hashes an OSN user into a succinct representation, where the hash values of similar users collide with high probability (w.h.p.). At a pre-processing

© Springer International Publishing AG 2016
L. Amsaleg et al. (Eds.): SISAP 2016, LNCS 9939, pp. 236–249, 2016.
DOI: 10.1007/978-3-319-46759-7_18

stage, LSH maps users into collections of objects called *buckets* based on common hashes. Upon receiving a query, LSH limits the search to buckets to which the query is mapped; these contain similar users w.h.p. We follow a variant of LSH, called MultiProb-LSH [20], which increases search quality by additionally searching *near buckets*, which are buckets that are similar to the query's bucket.

We present *NearBucket-LSH*, which integrates LSH into a P2P architecture. For our P2P overlay we use *Content Addressable Network (CAN)* [25], which is a good fit for a distributed LSH implementation. We use CAN to dynamically map and store LSH buckets within nodes, and refresh bucket contents once in a while in order to adjust to changes in the data. Upon search, we use CAN to locate the buckets to search in.

In P2P settings, searching additional buckets entails contacting additional nodes, which is a network-costly operation. We improve the network-efficiency when searching near buckets by exploiting the internals of CAN: We observe that in CAN, near buckets reside in a bucket's neighboring nodes, and thus contacting them incurs a low network cost. We further eliminate this network cost by caching near buckets in each CAN node. We show, both analytically and empirically, that the cache-based NearBucket-LSH provides the greatest search quality for a given network cost, compared to other approaches.

2   Model and Problem Definition

In this section we detail the model we consider. We formally define the notion of similarity search (Sect. 2.1), and provide details about P2P OSNs (Sect. 2.2).

2.1   User Similarity Search in OSN

An OSN user exposes an *interest profile*, which we represent as a non-negative weighted feature vector in a high d-dimensional vector space $V = (\mathbb{R}_0^+)^d$. The interests-weighting scheme may be arbitrary. A *similarity function* [8] measures the similarity between two user vectors. It returns a *similarity value* within the range $[0,1]$, where a similarity value of 1 denotes complete similarity, and 0 denotes no similarity.

An *m-similarity search* algorithm accepts as an input a *query* vector $q \in V$. It returns a unique *ideal result set* of m user vectors that are most similar to q, according to the given similarity function. An *approximate m-similarity search* algorithm trades-off efficiency with accuracy. Given a query q, it returns an *approximate result set* of m user vectors, which may differ from q's ideal result set. We consider the commonly used *cosine similarity* function [22], also proposed in the context of similarity between OSN users [3].

2.2   P2P OSN and CAN

P2P networks are distributed systems organized as overlay networks with no central management. Nodes (also called *peers*) are autonomous entities that may join or leave at any time. P2P networks provide massive scalability, fault tolerance, privacy, anonymity, and load balancing (see [18] for a survey). We consider a P2P Online Social Network [6,9,21,24], in which users' content is distributed among nodes. Any node in the P2P OSN may initiate a similarity search query.

In our algorithm, we use CAN [25] as our overlay, which naturally fits a distributed LSH implementation, as we later show. CAN implements a self-organizing P2P network representing a virtual c-dimensional Cartesian coordinate space on a c-torus. The Cartesian space is dynamically partitioned into *zones*, which are distributed among CAN nodes. CAN implements a *Distributed Hash Table (DHT)* abstraction, which provides a distributed *lookup* operation that accepts a vector as key, and returns a node that owns the zone to which the vector belongs. Each node maintains a table of *neighbors*, which are nodes that own zones adjacent to its own. These tables are used for routing messages within CAN.

3   Background and Previous Work

Before diving into our algorithm, we provide essential background and overview previous work. In Sect. 3.1, we overview LSH [14,16] and its space-efficient variant, MultiProb-LSH [20]. Section 3.2 discusses Layered-LSH [4,15], which is a distributed LSH implementation that optimizes network cost. In addition to distributed solutions, there are also parallel LSH variants, e.g. [26]. However, these do not focus on improving network-efficiency, which is not of essence in a parallel setting. Other P2P similarity search methods have been proposed [5], in particular, Falchi et al. [12] use CAN as their overlay. However, these methods are not based on LSH, which is the focus of our work.

3.1   Locality Sensitive Hashing

Locality sensitive hashing [14,16] is a widely used approximate similarity search algorithm for high-dimensional spaces, with sub-linear search time complexity. LSH limits the search to vectors that are similar to the query vector w.h.p. instead of linearly searching over all vectors. This reduces the search time complexity at the cost of missing similar vectors with a some probability.

LSH uses hash functions that map a vector in the high dimensional input space $(\mathbb{R}_0^+)^d$ into a representation in a lower dimension $k << d$, so that the hashes of similar vectors are likely to collide. LSH executes a pre-processing (index building) stage, where it assigns vectors into buckets according to their hash values. Then, given a query vector, the similarity search algorithm computes its hashes and searches vectors in the corresponding buckets. The LSH algorithm is parametrized by k and L, where k is the hashed domain's dimension, and L is the number of hash functions used. Formally [7]: a *locality sensitive hashing* with similarity function sim is a distribution on a family $\mathcal{H}$ of hash functions on a collection of vectors, $h : V \to \{0, 1\}$, such that for two vectors u, v,

$$Pr_{h \in \mathcal{H}}[h(u) = h(v)] = sim(u, v). \tag{1}$$

We use here a hash family $\mathcal{H}$ for angular similarity [7], which fits cosine-based similarity search [7]. In order to increase the probability that similar vectors are mapped to the same bucket, the algorithm defines a family $\mathcal{G}$ of hash functions, where each $g(v) \in \mathcal{G}$ is a concatenation of k functions chosen randomly and independently from $\mathcal{H}$. In the case of angular similarity, $g : V \rightarrow \{0,1\}^k$, i.e., g hashes v into a binary *sketch vector*, which encodes v in a lower dimension k. For two vectors u, v, $Pr_{g \in \mathcal{G}}[g(u) = g(v)] = (sim(u, v))^k$, for any randomly selected $g \in \mathcal{G}$. The larger k is, the higher the precision.

In order to mitigate the probability to miss similar items, the *LSH* algorithm selects L functions randomly and independently from $\mathcal{G}$. The item vectors are now replicated in L hash tables, where each vector is mapped to L buckets. Upon query, search is performed in L buckets. This increases the recall at the cost of additional storage and processing. In order to reduce the storage cost, MultiProb-LSH [20] additionally searches in *near buckets* within the same hash table, which are buckets that slightly differ from the query's *exact bucket* $g(q)$, and have a high probability to contain vectors similar to the query.

3.2   Layered LSH

In P2P networks, buckets are distributed over the overlay nodes. Contacting a near bucket involves performing a DHT lookup of its node, which incurs high network cost. Prior art [4,15] suggests *Layered-LSH*, which maps buckets to nodes using a second LSH, such that near buckets are assigned to the same node w.h.p. Queries now access a single node holding the desired buckets, which reduces the network cost. In Sect. 5.1, we show that in the case of cosine similarity, Layered-LSH is equivalent to the basic LSH for an appropriate choice of k.

4   Algorithm

We describe NearBucket-LSH, our network-efficient P2P user similarity search that is based on MultiProb-LSH. We construct a dedicated overlay above the CAN infrastructure, and exploit its internals for reducing search network cost when searching near buckets.

The Overlay. We use a k-dimensional CAN (i.e., $c = k$) to store and lookup LSH buckets in a decentralized manner. For simplicity, we assume that $N = 2^k$, where N denotes the number of CAN nodes. Each CAN node owns the zone of a single k-dimensional binary vector v representing some LSH sketch vector, and maintains the bucket of user vectors that are mapped to v by some hash function $g \in \mathcal{G}$. We name such a node the *bucket node* of v. The bucket node provides a local similarity search facility over its locally stored user vectors. The local search time is typically proportional to the searched bucket size [14]. The internal bucket data-structure and local search implementation are orthogonal to this research.

Each CAN node in our overlay has k neighbors; the i-th neighbor of node v owns a vector u that differs from v in the i-th entry only. Routing a message from node v to one of its neighbors requires a single hop, i.e., a single message. Routing a message from an arbitrary source node v to an arbitrary target node u, entails modifying the binary vector entries that differ between u and v. Two vectors of length k, differ in $k/2$ entries in expectation, and thus, the expected path length is $k/2$ hops[1].

The L hash functions $g = \{g_1, \cdots, g_L\}$ are randomly selected from $\mathcal{G}$ a priori. They are given to the distributed algorithm as a configuration parameter, and are known to all bucket nodes. CAN supports multiple hash functions [25], which we use for supporting multiple g_i's and mapping each user vector into L bucket nodes.

Bucket Maintenance. Our algorithm constructs and refreshes the buckets continuously, in a decentralized manner. Each user periodically re-hashes its vector using LSH into L sketch vectors. It then performs DHT lookups to locate the corresponding bucket nodes, and sends them the fresh user vector. Note that the user vector may or may not have changed since the previous update message.

We do not construct buckets a priori. Rather, bucket construction is triggered by vector update messages. A CAN node becomes an active bucket node when it first receives a notification of some user vector. Since user vectors change dynamically, their hashes change accordingly. Obsolete vectors that are not refreshed for a certain predefined length of time are garbage-collected from bucket nodes.

Query Processing. Algorithm 1 depicts NearBucket-LSH query processing procedure: Each P2P node may trigger an m-similarity search request for an input query q. The initiating node, denoted n, hashes q into L sketch vectors $g_i(q)$, looks-up the corresponding exact bucket nodes, and sends them m-similarity search requests in parallel. Once a query request reaches some exact bucket node $g_i(q)$, the node performs a local similarity search in its own bucket, and also forwards the request to the k near bucket nodes that differ from $g_i(q)$ in exactly one entry. All (exact and near) bucket nodes send back a set of up to m results to node n. Node n merges the result sets and returns a final m-result set to the caller.

As a CAN node maintains a table of k neighbors that differ from it in exactly one entry, these neighbors hold the desired near buckets. Thus, contacting a near bucket node costs a single message, and a total of kL messages per query. We further eliminate these additional messages by caching k near buckets at each CAN node. In order to maintain fresh caches, each node periodically sends its bucket to its neighbors. The cache requires an additional storage of size kB at each node, where B is the average bucket size.

It is possible to cache all k near buckets or any subset of them. For the purpose of the analysis and evaluation in the next sections, we refer to the following two

---

[1] Note that in a general c-dimensional CAN of N nodes, the expected routing length is $c/4 \left(N^{1/c} \right)$ [25], which equals $k/2$ for $c = k$ and $N = 2^k$.

extremes: we name NB-LSH a NearBucket-LSH that does not use caching at all, and CNB-LSH a NearBucket-LSH that caches all k near buckets. In addition, we refer in LSH to the basic LSH algorithm, which completely avoids searching near buckets.

Algorithm 1. NearBucket-LSH Query Processing

1:	**function** QUERY(q)	▷ At the query node
2:	**pforeach** $g_i \in g$ **do**	▷ A parallel foreach
3:	$v_i \leftarrow g_i(q)$	
4:	$n_i \leftarrow DHT.\text{LOOKUP}(v_i)$	▷ Lookup bucket node
5:	$n_i.\text{SENDREQ}(SimSearchNB, q, n)$	▷ Send request
6:	**end pforeach**	
7:	$hits \leftarrow$ collect results from bucket nodes	
8:	**return** top m hits	▷ Rank and return top m
9:	**end function**	
10:	**function** SIMSEARCHNB(q, n)	▷ Query q from n
11:	$res \leftarrow Bucket.\text{LOCALSIMSEARCH}(q)$	▷ Local search
12:	$n.\text{SENDRES}(res)$	▷ Send back result
13:	**pforeach** $j \in \{1, \cdots, k\}$ **do**	▷ A parallel foreach
14:	$n_j \leftarrow Neighbors.j$	▷ Extract the j-th neighbor
15:	**if** $Bucket_j$ is cached **then**	▷ Neighbor's bucket is cached
16:	$res \leftarrow Bucket_j.\text{LOCALSIMSEARCH}(q)$	▷ Local search
17:	$n.\text{SENDRES}(res)$	▷ Send back result
18:	**else**	
19:	$n_j.\text{SENDREQ}(SimSearch, q, n)$	▷ Forward request
20:	**end if**	
21:	**end pforeach**	
22:	**end function**	
23:	**function** SIMSEARCH(q, n)	▷ Query q from n
24:	$res \leftarrow Bucket.\text{LOCALSIMSEARCH}(q)$	▷ Local search
25:	$n.\text{SENDRES}(res)$	▷ Send back result
26:	**end function**	

---

5   Theoretical Analysis

We theoretically analyze an algorithm's capability of retrieving similar objects, and show the superiority of NearBucket-LSH to successfully retrieve similar objects for a given network cost.

5.1   Success Probability Formulation

The basic building block in our analysis is the *success probability* [20] of an algorithm A to find object y that has a similarity value s to query object q, under a random selection of $g \in \mathcal{G}$. We denote this success probability by $SP(A, s)$.

LSH. Let $LSH(k, L)$ denote the angular-LSH algorithm with parameters k and L, and let s denote the angular similarity between query q and searched object y. According to the LSH theory [7], for a randomly selected $h \in \mathcal{H}$:

$$Pr_{h \in \mathcal{H}}[h(q) = h(y)] = s, \text{ and } Pr_{h \in \mathcal{H}}[h(q) \neq h(y)] = (1 - s). \tag{2}$$

$LSH(k, L)$ searches in L exact buckets independently, thus, it finds y in any of these buckets with probability:

Proposition 1.

$$SP(LSH(k, L), s) = 1 - (1 - s^k)^L.$$

NearBucket-LSH. We define *b-near buckets* to be buckets that differ from an exact bucket in $0 \leq b \leq k$ entries (note that a 0-near bucket is an exact bucket). The success probability of finding y in a b-near bucket of $g(q)$ is:

$$s^{k-b}(1 - s)^b. \tag{3}$$

As our vectors are non-negative, their angular similarities s satisfy that $s \in [0.5, 1]$. This implies that $\forall s, (1 - s) \leq s$, and therefore, for $0 \leq b_1 < b_2 \leq k$, $s^{k-b_2}(1 - s)^{b_2} \leq s^{k-b_1}(1 - s)^{b_1}$, thus:

Proposition 2. *The success probability when searching in a b_1-near bucket is greater or equal to the success probability when searching in a b_2-near bucket, for any $0 \leq b_1 < b_2 \leq k$. Hence, NearBucket-LSH's selection of k 1-near buckets is optimal, with respect to any other k buckets selected for search, in addition to the exact bucket.*

The exact bucket and its near buckets are disjoint, as an object is mapped to exactly one bucket according to a specific g. NearBucket-LSH searches in L exact buckets each along with its k 1-near buckets. Thus,

Proposition 3.

$$SP(NearBucket\text{-}LSH(k, L), s) = 1 - (1 - (s^k + ks^{k-1}(1 - s)))^L.$$

Layered-LSH. We show that for the angular similarity, Layered-LSH is equivalent to the basic LSH. Layered-LSH maps near buckets to the same node w.h.p., which can be achieved by using Hamming-based LSH [8,14] as follows. Let g_{ang} be the angular-LSH used for mapping vectors to buckets. By definition, g_{ang} is a concatenation of h_i angular-LSH functions. Let g_{ham} be the Hamming-LSH used for mapping buckets to nodes. Hamming-based LSH hashes a binary vector to another binary vector of a lower dimension k, by randomly and independently selecting k entries of the input vector. In our case, this resorts to randomly and independently selecting k entries from $g_{ang}(v)$, each of which corresponds to some $h_i \in \mathcal{H}$. We get that $g_{ham}(g_{ang}(v))$ maps v to a node according to k randomly selected $h \in \mathcal{H}$ functions, which is equivalent to using the angular-LSH with parameter k.

5.2 Success Probability Comparison

We use Propositions 1 and 3 to compare the success probabilities of LSH, Layered-LSH, and NearBucket-LSH. We compute an algorithm's success probability as a function of the cosine similarity between the query and the searched object[2]. As Layered-LSH is equivalent to LSH, we refer to both as LSH in this discussion. For the purpose of the demonstration, we present graphs for selected k and L values. Note however that we observed the same trend for other k and L values; we omit the respective graphs from this text.

Constant Number of Hash Functions. We compare LSH and NearBucket-LSH for a constant L. Figure 1 depicts their success probabilities for $k = 12$ and for increasing L values of 1, 10, and 100. As the graphs demonstrate, the success probability of NearBucket-LSH is greater than or equal to the success probability of LSH for all similarities, for a constant L This stems from the fact that NearBucket-LSH searches in kL additional near buckets, which increases its success probability.

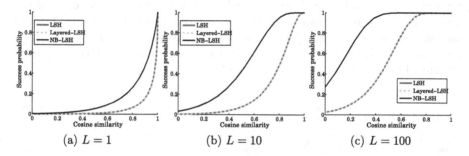

(a) $L = 1$ (b) $L = 10$ (c) $L = 100$

Fig. 1. Analytical success probability as a function of L ($k = 12$). NearBucket-LSH guarantees a greater or equal success probability compared to LSH and Layered-LSH, as it searches in more buckets (namely, near buckets). The gap increases as L increases.

Network Efficiency. As we have seen, for a constant L, NearBucket-LSH increases the success probability of LSH at the cost of contacting additional buckets. We proceed to analyzing the success probability as a function of the network cost. We measure the network cost by the average number of messages per query. We distinguish between the cached (CNB-LSH) and non-cached (NB-LSH) versions of NearBucket-LSH.

The first column of Table 1 summarizes the number of bucket nodes contacted (and searched) by each of the algorithms, for given k and L parameters. Looking up an exact bucket node requires an average of $k/2$ routing hops, and contacting a neighbor node costs one message. The second column in Table 1 summarizes the average number of messages per query, for given k and L parameters.

---

[2] We transform cosine similarity into angular similarity and then apply the success probability formulas.

Table 1. Summary of costs of similarity search in CAN-based LSH variants for given k, L LSH parameters.

	Number of nodes contacted per query	Average number of messages per query	Number of vectors stored in a node	Number of vectors searched per query
LSH	L	$\frac{1}{2}kL$	B	LB
Layered-LSH	L	$\frac{1}{2}kL$	B	LB
NB-LSH	L(1+k)	$1\frac{1}{2}kL$	B	$L(k+1)B$
CNB-LSH	L	$\frac{1}{2}kL$	$(k+1)B$	$L(k+1)B$

Figure 2 depicts success probability for $k = 12$ and an increasing network costs of 18, 180, and 1800 average number of messages. The graphs illustrates that, thanks to the low network cost of searching near buckets, NearBucket-LSH, (and more notably CNB-LSH), improves LSH's success probability for all similarity values, for a constant average number of messages. Note that one could further extend NearBucket-LSH to search in near buckets that differ from the query's bucket in more than one entry. The success probability of such buckets decreases (Proposition 2), whereas the network cost in NB-LSH and the storage cost in CNB-LSH increases compared to 1-near buckets. Thus, searching additional buckets is expected to be less effective.

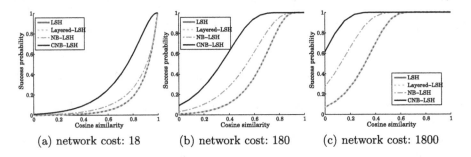

(a) network cost: 18   (b) network cost: 180   (c) network cost: 1800

Fig. 2. Analytical success probability as a function of network cost for $k = 12$. NB-LSH exploits the low lookup cost of near buckets in CAN, and increases LSH's and Layered-LSH's success probability for a given network cost. CNB-LSH further saves messages by caching near buckets, and achieves the greatest success probability for a given network cost.

Other Considerations. Our work focuses on minimizing the network cost, which is a dominant cost in P2P networks. For completeness, we present in the third and fourth columns of Table 1 other costs which tradeoff with network-efficiency. We denote the average bucket size by B. In terms of storage capacity, NB-LSH preserves the same space complexity as LSH and Layered-LSH. CNB-LSH increases the space complexity due to caching, while being more network-efficient than NB-LSH. Both NearBucket-LSH variants search over a larger number of vectors than LSH, implying more processing work per query. As our algorithm searches the buckets in parallel, and the average bucket size is equal in all algorithms, this does not affect the query latency.

6   Evaluation

We empirically evaluate our algorithm on three real world OSN datasets of varying sizes, and demonstrate the superiority of CNB-LSH over other approaches.

6.1   Search Quality Measures

Recall. (at m) is defined as follows [20]:

Definition 1 (recall at m). Given a query q, let $I_m(q)$ denote its ideal m-result set. Let $A_m(q)$ denote the approximate m-result set of q returned by some algorithm A. An algorithm's recall for query q is the fraction of results from q's m-ideal result set that are returned by A:

$$recall@m(A, q) = \frac{|A_m(q) \cap I_m(q)|}{|I_m(q)|}. \tag{4}$$

An algorithm's recall, $recall@m\,(A)$, is the mean of the queries' recall averaged over a query set Q.

Normalized Cumulative Similarity. We measure an algorithm's precision by comparing the similarity scores of its m-result set to those of the ideal m-result set. We define the following ratio, which we name the *normalized cumulative similarity (NCS)*:

Definition 2 (NCS at m). Given a query q, let $CumSim(I_m, q)$ denote the sum of the similarity values to q of the results in q's ideal m-result set.
    Let $CumSim(A_m, q)$ denote the sum of the similarity values to q of the results in q's m-result set of a given algorithm A. Then,

$$NCS@m(A, q) = \frac{CumSim(A_m, q)}{CumSim(I_m, q)} \tag{5}$$

We measure the NCS of an algorithm, $NCS@m(A)$, by averaging it over the query set Q.

Note that $CumSim(I_m, q) \geq CumSim(A_m, q)$, and both are positive. Therefore, $NCS@m(A) \in [0, 1]$.

6.2 Methodology

Datasets. We use three real-world publically-available datasets of OSNs [29]:

- *DBLP* [11], the computer science bibliography database: Authors are users, and venues are interests. We use a crawl of 13,477 interests, and 260,998 users that have at least one interest.
- *LiveJournal* [17] blogging-based OSN: Users publish blogs and form interest groups, which users can join. The LiveJournal crawl consists of 664,414 such groups, which we consider as user interests. There are 1,147,948 users with at least one interest.
- *Friendster* [13] online gaming network: Similarly to LiveJournal, Friendster allows users to form interest groups, which we consider as interests. The dataset consists of 1,620,991 interest groups, and 7,944,949 users with at least one interest.

All datasets contain anonymous user ids and interest information. We filtered out users having no interest.

Parameters. We set $k = 10$ in DBLP, $k = 12$ in LiveJournal and $k = 15$ in Friendster. We follow previous art [4,15] that uses k values between 10 and 20, and bucket sizes of a few hundreds [14]. Thus, we have 1,024 buckets in DBLP, 4,096 in LiveJournal, and 32,768 in Friendster. The average bucket size is approximately 250 vectors in all datasets. We set m, the number of search results, to 10.

Creating Sketch Vectors. We construct users' weighted interest vectors according to the dataset at hand. We weight each interest I based on its inverse frequency in user vectors [1]: $w(I) = ln(\frac{N_u}{N_I+1}) + 1$, where N_u denotes the total number of users, and N_I denotes the number of users having interest I. The user vector entry v_i is zero or $w(I)$ according to whether the user is associated with specific interest I. We use TarsosLSH's [27] for mapping vectors into LSH buckets.

Simulator. We implement a simulator of our CAN-based overlay using Apache Lucene 4.3.0 [19] centralized search index. We simulate distributing user vectors in bucket nodes by indexing vectors by their hash values (sketch vectors). The hash is then used for looking up a specific bucket node, and local similarity search is performed by limiting the search to the selected bucket (using Lucene's Filter mechanism). We additionally use Lucene to compute the ideal result set of a given query, by executing the query over the whole dataset. We score results according to the cosine similarity.

Evaluation Set. We construct a query set of 3,000 randomly sampled users. For each query q, we retrieve its ideal result set, as well as the result sets according to the algorithms we compare. For each dataset, we measure recall and precision over the query set in use.

6.3   Search Quality Results

Figure 3 illustrates our experimental results as a function of network cost. As in Sect. 5.2, we measure the network cost by the average number of messages per query according to Table 1. We increase the network cost by gradually increasing L, which increases search quality for all datasets as expected. We use larger values of k for larger datasets in order to preserve a common average bucket size. This ensures that local search takes the same time, and the cache sizes are identical. The larger k is, the lower the success probability is, thus, we expect a decrease in search quality when the dataset size increases, which is indeed demonstrated in the graphs.

The three datasets show a similar trend. Layered-LSH's search quality equals that of the basic LSH as expected. NearBucket-LSH (both cached and non-cached) demonstrates an increase in search quality compared to LSH and Layered-LSH, which is achieved by searching in additional near buckets stored at neighboring nodes or the node itself. For example, in LiveJournal (second column), LSH requires an average of 96 messages per query in order to achieve 0.59 precision, whereas CNB-LSH achieves a precision of 0.57 using only 12 messages. CNB-LSH also improves recall significantly, for example, achieving a 0.59 recall using 72 queries, compared to a recall of 0.35 for LSH. In all cases, NB-LSH is between LSH and CNB-LSH.

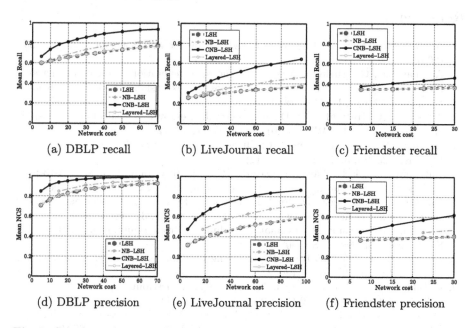

(a) DBLP recall   (b) LiveJournal recall   (c) Friendster recall

(d) DBLP precision   (e) LiveJournal precision   (f) Friendster precision

Fig. 3. Search quality as a function of the average number of messages per query, for three real world datasets: DBLP, LiveJournal, and Friendster ($k = 10$, $k = 12$, $k = 15$, respectively). For all datasets, CNB-LSH provides the greatest search quality as a function of the network cost, according to two metrics: recall and precision.

7   Conclusions and Future Work

We presented NearBucket-LSH, a network-efficient LSH algorithm for P2P OSNs, which provides good search quality. We first analytically showed that, for angular similarity, our choice of searched near buckets is optimal, that is, near buckets that differ in a single entry from the query's bucket are more likely to contain similar vectors than other near buckets. We then showed, both mathematically and empirically, that one may dramatically lower the additional network cost for searching in these buckets by exploiting CAN's internal structure and judicious caching.

Our proposed overlay focuses on angular-LSH, which fits OSN similarity search. It would be of an interest to extend our overlay to support other LSH families such as l_p-LSH, which map vectors to hashes in $\mathbb{Z}^k$ [10]. We expect such an extension to naturally fit our CAN overlay: According to l_p-LSH, Near buckets are computed by adding $\{-1, +1\}$ to an entry of a given hash vector [20]. Thus, when constructing a CAN over $\mathbb{Z}^k$, a near bucket's node resorts to the current node or its neighbor [25], which follows our design.

Acknowledgments. Naama Kraus is grateful to the Hasso-Plattner-Institut (HPI) for the scholarship for doctoral studies.

References

1. Adamic, L.A., Adar, E.: Friends and neighbors on the web. Soc. Netw. **25**, 211–230 (2001)
2. Adomavicius, G., Tuzhilin, A.: Toward the next generation of recommender systems: a survey of the state-of-the-art and possible extensions. IEEE Trans. Knowl. Data Eng. **17**(6), 734–749 (2005)
3. Anderson, A., Huttenlocher, D., Kleinberg, J., Leskovec, J.: Effects of user similarity in social media. WSDM 2012, pp. 703–712 (2012)
4. Bahmani, B., Goel, A., Shinde, R.: Efficient distributed locality sensitive hashing. In: CIKM 2012, pp. 2174–2178 (2012)
5. Batko, M., Novak, D., Falchi, F., Zezula, P.: Scalability comparison of peer-to-peer similarity search structures. Future Gener. Comp. Syst **24**(8), 834–848 (2008)
6. Buchegger, S., Schiöberg, D., Vu, L.H., Datta, A.: PeerSoN: P2P social networking - early experiences and insights. In: SNS 2009, pp. 46–52, 31 March 2009
7. Charikar, M.S.: Similarity estimation techniques from rounding algorithms. In: STOC 2002, pp. 380–388 (2002)
8. Chierichetti, F., Kumar, R.: LSH-preserving functions and their applications. In: SODA 2012, pp. 1078–1094 (2012)
9. Cutillo, L.A., Molva, R., Önen, M., Safebook: a distributed privacy preserving online social network. In: WOWMOM, pp. 1–3 (2011)
10. Datar, M., Immorlica, N., Indyk, P., Mirrokni, V.S.: Locality-sensitive hashing scheme based on p-stable distributions. In: SCG 2004, pp. 253–262 (2004)
11. DBLP. http://www.informatik.uni-trier.de/ley/db/
12. Falchi, F., Gennaro, C., Zezula, P.: A content–addressable network for similarity search in metric spaces. In: Moro, G., Bergamaschi, S., Joseph, S., Morin, J.-H., Ouksel, A.M. (eds.) DBISP2P 2005-2006. LNCS, vol. 4125, pp. 98–110. Springer, Heidelberg (2007). doi:10.1007/978-3-540-71661-7_9

13. Friendster. http://www.friendster.com/
14. Gionis, A., Indyk, P., Motwani, R.: Similarity search in high dimensions via hashing. In: VLDB 1999, pp. 518–529 (1999)
15. Haghani, P., Michel, S., Aberer, K.: Distributed similarity search in high dimensions using locality sensitive hashing. In EDBT 2009, pp. 744–755 (2009)
16. Indyk, P., Motwani, R.: Approximate nearest neighbors: towards removing the curse of dimensionality. In: STOC 1998, pp. 604–613 (1998)
17. Livejournal. http://www.livejournal.com/
18. Lua, E.K., Crowcroft, J., Pias, M., Sharma, R., Lim, S.: A survey and comparison of peer-to-peer overlay network schemes. IEEE Commun. Surv. Tutorials **7**, 72–93 (2005)
19. Lucene. http://lucene.apache.org/core/
20. Lv, Q., Josephson, W., Wang, Z., Charikar, M., Li, K.: Multi-probe LSH: efficient indexing for high-dimensional similarity search. In: VLDB 2007, pp. 950–961 (2007)
21. Mani, M., Nguyen, A.-M., Crespi, N.: Scope: a prototype for spontaneous P2P social networking. In: PerCom Workshops, pp. 220–225 (2010)
22. Manning, C.D., Raghavan, P., Schütze, H.: Introduction to Information Retrieval. Cambridge University Press, Cambridge (2008)
23. McPherson, M., Smith-Lovin, L., Cook, J.M.: Birds of a feather: homophily in social networks. Ann. Rev. Sociol. **27**, 415–444 (2001)
24. Narendula, R., Papaioannou, T.G., Aberer, K.: Towards the realization of decentralized online social networks: an empirical study. In: ICDCS Workshops, pp. 155–162 (2012)
25. Ratnasamy, S., Francis, P., Handley, M., Karp, R., Shenker, S.: A scalable content-addressable network. In: SIGCOMM 2001, pp. 161–172, New York, NY, USA (2001)
26. Sundaram, N., Turmukhametova, A., Satish, N., Mostak, T., Indyk, P., Madden, S., Dubey, P.: Streaming similarity search over one billion tweets using parallel locality-sensitive hashing. Proc. VLDB Endow. **6**(14), 1930–1941 (2013)
27. TarsosLSH. https://github.com/jorensix/tarsoslsh
28. Xiang, R., Neville, J., Rogati, M.: Modeling relationship strength in online social networks. In: WWW 2010, pp. 981–990 (2010)
29. Yang, J., Leskovec, J.: Defining, evaluating network communities based on ground-truth. In: MDS 2012, pp. 3: 1–3: 8 (2012)

Speeding up Similarity Search by Sketches

Vladimir Mic$^{(\boxtimes)}$, David Novak, and Pavel Zezula

Masaryk University, Brno, Czech Republic
`xmic@fi.muni.cz`

Abstract. Efficient object retrieval based on a generic similarity is one of the fundamental tasks in the area of information retrieval. We propose an enhancement for techniques that use the distance-based model of similarity. This enhancement is based on sketches–compact bit strings compared by the Hamming distance which represent data objects from the original space. The sketches form an additional filter that reduce the number of accessed data objects while practically preserving the search quality. For a certain class of state-of-the-art techniques, we can create the sketches using already known information, thus the time overhead is negligible and the memory overhead is subtle. According to the presented experiments, the sketch filtering can reduce the number of accessed data objects by 60–80 % in case of *M-Index*, and 30 % in case of *PPP-Codes* index while hurting the recall by less than 0.4 % on 10-NN search.

1 Introduction

Similarity retrieval represents a fundamental challenge of modern data processing. Handling objects according to their mutual similarity closely corresponds to the human perception of reality [6], therefore similarity retrieval can provide a natural way to access various types of data. Independently of a specific measure of similarity, we focus on the efficient similarity-based retrieval in large data collections. We adopt the broad model of the *metric space* that considers a data domain D together with a distance function $d : D \times D \mapsto \mathbb{R}$ to express the dissimilarity of two data objects from domain D. The distance function must satisfy properties of non-negativity, identity, symmetry and triangle inequality [19]; the triangle inequality is not explicitly utilized by the proposed technique. We assume an approximate evaluation of k-NN queries. The approximation quality is measured by *recall*, i.e. the relative size of the intersection of the approximate k-NN answer with the precise one. In our case the recall is equal to precision. We focus on speeding up existing distance-based similarity indexes. Majority of these indexes break the data collection down into disjoint partitions that are supposed to contain data objects mutually similar. Given a query object $q \in D$, the most promising partitions are identified; the union of data objects in these partitions forms a *candidate set* for the query q. Data objects x from the candidate set are accessed and distances $d(q, x)$ are evaluated to return the k most similar objects; this phase is further denoted as *refinement* of the candidate set. In high dimensional spaces, majority of data objects in the candidate set

© Springer International Publishing AG 2016
L. Amsaleg et al. (Eds.): SISAP 2016, LNCS 9939, pp. 250–258, 2016.
DOI: 10.1007/978-3-319-46759-7_19

are usually non-relevant [15], but these are uncovered only during the generally expensive refinement phase.

Objectives, Approach and Related Work

We propose to enrich a generic indexing technique with compact bit strings, called *sketches*, for all data objects and use them to reduce the candidate set. In order to be effective, this additional filter is supposed to work on a different space partitioning principle. Experiments with two indexes on two high-dimensional datasets show that the effect of the sketch filtering can be radical.

The basic reasoning behind our proposal is the following:

- The quality and efficiency of the approximate similarity search is closely related to the candidate set, i.e. the set should contain all relevant data objects and be as small as possible. By filtering out non-relevant data objects from the set, we can save the evaluations of the distance function d (which can be expensive) during the refinement phase and, if the data objects are stored on the disk, reduce the I/O costs.
- Many successful indexing techniques use a static set of reference objects (*pivots*) and the distances between each data object o and each pivot are evaluated during the preprocessing phase [1,5,12,15,17]. In these cases, we can create sketches using the already known object-pivot distances (see Sect. 2 for details). Then the time overhead introduced by the additional sketch filtering is negligible and the memory overhead subtle in comparison with the potential gain.

Combining different space partitionings is not a new idea and it seems to be a viable approach to fight the curse of dimensionality. A typical approach is to apply the same space partitioning principle but with different randomized parameters (e.g. sets of pivots) [5,10,13,15]. The data objects are either kept in memory and only accessed "from different points of view" [10], replicated on the disk [5,13], or kept separately from the memory indexes on an SSD disk [15]. There are also techniques that combine completely different principles of space partitioning into a single index, for instance the Pivoting M-Tree [16]. We do not know about any work that would propose a secondary filtering for approximate similarity search that would use the same pivot set for a different partitioning.

Section 2 of this paper describes the proposed approach and its properties in detail. Section 3 contains evaluation of the effectiveness of the sketch filtering on two indexes M-Index and PPP-Codes (Sect. 3.1) and on two datasets (CoPhIR and DeCAF – Sect. 3.2). The paper is concluded in Sect. 4.

2   Bit-String Sketches for Candidate Set Reduction

The sketch of data object $o \in D$ is a bit string in Hamming space which approximates the location of o in the original space (D, d). In particular, each bit value of $sketch(o)$ limits a subspace in (D, d) where the object o is located by means of

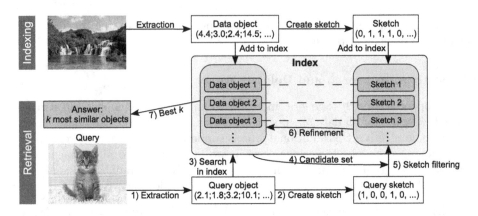

Fig. 1. Similarity search with additional sketch based filtering

a certain space partitioning. A typical example is a *generalized hyperplane partitioning* (GHP) which divides the data objects to two parts by means of their distances from two selected pivots $p_1, p_2 \in D$ [8,11,19]. Several authors study the similarity search based on sketches [4,8,11,18], and they usually achieve promising results for different data types, data dimensions, and distance functions.

In our previous paper [8], we analysed three properties of sketches desirable for efficient similarity search purely based on sketches. Having a set of sketches created for a given dataset, we have shown that (1) bit values of each *sketch(o)* should strongly depend on the position of object o in space (D, d), (2) the sketches should have *balanced bits*, i.e. each bit should be set to 1 in one half of the sketches, and (3) bits of sketches should be mutually as low correlated as possible. The first requirement is satisfied e.g. by sketches created using the GHP partitioning and other requirements can be satisfied by proper selection of specific pivots for GHP.

In this paper, we propose to use such sketches to enrich practically any distance-based search indexing technique. In particular, we propose to maintain a *sketch(o)* for each indexed data object o and use these sketches to reduce the candidate set to be refined (see Fig. 1). Given a query object q, the *sketch(q)* is created and we filter out some candidate objects o based on Hamming distance between *sketch(q)* and *sketch(o)*. Either, we can remove all data objects o with the sketch distance higher than some threshold, or we can filter out a given percentage of the candidate objects with the highest sketch distances. The distance threshold approach can be applied already during the primary candidate set generation. On the other hand, the percentage to be filtered can be determined without any knowledge about the Hamming space. According to our rigorous testing, the results of these approaches are comparable and thus we present only results of the second one which are slightly better.

The expenses of the sketch filtering are the following: (1) CPU time to investigate pivot pairs suitable for the sketches; this step is performed during the preprocessing. (2) CPU time to obtain *sketch(o)* for each object o and query sketch $sketch(q)$; using the GHP partitioning, this means evaluation of $2 \cdot b$ distances to pivots, where b is the sketch length; in indexes with a fixed set of pivots all the object-pivot distances are often known. In these cases sketches are created practically for free. (3) CPU time to evaluate the Hamming distances; this operation is very efficient on modern CPUs in comparison with often expensive evaluations of distance d. (4) Memory overhead of keeping $sketch(o)$ for each data object o; we show that even short sketches with lengths $b = 32$ or $b = 64$ can be very effective.

In general, the additional filtering can be employed in the following two ways. Either we can (1) reduce the number of refined data objects while, in ideal case, preserving the search quality (recall), (2) or we can preserve the number of refined data objects by adding more data objects instead of the ones filtered out by sketches; in this way, we can improve the search quality. Results presented in Sect. 3 can be interpreted in both ways.

3   Evaluation

In this section, we present evaluation of the similarity search with the additional sketch filtering. We conducted experiments using two different indexing techniques (see Sect. 3.1) and two real-life datasets (see Sect. 3.2). The sketch filtering was evaluated using sketches of lengths 32 and 64 bits and the results are presented and discussed in Sect. 3.3.

3.1   Similarity Indexes

The first index structure is the M-Index [12]. It uses a fixed set of pivots to perform Voronoi partitioning and each Voronoi cell is recursively partitioned by the same principle. The depth of this partitioning is determined dynamically according to occupation of the leaf partitions. Each data object is then stored according to its several closest pivots (according to prefix of a *pivot permutation*). Given a query object, the M-Index uses the query-pivot distances to determine the most promising partitions to contain query-relevant data objects [12].

The PPP-Codes search structure [14,15] has a slightly different architecture. It maintains a memory index which determines the set of candidate objects by their unique IDs and this set is retrieved from a disk storage (SSD) and refined. The space partitioning is also based on pivot permutation prefixes (PPP) but the candidate set can be an order of magnitude smaller than for M-Index [15]. The effect is achieved by decomposing the pivot set into several sets and thus creating several PPPs for each data object. Given a query object, the candidate set is determined by a selective combination of candidate sets from individual pivot spaces. In this way, PPP-Codes can appropriately filter out objects that seem to be relevant in one pivot space but the other spaces indicate the opposite. The process of candidate set formation is more computationally demanding.

3.2   Testing Data

The experiments are conducted on two real-life data collections, both consisting of visual descriptors extracted from images. The first set is formed by *DeCAF* descriptors [3] – 4096-dimensional vectors taken as an output from the last hidden layer of a deep convolutional neural network [7]. These descriptors were extracted from a 1M subset of the *Profiset collection*[1]. The DeCAF descriptors are compared by the Euclidean distance to form the metric space (D, d).

The second dataset consists of a combination of five MPEG-7 visual descriptors [9] as provided by the *CoPhIR*[2] data collection [2]. Each of these descriptors is accompanied with a suitable distance function [9] and the descriptors extracted from each image are combined into a single metric space (D, d) by a weighted sum of individual distances [2]. In total, this representation can be viewed as a 280-dimensional vector. We take a 1M subset of CoPhIR.

For each dataset, we have randomly selected a set of 512 pivots that are used by M-Index and PPP-Codes for indexing. In order to create sketches, we investigate all $\binom{512}{2}$ pivot pairs. Each pair has been used to partition a random subset of 100,000 data objects by GHP and in this way we have identified those pairs that divide the data into parts balanced at least 55 % to 45 %. From these balanced pivot pairs ($\approx$8,000) we further select those producing sketches with low correlated bits using a heuristics described in [8]. Both 1M subsets of datasets, query objects and pivots were selected randomly and are publicly available.

3.3   Results

In this section, we evaluate the ability of the sketches to filter out non-relevant data objects from the candidate set. In the first set of experiments, we let the M-Index identify the candidate set with the 50,000 most promising data objects for a representative query object q; such set is denoted *M50K*. Let us first observe the distribution of distances $d(q, x)$ for all 50 K data objects x in this candidate set. This distribution is denoted dd_{M50K} and is depicted by the black curve in Fig. 2. Further, we denote *M50K,S50 %* the same candidate set with 50 % data objects filtered out by sketches (according to Hamming distances from $sketch(q)$); the respective query-object distance distribution $dd_{M50K,S50\%}$ is also depicted in Fig. 2. Please, note that these plots are taken for a single query on the DeCAF dataset and that both plots are the same, just the right one uses a logarithmic scale of axis y. The y axis expresses the frequency, formally:

$$\int_{-\infty}^{\infty} dd_{M50K}(x)\, dx = 1, \text{ therefore: } \int_{-\infty}^{\infty} dd_{M50K,S50\%}(x)\, dx = 0.5.$$

This example illustrates the ability of the sketches to preserve relevant data objects, since the beginnings of both curves are the same (see marked point from which these curves differs). In particular 499 out of 500 data objects with the smallest distances to the query object are preserved in this example.

---

[1] http://disa.fi.muni.cz/profiset/.
[2] http://cophir.isti.cnr.it/.

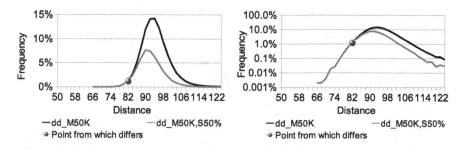

Fig. 2. Distance distributions on data objects M50K and M50K,S50 %, left plot with linear scale and right plot with logarithmic scale

In the rest of this section, we present results averaged over 1,000 randomly selected queries. We focus on the probability of false negatives, i.e. that the actual kth nearest neighbour from the candidate set is filtered out by sketches. Figure 3 depicts these probabilities for the DeCAF dataset, M50K data objects and filtering out 50 % of these data objects by sketches. Results for 32 bit and 64 bit sketches are presented. The genuine results are depicted in a gray color and the trend curves are in black. The results show significantly better ability of 64 bit sketches over 32 bit to preserve the most similar data objects in an answer.

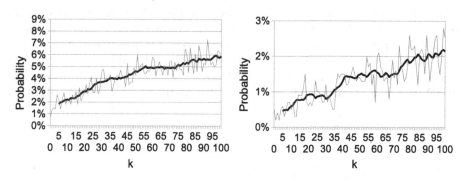

Fig. 3. Chances of omitting the kth most similar object by additional sketch filtering, left plot with 32 bit sketches and right plot with 64 bit sketches

In the final set of experiments, we use 64 bit sketches and we focus on overall k-NN recall for $k = 10$ (denoted as recall@10). Figures 4 and 5 show results for CoPhIR and DeCAF datasets, respectively, always for candidate sets from both M-Index and PPP-Codes. The vertical axes represent the average recall@10 after the sketch filtering. The horizontal axes show the relative size of the candidate set with respect to the size of dataset (which is 1 million). Individual curves correspond to percentages of this candidate set filtered out by the sketches (curves *sketch filter 0 %* show original results of M-Index and PPP-Codes).

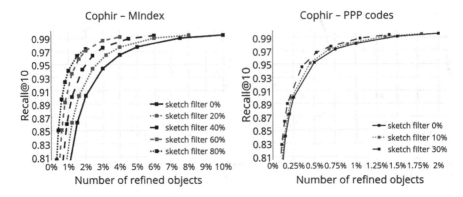

Fig. 4. Recall of double filter and refine similarity retrieval on CoPhIR dataset

We can read these graphs in two ways: (1) For a given number of refined objects, we can observe which combination of original candidate set size and percentage of sketch filtering gives the highest recall, and (2) for a required recall level, we can look for parameters leading to the smallest number of refined objects. For instance, for recall below 0.95, the most efficient is letting the sketches filter out even 80 % of the M-Index candidates and over 30 % in case of PPP-Codes. For a highly accurate retrieval with recall about 0.99, it is better to use a bigger original candidate set and filter out about 50–60 % in case of M-Index and 30 % objects for PPP-Codes. Let us observe specific selected values for DeCAF (Fig. 5): The pure M-Index must refine 100,000 objects to achieve recall 97.57, while the sketches can filter out all but 40,000 objects and preserve average recall of 97.21. In case of PPP-Codes, only 14,000 objects instead of 20,000 have to be refined while the recall value decreases from 97.32 to 97.13.

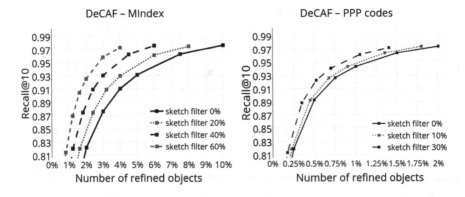

Fig. 5. Recall of double filter and refine similarity retrieval on DeCAF dataset

4 Conclusions

We have proposed and evaluated an enhancement of traditional similarity search techniques with an additional sketch-based filtering. Sketches, compact binary strings, can be created practically for free for techniques that use a static set of pivots. Their contribution to the quality of filtering can be huge, as shown on two state-of-the-art indexes M-Index and PPP-Codes and two real-life datasets. We have demonstrated the ability of the sketches to filter out many non-relevant data objects while preserving almost all relevant ones. In case of M-Index, the number of refined data objects can be reduced by 60–80 % and in case of PPP-Codes by 30 % while the decrease of the recall@10 was only negligible.

Acknowledgements. This work was supported by the Czech Science Foundation project GA16-18889S.

References

1. Amato, G., Gennaro, C., Savino, P.: MI-File: using inverted files for scalable approximate similarity search. Multimedia Tools Appl. **71**(3), 1333–1362 (2014)
2. Batko, M., Falchi, F., Lucchese, C., Novak, D., Perego, R., Rabitti, F., Sedmidub-sky, J., Zezula, P.: Building a web-scale image similarity search system. Multimedia Tools Appl. **47**(3), 599–629 (2010)
3. Donahue, J., Jia, Y., Vinyals, O., Hoffman, J., Zhang, N., Tzeng, E., Darrell, T.: DeCAF: a deep convolutional activation feature for generic visual recognition. arXiv preprint arXiv:1310.1531 (2013)
4. Dong, W., Charikar, M., Li, K.: Asymmetric distance estimation with sketches for similarity search in high-dimensional spaces. In: Proceedings of ACM SIGIR 2008, pp. 123–130. ACM (2008)
5. Esuli, A.: Use of permutation prefixes for efficient and scalable approximate simi-larity search. Inf. Process. Manage. **48**(5), 889–902 (2012)
6. Kemler, D.G.: Classification in young and retarded children: the primacy of overall similarity relations. Child Dev. **53**(3), 768–779 (1982)
7. Krizhevsky, A., Sutskever, I., Hinton, G.E.: Imagenet classification with deep con-volutional neural networks. In: Advances in neural information processing systems, pp. 1097–1105 (2012)
8. Mic, V., Novak, D., Zezula, P.: Improving sketches for similarity search. In: Pro-ceedings of MEMICS 2015, pp. 45–57 (2015)
9. MPEG7: Multimedia content description interfaces. part 3: Visual (2002)
10. Muja, M., Lowe, D.G.: Scalable nearest neighbour algorithms for high dimensional data. IEEE Trans. Pattern Anal. Mach. Intell. **36**(11), 1–14 (2014)
11. Muller-Molina, A.J., Shinohara, T.: Efficient similarity search by reducing i/o with compressed sketches. In: Proceedings of SISAP 2009, pp. 30–38. IEEE Computer Society (2009)
12. Novak, D., Batko, M., Zezula, P.: Metric index: an efficient and scalable solution for precise and approximate similarity search. Inf. Syst. **36**(4), 721–733 (2011)
13. Novak, D., Zezula, P.: Performance study of independent anchor spaces for simi-larity searching. Comput. J. **57**(11), 1741–1755 (2014)

14. Novak, D., Zezula, P.: Rank aggregation of candidate sets for efficient similarity search. In: Decker, H., Lhotská, L., Link, S., Spies, M., Wagner, R.R. (eds.) DEXA 2014. LNCS, vol. 8645, pp. 42–58. Springer, Heidelberg (2014). doi:10.1007/978-3-319-10085-2_4

15. Novak, D., Zezula, P.: PPP-codes for large-scale similarity searching. In: Hameurlain, A. (ed.) TLDKS XXIV. LNCS, vol. 9510, pp. 61–87. Springer, Heidelberg (2016). doi:10.1007/978-3-662-49214-7_2

16. Skopal, T., Pokorny, J., Snasel, V.: PM-Tree: pivoting metric tree for similarity search in multimedia databases. In: Proceedings of ADBIS 2004, pp. 99–114 (2004)

17. Tellez, E.S., Chavez, E., Navarro, G.: Succinct nearest neighbor search. Inf. Syst. **38**(7), 1019–1030 (2013)

18. Wang, Z., Dong, W., Josephson, W., Lv, Q., Charikar, M., Li, K.: Sizing sketches: a rank-based analysis for similarity search. SIGMETRICS Perform. Eval. Rev. **35**(1), 157–168 (2007)

19. Zezula, P., Amato, G., Dohnal, V., Batko, M.: Similarity Search: the Metric Space Approach. Advances in Database Systems, vol. 32. Springer Science & Business Media, New York (2006)

Fast Hilbert Sort Algorithm Without Using Hilbert Indices

Yasunobu Imamura[1($\boxtimes$)], Takeshi Shinohara[1], Kouichi Hirata[1], and Tetsuji Kuboyama[2]

[1] Department of Artificial Intelligence,
Kyushu Institute of Technology, Kitakyushu, Japan
imamura.kit@gmail.com
[2] Computer Centre, Gakushuin University,
Toshima, Japan

Abstract. *Hilbert sort* arranges given points of a high-dimensional space with integer coordinates along a Hilbert curve. A naïve method first draws a Hilbert curve of a sufficient resolution to separate all the points, associates integers called *Hilbert indices* representing the orders along the Hilbert curve to points, and then, sorts the pairs of points and indices. Such a method requires an exponentially large cost with respect to both the dimensionality n of the space and the order m of the Hilbert curve even if obtaining Hilbert indices. A known improved method computes the Hilbert index for each point in $O(mn)$ time. In this paper, we propose an algorithm which directly sorts N points along a Hilbert curve in $O(mnN)$ time without using Hilbert indices. This algorithm has the following three advantages; (1) it requires no extra space for Hilbert indices, (2) it handles simultaneously multiple points, and (3) it simulates the Hilbert curve in heterogeneous resolution, that is, in lower order for sparse space and higher order for dense space. It, therefore, runs much faster on random data in $O(N\log N)$ time. Furthermore, it can be expected to run very fast on practical data, such as high-dimensional features of multimedia data.

1 Introduction

Hilbert curve [4], one of the space-filling curves, can be defined in any dimensionality. In Fig. 1 Hilbert curves in the 2-dimensional space of the first to the third order are shown. It is known that any interval of sorted objects along a Hilbert curve forms a relatively good cluster. For example, R-tree, which is one of the hierarchical spatial index structures, exhibits high performance when constructing such clusters [6]. In this paper we consider Hilbert sort problem which sorts given objects in high-dimensional space with integer coordinate values along a Hilbert curve.

This work was partially supported by Grant-in-Aid for Scientific Research 16H02870, 26280090, 15K12102, 26280085 and 16H01743 from the Ministry of Education, Culture, Sports, Science and Technology, Japan.

L. Amsaleg et al. (Eds.): SISAP 2016, LNCS 9939, pp. 259–267, 2016.
DOI: 10.1007/978-3-319-46759-7_20

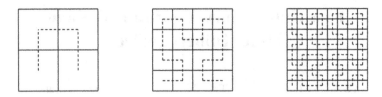

Fig. 1. Hilbert curves of the first order to the third order in two-dimensional space

A naïve method of Hilbert sort first draws the Hilbert curve of the m-th order with a sufficient resolution to separate all the points, associates integers called *Hilbert indices* representing the orders along the Hilbert curve to points, and then, sorts the pairs of points and indices. Such a method requires an exponential cost with respect to both the dimensionality n of the space and the order m of the Hilbert curve only for obtaining Hilbert indices [5]. A known improved method [1–3] computes the Hilbert index for each point in $O(mn)$ time. Thus, we have already known the Hilbert sort for N objects in n-dimensional space of m-th order can be solved in $O(mnN + mnN\log N) = O(mnN\log N)$ time.

In this paper, we propose an algorithm that directly sorts N points along a Hilbert curve in $O(mnN)$ time without using Hilbert indices, which was originally introduced by Tanaka [7][1]. This algorithm has three advantages; (1) it requires no extra space for Hilbert indices, (2) it handles simultaneously multiple points, and (3) it simulates the Hilbert curve in heterogeneous resolution, that is, in lower order for sparse space and higher order for dense space. As shown later, it can be observed to run very fast on practical data, such as multimedia data with high-dimensional features. It runs much faster on random data in $O(N\log N)$ time.

2   Outline of Proposed Algorithm

In this section, we explain the outline of the proposed algorithm by using an example. Let's consider 9 points in 2-dimensional space shown in Fig. 2. To separate all the points we have to draw the Hilbert curve of the third order, where the dimensionality n is 2, the order (resolution) m of Hilbert curve is 3, and the space is divided into $2^{mn} = 64$ subspaces. Without loss of generality we can deal with every point in the set represented by nm-bit unsigned integers. For example, the rightmost one in the 9 points is in the position (5, 6) represented by "101110".

Our algorithm simulates the Hilbert curve by dividing space into two in an axis. Here we trace it in a breadth-first manner, while it runs in a depth-first manner by a recursive call in a natural implementation.

We represent the start point, end point and crossing area of the Hilbert curve by an arc-like arrow. For example, Fig. 3 shows that the Hilbert curve crosses the square from the lower left corner to the lower right.

---

[1] We found that Tanaka's implementation has $O(mn^2N)$ running time.

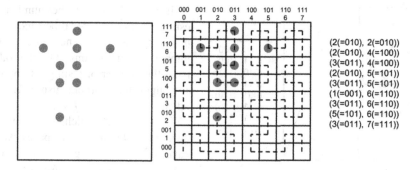

(2(=010), 2(=010))
(2(=010), 4(=100))
(3(=011), 4(=100))
(2(=010), 5(=101))
(3(=011), 5(=101))
(1(=001), 6(=110))
(3(=011), 6(=110))
(5(=101), 6(=110))
(3(=011), 7(=111))

Fig. 2. 9 points separated by Hilbert curve of the third order

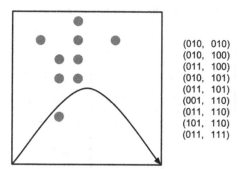

(010, 010)
(010, 100)
(011, 100)
(010, 101)
(011, 101)
(001, 110)
(011, 110)
(101, 110)
(011, 111)

Fig. 3. Arc-like arrow of Hilbert curve

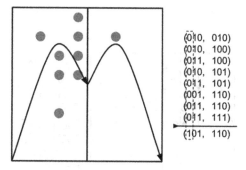

(010, 010)
(010, 100)
(011, 100)
(010, 101)
(011, 101)
(001, 110)
(011, 110)
(011, 111)
(101, 110)

Fig. 4. Division in the horizontal axis

First, we divide the space into two in the horizontal axis as in Fig. 4. Note that this division is done by tests on the first bits, which are enclosed by a dashed line in Fig. 4. At this stage, 8 points in the left subspace are decided to precede a point in the right subspace. Since just one point exists in the right side, no more division is necessary,

The next division in the vertical axis is applied only to the left subspace (Fig. 5). Then the simulation of the first order of the Hilbert curve is complete.

The simulation of the second order is applied only to the upper left subspace (Fig. 6). The simulation of the third order is shown in Fig. 7.

Finally, in this example, to sort 9 points, 9 space divisions are done to generate 10 subspaces.

If we adopt the algorithm in [2, 3] to compute Hilbert indices for 9 points, then the number of simulations at the smallest level amounts to $mnN = 54$, where $m = 3$ is the order of Hilbert curve, $n = 2$ is the dimensionality and $N = 9$ is the number of points. In contrast, our algorithm requires only 9 simulations, which is the same as the number of divisions. Thus, the proposed algorithm can reduce the cost to sort objects.

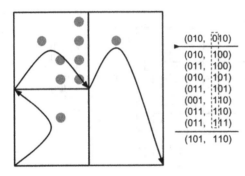

(010, 010)
(010, 100)
(011, 100)
(010, 101)
(011, 101)
(001, 110)
(011, 110)
(011, 111)

(101, 110)

Fig. 5. Division in the vertical axis

It is obvious that the number of testing bits to compute Hilbert indices is $mnN = 54$, which is the same as the number of simulations. On the other hand, the number of testing bits by our sort algorithm without using Hilbert indices is $9 + 8 + 7 + 6 + (2 + 4) + (2 + 2 + 2) = 42$, which is the sum of the number of points in subspaces except leaves as shown in Fig. 8. Here the difference is not large, because m is small and many points need to be simulated in the largest order. However, the larger the order m is, the larger the difference is.

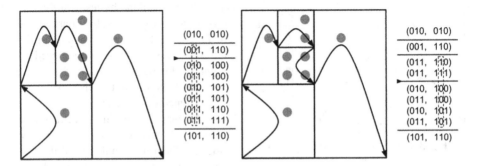

(010, 010)
(001, 110)
(010, 100)
(011, 100)
(010, 101)
(011, 101)
(011, 110)
(011, 111)
(101, 110)

(010, 010)
(001, 110)
(011, 110)
(011, 111)
(010, 100)
(011, 100)
(010, 101)
(011, 101)
(101, 110)

Fig. 6. Simulation of the second order

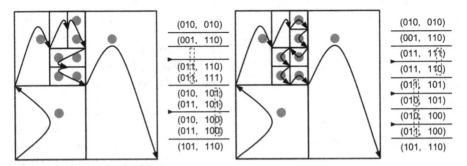

(010, 010)
(001, 110)

(011, 110)
(011, 111)
(010, 101)
(011, 101)
(010, 100)
(011, 100)
(101, 110)

(010, 010)
(001, 110)
(011, 111)
(011, 110)
(011, 101)
(010, 101)
(010, 100)
(011, 100)
(101, 110)

Fig. 7. Simulation of the third order

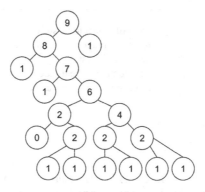

Fig. 8. The number of points in subspaces

3 Pseudo-code of Proposed Algorithm

In Fig. 9, we present C++ like pseudo-code of the proposed algorithm *Hilbert-Sort*, which consists of two functions `partition` and `HSort`. The constants m and n, which are the order of Hilbert curve and the dimensionality of data, are defined externally, such as macros. For clarity, here we use `bitset<m>` to represent coordinate values. In practice, we may use an integer type, such as `unsigned char` for $m = 8$, or `unsigned long` for $m = 32$. To sort N objects, we call `HSort` with parameters: `st = 0, en = N - 1, od = m - 1, c = 0, e = bitset<m>(), d = 0, di = false, cnt = 0`.

The function `partition` arranges data in a similar way as the famous partition function in quick sort. The function `HSort` is the main sorting function, which simulates Hilbert curve in almost the same way as in Hilbert-Index [2, 3]. Therefore we omit the formal proof of the correctness of simulation. The key difference of Hilbert-Sort from Hilbert-Index is the bit-wise simulation, which avoids redundant simulations to provide practical high speed.

4 Experiments

In this section, we demonstrate effectiveness of proposed algorithm Hilbert-Sort by running experiments. We use a library function sort in C++STL to sort using indices calculated by Hilbert-Index.

We use three kinds of data to be sorted, (1) random data, (2) random pair data, and (3) feature data extracted from images. Every bit of random data is expected to divide uniformly. A set of random pairs data is made by duplicating the half size random data, which is expected to derive the worst running time by the proposed algorithm because the deepest order simulation of Hilbert curve is necessary and the advantage of simultaneous simulations for multiple points is hard to preserve. For feature data, we prepare about 7 million image data extracted from video by processing grayscale transformation, down scaling and 2-dimensional FFT. Dimensionality of image features is 64. Each axis is represented by an 8 bits unsigned char.

```
int partition(
    bitset<m> *A[],        // array of points to be sorted
    int st, int en,        // represent range A[st], ... , A[en]
    int od,                // order of interest
    int ax,                // axis number to be divided
    bool di                // direction, false -> ascending, true -> descending
    )
{
  int i, j;
  i = st - 1;
  j = en + 1;
  while(true) {
    do i = i + 1; while(A[i][ax].test(od) == di);
    do j = j - 1; while(A[j][ax].test(od) != di);
    if(j < i) return i;      // partition is completed
    swap(A[i], A[j]);
  }
}

void HSort(
    bitset<m> * A[],    // array of points to be sorted
    int st, int en,     // represent range A[st], ... , A[en]
    int od,             // current order of Hilbert curve
    int c,              // axis counter in current order
    bitset<m> & e,      // start positions of axes
    int d,              // axis number of first division on current order
    bool di,            // direction of previous division:
                        // false -> forward,   true -> backward
    int cnt             // number of continuous occurrences of di at same order
    )
{
  int p, d2;
  if(en <= st) return;   // nothing to do for empty or singleton
  p = partition(A, st, en, od, (d + c) % n, e.test((d + c) % n));
  if(c == n - 1) {            // simulation done, goto next order
    if(b == 0) return;   // no more order to simulate
    d2 = (d + n + n - (di ? 2 : cnt + 2)) % n;
    e.flip(d2); e.flip((d + c) % n);
    HSort(A, st, p - 1, b - 1, 0, e, d2, false, 0);
    e.flip((d + c) % n); e.flip(d2);   // undo of flips (2 before line)
    d2 = (d + n + n - (di ? cnt + 2 : 2)) % n;
    HSort(A, p, en, b - 1, 0, e, d2, false, 0);
  } else {
    HSort(A, st, p - 1, b, c + 1, e, d, false, di ? 1 : cnt + 1);
    e.flip((d + c) % n); e.flip((d + c + 1) % n);
    HSort(A, p, en, b, c + 1, e, d, true, di ? cnt + 1 : 1);
    e.flip((d + c + 1) % n); e.flip((d + c) % n);   // undo of flips
  }
}
```

Fig. 9. Pseudo-code of Hilbert-Sort

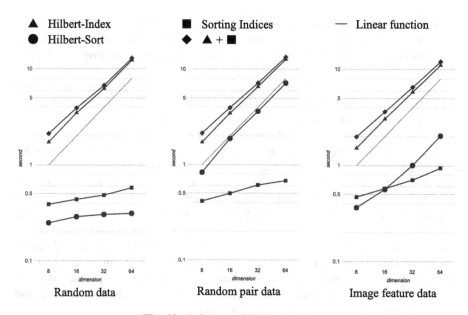

Fig. 10. Influence of dimensionality n

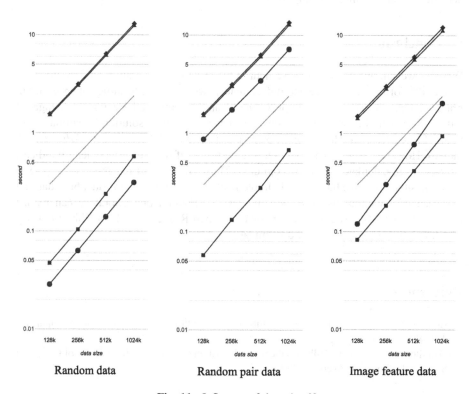

Fig. 11. Influence of data size N

In all the experiments, we fix $m = 8$, $n = 64$, $N = 1,048,576$ (=1024k) unless we explicitly vary the values. We present two graphs for each data to observe the effect of the dimensionality n and the number N of data. Every graph is plotted on a log-log scale.

As shown in Fig. 10, for random data, Hilbert-Sort runs extremely faster than Hilbert-Index in sublinear time with respect to n, whereas Hilbert-Index runs in just linear time. On the other hand, as shown in Fig. 11, for random data, both algorithms run in linear time with respect to the number N of data.

For random pair data, Hilbert-Sort has no advantage compared with Hilbert-Index in computational complexity, however, in our implementations, Hilbert-Sort runs about twice as fast as Hilbert-Index. (Figs. 10 and 11). For feature data, Hilbert-Sort runs about 5 times as fast as Hilbert-Index (Figs. 10 and 11).

From Fig. 11, the running time of Hilbert-Sort for feature data looks like worse than linear. The feature data are extracted from videos, which contain many similar data. Since we prepare the different sizes of images by random selection from 7 million data, the larger the size of data set is, the higher the probability of similar data is. Thus, the results for them are similar as ones for random data in smaller size, while they have many similar pairs like random pair data. Therefore, we can conclude that the running time of Hilbert-Sort is linear even for feature data.

Additional experiments on colors, one of SISAP databases, also show a similar behavior as on the image feature data.

5   Conclusion

Whereas we can observe no improvement in the order of computation time for practical feature data of images as expected unfortunately, we can achieve sufficient speed-up in practice. One of reasons for improvement can be explained by the advantage of Hilbert-Sort that simulates Hilbert curve in heterogeneous resolution. A similar technique for calculating Hilbert indices is possible with introducing *variable length* indices. However, when a new data set is added, even if we use variable length indices, we have to re-calculate indices for old data. Thus, Hilbert-Sort has more advantage in dynamic situations. We have already implemented Hilbert-Merge to add a bulk data set to sorted data without using Hilbert indices [8]. Finally, our laboratory can realize *compact Hilbert R-trees* as online versions of Hilbert R-tree [6] without Hilbert indices, which can also be used even in dynamical situations.

References

1. Butz, A.R.: Alternative algorithm for Hilbert's space-filling curve. IEEE Trans. Comput. **20**, 424–426 (1971)
2. Hamiltonm, C.: Compact Hilbert indices. Technical report CS-2006–07, Faculty of Computer Science, Dalhousie University (2006)

3. Hamilton, C.: Compact Hilbert indices: space-filling curves for domains with unequal side lengths. Inf. Process. Lett. **105**, 155–163 (2008)
4. Hilbert, D.: Uber die stetige Abbildung einer Linie auf ein Flachenstuck. Math. Ann. **38**, 459–460 (1891)
5. Kamata, S., Perez, A., Kawaguchi, E.: A computation of Hilbert's curves in N dimensional space. IEICE **J76-D-II**, 797–801 (1993)
6. Kamel, I., Faloutsos, C.: Hilbert R-tree: an improved R-tree using fractals. In: The 20th International Conference on Very Large Data Bases (VLDB), pp. 500–509 (1994)
7. Tanaka, A.: Study on a fast ordering of high dimensional data to spatial index. Master thesis, Kyushu Institute of Technology (2001). (in Japanese)
8. Tashima, K.: Study on efficient method of insertion for spatial index structure by using Hilbert sort. Master thesis, Kyushu Institute of Technology (2011). (in Japanese)

Time-Evolving Data

Similarity Searching in Long Sequences of Motion Capture Data

Jan Sedmidubsky$^{(\boxtimes)}$, Petr Elias, and Pavel Zezula

Masaryk University, Brno, Czech Republic
xsedmid@fi.muni.cz

Abstract. Motion capture data digitally represent human movements by sequences of body configurations in time. Searching in such spatio-temporal data is difficult as query-relevant motions can vary in lengths and occur arbitrarily in the very long data sequence. There is also a strong requirement on effective similarity comparison as the specific motion can be performed by various actors in different ways, speeds or starting positions. To deal with these problems, we propose a new subsequence matching algorithm which uses a synergy of elastic similarity measure and multi-level segmentation. The idea is to generate a minimum number of overlapping data segments so that there is at least one segment matching an arbitrary subsequence. A non-partitioned query is then efficiently evaluated by searching for the most similar segments in a single level only, while guaranteeing a precise answer with respect to the similarity measure. The retrieval process is efficient and scalable which is confirmed by experiments executed on a real-life dataset.

1 Introduction

Current motion capturing technologies can accurately record a human motion at high spatial and temporal resolutions. The recorded motion is represented as an ordered sequence of *poses* that describe skeleton configurations in corresponding video frames. The skeleton configuration is represented by a set of 3D coordinates determining positions of the captured body *joints* in space. The recorded motion sequences can be used in a variety of applications, e.g., in sports to compare performance of athletes, in law-enforcement to detect suspicious events, in health care to determine the success of rehabilitative treatments, or in computer animation to synthesize and generate realistic human motions for production of high-quality games or movies.

These applications require efficient subsequence searching: Given a short *query sequence* and a long *data sequence*, search the data sequence and locate its subsequences that are the most similar to the query sequence. For example, find occurrences of acrobatic elements within a 5-minute dancing exercise. Locating such query-relevant subsequences constitutes a hard task since their lengths and positions (i.e., beginnings and endings) are not known in advance. Moreover, the query need not correspond to any semantic action, so textual-annotation-based

© Springer International Publishing AG 2016
L. Amsaleg et al. (Eds.): SISAP 2016, LNCS 9939, pp. 271–285, 2016.
DOI: 10.1007/978-3-319-46759-7_21

retrieval cannot be applied. To deal with these problems, a proper segmentation technique along with an effective similarity measure are needed.

The contribution of this paper is an efficient subsequence retrieval algorithm based on a new multi-level segmentation. The proposed multi-level structure produces a minimal number of segments with respect to the elasticity property of the used similarity measure. The elasticity allows segments to be shifted much larger than of a single frame, which speeds up the retrieval process.

2   Related Work

Subsequence retrieval methods for motion capture data generally require a (1) segmentation technique to partition a data sequence into meaningfully-long segments, (2) similarity measure to compare query and data segments, and (3) retrieval algorithm to efficiently localize query-relevant subsequences.

Segmentation. A segmentation technique partitions the data sequence [3] into short segments to be comparable with segment(s) of the query sequence. The segmentation can be done in a semantic way by localizing non-overlapping segments which correspond to the predefined actions (e.g., walking, kicking and jumping) [9,11]. Another kind of approaches identifies segments according to some property, such as changes in pose distribution or intrinsic dimensionality [1] or occurrences of repetitive movements [17]. However, these methods are not suitable for universal subsequence retrieval as they do not cope well with queries that search for motions that are not annotated or occur on the boundaries of segments. We avoid this problem by a multi-level overlapping segmentation to ensure that an arbitrary query-relevant data subsequence (bounded in length by the user) highly overlaps with at least one data segment of a similar length.

Similarity Measure. Motion data are usually represented as a sequence of poses representing how the specific features change in time, such as joint coordinates, angles and velocity. To determine similarity of two motion sequences, temporal alignment techniques are commonly used, such as Dynamic Time Warping (DTW) and its variants [2,8,16], Longest Common Subsequence (LCS) [14] and Smith-Waterman distance [19]. Sometimes, motion sequences are described via lower-dimensional feature representations such as 160-bit signatures in [18] that are compared by the Hamming distance. We also benefit from lower-dimension representations by extracting fixed-size feature vectors using the convolutional neural network and comparing them by the Euclidean distance.

Subsequence Retrieval. Having the segmentation technique and similarity measure defined, a subsequence retrieval algorithm is used to locate query-relevant parts within a long data sequence. A trie-based structure is used in [6] to efficiently access numerical features whose values are quantized into fixed-size intervals. Based on quantized intervals of the query features, the trie structure is traversed to identify query-relevant parts in the data sequence. However, the evaluation of search effectiveness on any standard dataset is not provided. Similarly in [18], the search accuracy is only commented on several query results evaluated on the perceptual level. In this paper, we analyze the search accuracy of our search algorithm on the largest annotated motion capture dataset and compare the results with a subsequence retrieval algorithm introduced in [15]. The algorithm in [15] partitions a query sequence into short fixed-size segments whose first poses are used to search for the most similar poses in the data sequence. The obtained ranked sets of candidate poses are post-processed in temporal order to identify query-relevant subsequences. We demonstrate that our approach is much more efficient because we do not split a query into segments.

Our Contributions. We present a new algorithm for subsequence matching in motion capture data. In particular, we propose a multi-level overlapping segmentation ensuring a traceability of an arbitrary query-relevant data subsequence which is bounded in its size. The number of segments in all levels is constructed to be minimal with respect to the ability of a similarity measure to compare slightly cropped/extended motions. We employ a similarity measure which is tolerant to slightly changed motions and also generates fixed-size feature vectors for motions of variable lengths. As the fixed-size features are compared by the Euclidean distance, the segments in each level can be very efficiently indexed. This makes our approach very efficient for subsequence searching, potentially in a 121-day long motion sequence within one second.

3   Similarity of Motion Data

To compare a pair of motion sequences, we employ an elastic similarity measure that attributes very convenient properties for subsequence matching because it (1) extracts very effective and fixed-size 4,096-dimensional feature vectors for motions of variable lengths, (2) compares these vectors by the efficient Euclidean distance, and (3) is able to tolerate a non-trivial degree of segmentation error when comparing similar motions.

3.1   An Elastic Similarity Measure

The used similarity measure firstly normalizes and transforms a motion sequence into a visual image representation [4], as illustrated in Fig. 1. This image is then processed by a neural network [7] to extract a 4,096-dimensional feature vector. The whole process is described in the following four steps.

1. **Normalization.** Spatio-temporal motion data are normalized in order to suppress the actor's absolute location in space, facing direction and sizes of limbs. This helps to unify motions that are perceived as the same but represented in different styles, e.g., "adult walking from left to right" and "infant walking from right to left" become the same "walking".
2. **Quantization.** Normalized 3D joint coordinates are then quantized into a discrete space of 256^3 bins. The quantization is very practical for the next visualization step, while introducing only millimeter errors in the joint coordinates with respect to the non-quantized data.

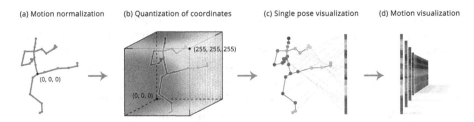

(a) Motion normalization   (b) Quantization of coordinates   (c) Single pose visualization   (d) Motion visualization

Fig. 1. Each pose is (a) normalized and (b) quantized into 256^3 space. The quantized positions of joints define a stripe of colors for each pose (c). The concatenation of these stripes forms a compact motion visualization (d). (Color figure online)

3. **Visualization.** Every joint position in the quantized space constitutes a triplet that assigns a given color in the RGB color space. All joints visualized one-by-one on the vertical axis of the resulting image change colors as their quantized positions change in time on the horizontal axis (see Fig. 1).
4. **Feature extraction.** Deep convolutional neural network [7] is used to discover inherent visual patterns in the diversely colorful motion images. The output of the last hidden layer of the network is a $4,096$-dimensional feature vector that describes a motion sequence with a high descriptive power.

The extracted feature vectors are then compared by the Euclidean distance to determine similarity of given motion sequences.

3.2 Properties of the Elastic Similarity Measure

The following paragraphs highlight important properties of the similarity measure and demonstrate its suitability for segment-based subsequence retrieval.

- **High descriptive power.** Effectiveness of the feature vectors can be even increased by fine-tuning the neural network for the specific application purpose. The features are also able to cluster similar images of categories on which the network has never been explicitly trained.
- **High efficiency.** Motions of variable lengths are described by fixed-size feature vectors that are compared by the Euclidean distance, which can be efficiently indexed to speed-up the retrieval process.
- **Compression of original data.** 5-second motion of 120-Hz frequency occupies 223 kB (55,800 floats for 31 joints) while the $4,096$-D vector only 16 kB. The features compress original motion data longer than 44 frames.
- **Elasticity.** The measure is robust when comparing motions that slightly differ in beginning and/or ending parts but are otherwise similar. Figure 2 shows how much different beginnings and endings influence effectiveness of the measure, which still works well for motions that lose as much as 10 % or carry as much as 20 % extra content with regards to their original length.

The elasticity property has a very positive impact to subsequence retrieval where automatic segmentation of motion data sequences introduces displacement

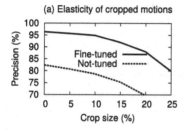 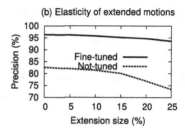

Fig. 2. Effectiveness of elasticity is measured by the precision using 1-nearest-neighbor search on a database of $2,345$ motions that are (a) cropped or (b) extended by 5, 10, 15, 20 and 25 % with respect to their original frame content. The features extracted using the fine-tuned neural network generally achieve a higher precision, but with similar trends as the features from the not-tuned network.

between the query and data segments. This property implies that the proposed measure can confidently match motions that are slightly different but overlapping in most of their content. For example, two subsequences of a similar length that are "slightly" shifted within a long motion sequence have a mutual distance close to 0. We quantify the maximum possible shift of consecutive overlapping segments as a covering factor cf and utilize it to segment data sequences.

4   Subsequence Retrieval by A Multi-level Segmentation

To search for query-relevant subsequences, the data sequence has to be partitioned into segments. Traditional methods [5] suggest partitioning the data sequence into disjoint (non-overlapping) segments, while the query sequence into overlapping segments using the sliding window principle (or vise-versa). Such partitioning facilitates locating relevant data segments that are similar to some query segments. Although the data segments can be indexed and efficiently retrieved, this concept has the following disadvantages:

- Longer queries are partitioned into a larger number of query segments. For each query segment an independent search (i.e., sub-query) has to be executed to retrieve the most similar data segments;
- The retrieved data segments of all sub-queries have to be intelligently merged respecting chronological order of query and retrieved segments to construct a set of relevant subsequences as the query result.

To overcome these problems, we propose to consider the query as a *single* segment. It means that only a single search is executed without the need of any other merging procedure. However, this would require sliding data segments for every potential query size, which results in a huge number of data segments. Such number can be dramatically reduced when the used similarity function can deal with a certain versatility in segmentation – sliding data segments can be then shifted much more than of a single frame only and can be constructed just for the specific sizes of queries.

Table 1. Table of symbols.

Symbol	Description
m	Data sequence
$\lvert m \rvert$	Length of data sequence in number of frames
$m[i:j]$	Subsequence of data sequence m starting at the i-th frame (inclusive) and ending at the j-th frame (exclusive), i.e., $\lvert m[i:j] \rvert = j - i$
m^Q	Query sequence
n	Number of segmentation levels
n^r	Number of segments within the r-th segmentation level
l^r	Length of segments at the r-th segmentation level
s_j^r	Starting frame of the j-th segment at the r-th segmentation level
rf	Replication factor
cf	Covering factor – used-defined parameter
l^{min}, l^{max}	Minimum/maximum query length – used-defined parameters

4.1 Problem Formalization

We partition the data sequence into segments in a way that an arbitrary data subsequence (bounded in length) overlaps with at least one segment in the majority of frames. Consequently, having a query as a single segment, each query-relevant data subsequence highly overlaps with at least one data segment. The high overlap ensures that relevant subsequences are always findable just by searching for similar segments. To quantify the high overlap between the specific subsequence and segment, we define *covering factor* $cf \in [0,1)$ which determines the maximum ratio between the number of their non-overlapping frames and the segment length. In other words, the covering factor denotes how much the similarity function is tolerant towards cropped/added content of two similar motions. The following definition defines the covering factor formally.

Definition 1. Given data sequence m and covering factor $cf \in [0,1)$: We say that any subsequence $m[i':j']$ is cf-covered by segment $m[i:j]$ if and only if $\frac{\lvert i'-i \rvert + \lvert j'-j \rvert}{j-i} \le cf$.

Our objective is to partition the data sequence into segments having optimal sizes and minimum possible overlaps with respect to the covering factor. To ensure that an arbitrary subsequence is covered by at least one segment, we need to restrict the subsequence length by the minimum $l^{min} \in \mathbb{N}$ and maximum $l^{max} \in \mathbb{N}$ value (the maximum length is supposed to be much shorter than the length of the data sequence). According to these limits (Table 1), we partition the data sequence according to the following objective.

Objective 1. *Given data sequence m and minimum l^{min} and maximum l^{max} subsequence length: Partition sequence m into a minimum number of segments* *so that an arbitrary subsequence $m[i : j]$ (bounded in length $l^{min} \leq j - i \leq l^{max}$) is cf-covered by at least one segment.*

4.2   Multi-level Segmentation

A query sequence m^Q is also restricted to a limited length in $[l^{min}, l^{max}]$ and always considered as a single segment. To partition the data sequence m according to Objective 1, we need to cover all potential query-relevant subsequences. Since positions (beginning and ending frames) of relevant hits are not known in advance, all possible data subsequences of the restricted length have to be covered.

Our idea is to define segmentation *levels* responsible for groups of queries in certain length intervals. Each level has its own size of segments that overlap by a fixed-size margin to cf-cover an arbitrary data subsequence. We naturally require to minimize the number of such levels as well as the size of overlaps between segments with respect to the predefined covering factor cf. These observations imply the following important lemma.

Lemma 1. *A single segmentation level with segments of fixed-size l can cf-cover the subsequences having their lengths in range $[l \cdot (1 - cf), l \cdot (1 + cf)]$.*

Proof. According to Lemma 1, a single level with segments of fixed-size l can cover only the subsequences which are maximally $l \cdot (1 + cf)$ long. Suppose that this statement is not true, then some segment $m[i, j]$ can also cover subsequence $m[i' : j']$ which is longer than $l \cdot (1 + cf) = (j - i) \cdot (1 + cf)$ frames:

$$j' - i' > (j - i) \cdot (1 + cf)$$
$$j' - i' > j - i + j \cdot cf - i \cdot cf$$
$$j' - i' > j - i + cf \cdot (j - i)$$
$$j' - i' - j + i > cf \cdot (j - i)$$
$$\frac{j' - i' - j + i}{j - i} > cf$$
$$\frac{|i - i'| + |j - j'|}{j - i} > cf,$$

which is in contradiction with Definition 1. Similarly, a segment of l frames can cover the subsequences which have minimally $l \cdot (1 - cf)$ frames. Due to the analogy with the previous case, the proof is omitted.   □

Lengths of Segments. To minimize the total number of segmentation levels (i.e., also the total number of segments), the individual levels have to cover subsequences of the possibly largest length interval. At the same time, the first level with segments of length l^1 needs to cover the shortest possible subsequences of length l^{min}:

$$l^{min} = l^1 \cdot (1 - cf) \quad \Leftrightarrow \quad l^1 = \frac{l^{min}}{1 - cf}$$

Consequently, this level also covers the subsequences which are up to $l^1 \cdot (1+cf)$ frames long (based on Lemma 1). The second level then covers the subsequences of at least $l^1 \cdot (1+cf)$ frames, so $l^2 = l^1 \cdot (1+cf)/(1-cf)$. Similarly as the first level, the second one covers maximally the subsequences of $l^2 \cdot (1+cf)$ frames. This continues until the n-th segmentation level covers the longest possible subsequences of l^{max} frames:

$$l^{n-1} \cdot (1+cf) < l^{max} \le l^n \cdot (1+cf) \quad \Leftrightarrow \quad l^n \ge \frac{l^{max}}{1+cf}.$$

Respecting these properties, the segment length is determined by constructing the individual levels. The fixed length l^r of segments at the r-th level ($r \in [1, n]$) can be recursively defined as:

$$l^1 = \frac{l^{min}}{1-cf} \qquad\qquad l^r = l^{r-1} \cdot \frac{1+cf}{1-cf}. \qquad (1)$$

Number of Segmentation Levels. The number n of segmentation levels can be calculated as $n = \lceil x \rceil + 1$, where x denotes the power parameter needed to skip to another level and is computed as:

$$\frac{l^{min}}{1-cf} \cdot \left(\frac{1+cf}{1-cf}\right)^x = \frac{l^{max}}{1+cf}$$

$$\left(\frac{1+cf}{1-cf}\right)^x = \frac{l^{max} \cdot (1-cf)}{l^{min} \cdot (1+cf)}$$

$$x \cdot \log\left(\frac{1+cf}{1-cf}\right) = \log \frac{l^{max} \cdot (1-cf)}{l^{min} \cdot (1+cf)}$$

$$x = \log_{\frac{1+cf}{1-cf}}\left(\frac{l^{max} \cdot (1-cf)}{l^{min} \cdot (1+cf)}\right)$$

$$\Rightarrow n = \left\lceil \log_{\frac{1+cf}{1-cf}}\left(\frac{l^{max} \cdot (1-cf)}{l^{min} \cdot (1+cf)}\right)\right\rceil + 1. \qquad (2)$$

Overlaps of Segments. The size of overlap among segments in each level is selected to minimize the number of segments while they all together cf-cover all possible subsequences bounded in length $[l^{min}, l^{max}]$. The lowest possible overlap we can afford corresponds exactly to the $100 \cdot (1-cf)\,\%$ frames with respect to the segment length. The initial position s_j^i of the j-th segment at the r-th level is then recursively defined as:

$$s_1^r = 1 \qquad\qquad s_j^r = s_{j-1}^r + l^r \cdot cf. \qquad (3)$$

Lemma 2. *The segments at the specific r-th level have to be maximally shifted by $l^r \cdot cf$ frames to cf-cover any subsequence of length in $[l^r \cdot (1-cf), l^r \cdot (1+cf)]$.*

Proof. Given the specific r-th segmentation level with segments of fixed length l^r and an arbitrary subsequence $m[i, j]$ belonging to this level, then:

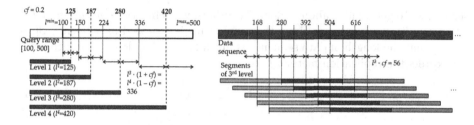

Fig. 3. Graphical illustration of segmentation: Based on covering factor $cf = 0.2$ and query length limits $l^{min} = 100$ and $l^{max} = 500$, the four segmentation levels are computed (left) and used to partition the data sequence (right), where only third-level segments are visualized. E.g., the third-level segments can 0.2-cover any data subsequence of length in $[224, 336]$.

1. We show that the shift between segments about $l^r \cdot cf$ frames is of a maximum possible size. Assume that the shift is higher, i.e., $l^r \cdot cf + 1$ frames, and there exist two consecutive segments $m'[i : i + l^r]$ and $m''[i + l^r \cdot cf + 1 : i + l^r \cdot cf + 1 + l^r]$ between which subsequence $m[i + \frac{l^r \cdot cf + 1}{2}, i + \frac{l^r \cdot cf + 1}{2} + l^r]$ of the same length l^r is located. Then this subsequence has the same number of non-overlapping frames with both the segments m' and m''. Considering m' and according to Definition 1, then:

$$\frac{\left| i + \frac{l^r \cdot cf + 1}{2} - i \right| + \left| i + \frac{l^r \cdot cf + 1}{2} + l^r - (i + l^r) \right|}{l^r} \leq cf \Rightarrow$$

$$\frac{\left| \frac{l^r \cdot cf + 1}{2} \right| + \left| \frac{l^r \cdot cf + 1}{2} \right|}{l^r} \leq cf \quad \Rightarrow \quad \frac{l^r \cdot cf + 1}{l^r} \leq cf \quad \Rightarrow \quad cf + \frac{1}{l^r} \leq cf,$$

   which is not valid for any (positive) length l^r of segment. □
2. It can be shown that an arbitrary subsequence of length in range $[l^r \cdot (1 - cf), l^r \cdot (1 + cf)]$ is covered by at least one segment. This can be proven via an induction step but the proof is omitted due to space limitations.

□

   Figure 3 illustrates the multi-level segmentation structure along with segments generated at the third level.

Number of Segments. When we partition the data sequence m of $|m|$ frames, the number n^r of segments at the r-th level is determined as:

$$n^r = 1 + \left\lfloor \frac{|m| - l^r}{l^r \cdot cf} \right\rfloor. \tag{4}$$

Replication Factor. The covering factor cf has the most important influence on the number of segmentation levels as well as the total number of generated segments. To get an idea about "global" overlaps, we define the *replication factor* rf that indicates how many times the same frame is repeated in segments. Supposing that the length of data sequence is much longer than lengths of segments, we express the replication factor rf as:

$$rf \approx \frac{n}{cf}, \tag{5}$$

where n stands for the number of segmentation levels. For example, having the covering factor $cf = 0.2$ and four segmentation levels, each frame of the original data sequence is involved twenty times in the specific segments.

4.3   Index Construction

The data sequence is preprocessed to be partitioned into the multi-level segmentation structure. Having specified covering factor cf and minimum l^{min} and maximum l^{max} query length, the number n of segmentation levels is calculated according to Eq. 2. In each r-th level ($r \in [1, n]$), the data sequence is partitioned into n^r segments of a fixed length of l^r frames by applying Eqs. 1 and 3. As the data sequence is partitioned, each segment is independently processed to extract the 4,096-dimensional feature vector using the deep convolutional neural network – see Sect. 3.1 for more detailed information.

The feature vectors within each level can be also independently indexed to speedup the retrieval process. As the feature vectors are compared by the Euclidean distance, any metric-based index structure can be utilized. We confront the naive sequential scan with the usage of indexing structure in the experimental evaluation in Sect. 5. In addition, the whole multi-level structure is dynamic because it simply enables adding the feature vectors of segments of new data sequences into each level.

4.4   Retrieval Algorithm

The objective of the retrieval phase is to search the data sequence and locate such subsequences that are the most similar to the query sequence, which is bounded in length $[l^{min}, l^{max}]$. Since each segmentation level is responsible for a certain interval of queries, only a single level is always searched for the most similar segments. Then the result contains segments of the same length that differs maximally about $100 \cdot cf\%$ with respect to the query length. Although the result segments need not be perfectly aligned with relevant subsequences, they should overlap in the majority of frames.

To search for similar segments, we evaluate a k-nearest-neighbor (k-NN) query. The query is specified by the number k of the most similar segments to be returned and sequence m^Q as the query object. The query sequence is firstly preprocessed to extract its 4,096-dimensional feature vector which is then compared to the feature vectors within the responsible segmentation level. Index

i of the responsible segmentation level for the query of $|m^Q|$ frames is determined by the following formula:

$$i = \begin{cases} 1 & |m^Q| \leq l^{min} \cdot \frac{1+cf}{1-cf} \\ 1 + \left\lceil \log_{\frac{1+cf}{1-cf}} \left(\frac{|m^Q| \cdot [1-cf]}{l^{min} \cdot [1+cf]} \right) \right\rceil & \text{otherwise.} \end{cases}$$ (6)

Considering the example in Fig. 3, the second segmentation level (i.e., $i = 2$) is responsible for a query ranging from 150 to 224 frames.

The advantage of retrieval is its high efficiency because of (1) comparing the fixed-size feature vectors, (2) evaluating only a single query segment, (3) accessing only a single segmentation level, and (4) presenting the results of the k-NN query directly without the need of any further post-processing. From the effectiveness point of view, the great advantage is the possibility to retrieve segments which are performed slowly/quickly or are not perfectly aligned with relevant subsequences up to $100 \cdot cf \%$ with respect to the query length. The actual effectiveness depends on the quality of the similarity measure and its tolerance to imprecise segmentation and accelerated/slowed motions.

5   Experimental Evaluation

The effectiveness and efficiency of the proposed subsequence retrieval algorithm is evaluated on the largest annotated motion capture dataset HDM05 [10]. This dataset contains 324 sequences performed by 5 different actors (with sampling frequency of 120 Hz). Similarly as in [11,15], we use a subset of 102 motion sequences (68 min in total) for which a ground truth is provided. This ground truth describes 1, 464 actions (subsequences within the 102 sequences) by 15 non-uniformly populated motion categories. The shortest action takes only 0.34 s (41 frames) while the longest one has 17.2 s (2, 063 frames).

5.1   Methodology

We concatenate 102 sequences into a single 68-minute data sequence and set the minimum $l^{min} = 41$ and maximum $l^{max} = 2,063$ query length according to the shortest and longest ground-truth actions. We evaluate search effectiveness and efficiency in 5 settings of the covering factor $cf \in \{0.1, 0.2, 0.3, 0.4, 0.5\}$. These settings reflect the ability of the similarity measure to deal with variously-size crops or extensions of motion content (see results in Fig. 2). All the settings follow the common methodology:

- For each cf setting, the multi-level segmentation structure is built and 4096-dimensional feature vectors are extracted using by a fine-tuned convolutional neural network model. The model is trained on a completely different HDM05 subset of 130 motion categories, which is also provided with the dataset;

Table 2. Effectiveness and efficiency evaluation of the preprocessing and retrieval phase of the 68-minute motion sequence for different settings of covering factor cf.

cf	# of levels	# of segments		rf	Feature ext. [min]	Sequential search [ms]	Precision	
		Total	1st level				$k = 1$	$k = 5$
0.1	18	631,746	111,774	180.0	263.2	447	87.30	84.37
0.2	9	150,971	51,230	45.0	62.9	205	86.75	84.13
0.3	6	66,972	31,526	20.0	27.9	126	86.89	82.98
0.4	5	37,345	21,955	12.5	15.6	88	85.79	82.65
0.5	4	23,669	16,393	8.0	9.9	66	84.43	81.99

- A k-NN query is constructed for each of $1,464$ ground-truth subsequences, that are used as query objects. Each query is then evaluated against the multi-level structure to obtain the k most relevant segments, excluding the segments that overlap with the query-object subsequence. We also exclude less-relevant segments that overlap with more relevant ones to finally obtain the k non-overlapping segments as the query result.

5.2   Analysis of Effectiveness

The query effectiveness is measured by *precision* as a ratio between positively retrieved segments and all retrieved segments (i.e., the number k). The segment is marked as positive if it overlaps with some ground-truth subsequence that is labeled with the same category as the query object. The global precision is then averaged over all $1,464$ queries. As the sparsest category in the dataset has only 6 motion instances, we use $k = 5$ to evaluate the search effectiveness. The precision @5 is above 84% as seen in the last column of Table 2 and decreases with an increasing covering factor cf. The "finer-grained" segmentation slightly improves the search accuracy, however, it brings a non-linear increase in the number of generated segments, i.e., efficiency is much worse.

To compare how much error is introduced by the proposed subsequence retrieval algorithm, we classify the same $1,464$ actions by comparing them each other using the 1-nearest neighbor search with the 93.9% accuracy. Based on the results in Table 2, we can reach a very high precision of 87.3% which is only about 6.6% worse. It is important to realize that our search space is much bigger, e.g., $111,774$ segments are generated for $cf = 0.1$, while the annotations constitute only $1,464$ objects.

5.3   Analysis of Efficiency

Efficiency of the preprocessing and retrieval phase depends on the total number of generated data segments, which is primarily influenced by the setting of cf.

Preprocessing phase. The bottleneck of the preprocessing phase is the extraction of the $4,096$-dimensional feature vector for each segment, which takes $25\,\mathrm{ms}$ using the GPU implementation – see Table 2 for total extraction times. However, we can still process the whole 68-minute sequence in real time using $cf = 0.2$ or higher. During the extraction, we can also simultaneously index the feature vectors within each segmentation level. For example, by employing the PPP-Codes [13] structure with $1,000$ pivots, we need only about $10\,\mathrm{min}$ to index all $150,971$ features (for $cf = 0.2$) using a single CPU (i7 960 at $3.2\,\mathrm{GHz}$).

Retrieval phase. As the retrieval algorithm accesses only a single segmentation level, the most-populated first level is considered for evaluating the upper bound on search performance. Without any indexing structure, the first-level segments can be stored in main memory and sequentially accessed by a single CPU that is able to perform approximately $250,000$ distance computations per second. The actual search times presented in Table 2 range from 66 to $447\,\mathrm{ms}$ per query based on the setting of the covering factor, while the approach in [2] needs about $36\,\mathrm{s}$ to search the 100k motion database. Importantly, our search times can be decreased by two orders of magnitude by applying the PPP-Codes [13] structure that has already proved to retrieve a similar kind of $4,096$-dimensional image features in a collection of 50 million images up to $1\,\mathrm{second}$ [12] (this time is measured by approximate search with the $90\,\%$ recall when the feature vectors are stored on SSD disk). In this way, we would possibly search online in a sequence of 121-day long with the same setting as in Fig. 3.

The proposed algorithm employing PPP-Codes indexing has search complexity $\log(|m|/(|l^Q| \cdot cf))$, where $|m|$ and $|l^Q|$ denote lengths of the data and average query sequence. This is much more efficient compared to existing approaches whose complexity increases much with the query length, such as $|m| \cdot |l^Q|$ in [15].

6   Conclusions

We propose a new subsequence matching algorithm which uses a synergy of elastic similarity measure and multi-level segmentation. The search space comprises overlapping segments of various sizes that ensure the bounded coverage of arbitrary parts within very long motion sequences. The size of overlaps and the total number of generated segments are bounded to be formally minimal with respect to the covering factor parameter, which reflects the versatility and effectiveness of the used similarity measure. Due to the efficient comparison of $4,096$-dimensional segment features by the Euclidean distance, the retrieval process is also very efficient, e.g., sequential search within the 68-minute motion sequence takes only $126\,\mathrm{ms}$ and has the accuracy of $87\,\%$. The advantage is that the segmentation levels can be processed independently and additionally indexed to speedup retrieval by two orders of magnitude. The segments of new data sequences can be also dynamically added.

Acknowledgements. This research was supported by GBP103/12/G084.

References

1. Barbič, J., Safonova, A., Pan, J.Y., Faloutsos, C., Hodgins, J.K., Pollard, N.S.: Segmenting motion capture data into distinct behaviors. In: Graphics Interface, pp. 185–194. Canadian Human-Computer Communications Society (2004)
2. Beecks, C., Hassani, M., Obeloer, F., Seidl, T.: Efficient query processing in 3D motion capture databases via lower bound approximation of the gesture matching distance. In: 2015 IEEE International Symposium on Multimedia (ISM 2015), pp. 148–153 (2015)
3. Bouchard, D., Badler, N.I.: Semantic segmentation of motion capture using Laban movement analysis. In: Pelachaud, C., Martin, J.-C., André, E., Chollet, G., Karpouzis, K., Pelé, D. (eds.) IVA 2007. LNCS (LNAI), vol. 4722, pp. 37–44. Springer, Heidelberg (2007)
4. Elias, P., Sedmidubsky, J., Zezula, P.: Motion images: an effective representation of motion capture data for similarity search. In: Amato, G., et al. (eds.) SISAP 2015. LNCS, vol. 9371, pp. 250–255. Springer, Heidelberg (2015). doi:10.1007/978-3-319-25087-8_24
5. Faloutsos, C., Ranganathan, M., Manolopoulos, Y.: Fast subsequence matching in time-series databases. SIGMOD Rec. **23**(2), 419–429 (1994)
6. Kapadia, M., Chiang, I.K., Thomas, T., Badler, N.I., Kider Jr., J.T.: Efficient motion retrieval in large motion databases. In: ACM SIGGRAPH Symposium on Interactive 3D Graphics and Games (I3D 2013), pp. 19–28. ACM (2013)
7. Krizhevsky, A., Sutskever, I., Hinton, G.E.: Imagenet classification with deep convolutional neural networks. In: Advances in Neural Information Processing Systems, vol. 25, pp. 1097–1105. Curran Associates, Inc. (2012)
8. Krüger, B., Tautges, J., Weber, A., Zinke, A.: Fast local and global similarity searches in large motion capture databases. In: ACM Symposium on Computer Animation, SCA 2010, pp. 1–10. Eurographics Association (2010)
9. Lan, R., Sun, H.: Automated human motion segmentation via motion regularities. Vis. Comput. **31**(1), 35–53 (2015)
10. Müller, M., Röder, T., Clausen, M., Eberhardt, B., Krüger, B., Weber, A.: Documentation Mocap database HDM05. Technical report CG-2007-2. Universität Bonn (2007)
11. Müller, M., Baak, A., Seidel, H.P.: Efficient and Robust annotation of motion capture data. In: ACM Symposium on Computer Animation (SCA 2009), p. 10. ACM Press (2009)
12. Novak, D., Cech, J., Zezula, P.: Efficient image search with neural net features. In: Amato, G., et al. (eds.) SISAP 2015. LNCS, vol. 9371, pp. 237–243. Springer, Heidelberg (2015). doi:10.1007/978-3-319-25087-8_22
13. Novak, D., Zezula, P.: Rank aggregation of candidate sets for efficient similarity search. In: Decker, H., Lhotská, L., Link, S., Spies, M., Wagner, R.R. (eds.) DEXA 2014, Part II. LNCS, vol. 8645, pp. 42–58. Springer, Heidelberg (2014)
14. Ren, C., Lei, X., Zhang, G.: Motion data retrieval from very large motion databases. In: International Conference on Virtual Reality and Visualization (ICVRV 2011), pp. 70–77 (2011)
15. Sedmidubsky, J., Valcik, J., Zezula, P.: A key-pose similarity algorithm for motion data retrieval. In: Blanc-Talon, J., Kasinski, A., Philips, W., Popescu, D., Scheunders, P. (eds.) ACIVS 2013. LNCS, vol. 8192, pp. 669–681. Springer, Heidelberg (2013)

16. Valcik, J., Sedmidubsky, J., Zezula, P.: Assessing similarity models for human-motion retrieval applications. Computer Animation and Virtual Worlds (2015)
17. Vögele, A., Krüger, B., Klein, R.: Efficient unsupervised temporal segmentation of human motion. In: ACM Symposium on Computer Animation (2014)
18. Wang, Y., Neff, M.: Deep signatures for indexing and retrieval in large motion databases. In: 8th ACM Conference on Motion in Games, pp. 37–45. ACM (2015)
19. Wu, S., Wang, Z., Xia, S.: Indexing and retrieval of human motion data by a hierarchical tree. In: 16th ACM Symposium on Virtual Reality Software and Technology (VRST 2009), pp. 207–214. ACM, New York (2009)

Music Outlier Detection Using Multiple Sequence Alignment and Independent Ensembles

Dimitrios Bountouridis[(✉)], Hendrik Vincent Koops, Frans Wiering, and Remco C. Veltkamp

Department of Information and Computing Sciences,
Utrecht University, Utrecht, Netherlands
d.bountouridis@uu.nl

Abstract. The automated retrieval of related music documents, such as cover songs or folk melodies belonging to the same tune, has been an important task in the field of Music Information Retrieval (MIR). Yet outlier detection, the process of identifying those documents that deviate significantly from the norm, has remained a rather unexplored topic. Pairwise comparison of music sequences (e.g. chord transcriptions, melodies), from which outlier detection can potentially emerge, has been always in the center of MIR research but the connection has remained uninvestigated. In this paper we firstly argue that for the analysis of musical collections of sequential data, outlier detection can benefit immensely from the advantages of Multiple Sequence Alignment (MSA). We show that certain MSA-based similarity methods can better separate inliers and outliers than the typical similarity based on pairwise comparisons. Secondly, aiming towards an unsupervised outlier detection method that is data-driven and robust enough to be generalizable across different music datasets, we show that ensemble approaches using an entropy-based diversity measure can outperform supervised alternatives.

1   Introduction

The World Wide Web (WWW) has revolutionized music, from its creation, production, distribution to the way people currently listen to it. Due to its open nature, WWW has put users in the center of content generation. Popular websites such as *Chordify*[1], *UltimateGuitar*[2], *WhoSampled*[3] or *Midomi*[4], allow users to submit their own chord transcriptions, tabs, discovered covers or sung interpretations of a song. However, this pleasant development does not come without shortcomings; it stands to reason that user content is not always trustworthy due to human error, malicious editing and so on.

---

[1] www.chordify.net.
[2] www.ultimate-guitar.com.
[3] www.whosampled.com.
[4] www.midomi.com.

© Springer International Publishing AG 2016
L. Amsaleg et al. (Eds.): SISAP 2016, LNCS 9939, pp. 286–300, 2016.
DOI: 10.1007/978-3-319-46759-7_22

Identifying and filtering out untrusted documents is of major importance for these content providing services, therefore techniques such as user ratings (e.g. star rating, social media "shares") have been frequently employed. However, ratings not only require large amount of users but, at least for chord transcriptions, have been shown to be uncorrelated to the quality of the document [18].

Outlier detection, the general task of locating those observations that *"... deviate so much from the other observations as to arouse suspicions that they were generated by a different mechanism"* [14], has found major applications in bio-informatics, fraud detection, medical diagnosis and other fields. It is therefore surprising that it has remained rather unexplored in the field of Music Information Retrieval (MIR). This can be attributed to the nature of outlier detection algorithms which by definition have two components [6]: the scores indicating level of "outliernes" of each sample, and their conversion to a binary decision by imposing thresholds based on their statistical distribution. "Outliernes" in MIR however, has been considered a welcome byproduct of music similarity. A strong music similarity model would assign low similarity scores to any documents that "do not belong." Therefore, outlier detection would be practically rendered obsolete.

In this paper we argue that, since music similarity is inherently ambiguous, it *(a)* has remained largely an unsolved problem *(b)* has been typically based on music heuristics or learning, thus becoming very task- and domain-specific and *(c)* relies on pairwise comparisons which can be time-consuming and unrealistic for large collections. Even if a strong theoretical model of similarity had been established, outlier detection in practice would be still non-trivial due to the following: first, the number of samples is usually small, and therefore an underlying "normal" model cannot be assumed or learned. Secondly, music is inherently pattern-based: therefore songs that do not belong to the reference set might share commonalities (e.g. similar chord progressions) and thus might not be deviating significantly. Consequently, the boundary between "normal" and anomalies becomes fuzzy.

Contribution: This paper's contribution is twofold. We firstly exploit the sequential nature of certain kinds of musical content, such as melodies and chord progressions, and the advantages of Multiple Sequence Alignment (MSA) in terms of sequence analysis. A sequential, music-agnostic representation of music allows for the development of tools that generalize across collections. The extensive work on MSA, mostly in the field of computational biology, allows for the adoption of tools that can separate better outlier sequences from the rest thus limiting the undesirable shortcomings of pairwise comparisons. Secondly, we present an almost settings-free outlier detection method that can find robust application to any form of music sequences. We use ensembles of different outlier solutions to form more informative decisions that avoid dependencies on specific artifacts related to a particular similarity method or data set.

Summary: The rest of the paper is organised as follows. Section 2 is a brief overview of related outlier detection approaches in MIR. Sections 3 and 4 describe the music datasets and the basic outlier detection method considered in our work. Sections 5 and 6 break down the outlier detection into two components (similarity methods and extreme value analysis) and further analyse them. Section 7 introduces the independent ensemble approach for outlier detection, which is evaluated in Sect. 8.

2   Related Work in MIR

There are only a few published approaches that explicitly aim to tackle the task of outlier detection in music. Panteli *et al.* [20] specifically focus on world music and use data mining techniques on audio-features (e.g. rhythmic, melodic, harmonic) to detect outliers. Lukashevich and Dittmar [17], aiming towards improving a GMM mood classifier, use a Support Vector Machine (SVM) classifier as a preliminary stage to filter out outlying samples. Livshin and Rodet [16] focus on automatic removal of bad samples from an instrument music library. Their algorithms are supervised variations of the Interquartile Range. Hansen *et al.* [13] use a combination of supervised and unsupervised learning to clean-up large-scale databases that included metadata (e.g. genre information). They model the relation between metadata and audio features by training conditional densities. Unconditional densities are modeled for spotting unlikely music features. Tangential to the outlier detection problem, with common methodologies [19], is novelty detection, the automatic differentiation between known and unknown object information during testing. Most notably, Flexer *et al.* compared different rejection rules and novelty detection methods in a genre classification context [7,8]. In general, all the published approaches are either supervised and domain specific or dependent on abundant samples.

3   Music Datasets

Before going into detail about the outlier detection problem, we should describe the different music datasets investigated and how they were represented as sequences in our work (available online[5]). We use four datasets of varying size and nature, ranging from expert-annotated melodies to non-expert chord transcriptions found online. This allows us to generalize any observations, derived from this work, to almost any music dataset of sequential nature. The data sets are further explained below, while summary statistics are presented in Table 1.

The Annotated Corpus of the Meertens Tune Collections [32] is a set of 360 Dutch folk songs grouped into 26 "tune families." Each contains a group of melody variations related through an oral transmission process. For this TuneFam-26 data set, expert annotators assessed the perceived similarity of every melody over a set of dimensions (contour, rhythm, lyrics, etc.) to a set of 26 prototype "reference melodies." The Cover Song Variation data set [2], or Csv-60, on the other hand is a set of expert-annotated, symbolically-represented

---

[5] www.projects.science.uu.nl/COGITCH/outlier.

Table 1. Summary statistics for the four datasets of our experiments.

	TuneFam-26	Csv-60	Shs-50	Beatles
Number of cliques	26	60	50	174
Number of sequences	360	243	467	948
Avg. cliques size	13.0 *(4.0)*	4.0 *(1.1)*	9.34 *(3.5)*	5.66 *(3.61)*
Avg. sequence length	43.0 *(14.9)*	53.0 *(21.4)*	130 *(78.58)*	74.17 *(58.20)*

vocal melodies derived from matching structural segments (such as verses and choruses) of different renditions of sixty pop and rock songs. Melodies in both datasets are represented as *pitch contours*, a series of relative pitch transitions constrained to the region between $+11$ and -11 semitones.

The Second Hand Song dataset[6] contains metadata for around 18,000 cover songs grouped into 6,000 cliques. In order to keep computations to a practicable level, we randomly picked 50 cliques (of more than 6 songs per cliques). We denote this subset Shs-50. Songs are represented as sequences of major-minor chords by finding the shift that maximises the correlation between the Krumhansl-Kessler profiles [26] and the chroma vector of each beat-frame (as extracted by the Echonest[7] API).

From *UltimateGuitar* we web-mined 948 user chord transcriptions corresponding to the 174 songs of the complete Beatles discography as provided by the famous Beatles dataset[8]. All transcriptions in the Beatles dataset are key-normalised and reduced to major-minor chord sequences.

It should be pointed out that each group of related music sequences in the datasets is not guaranteed to contain any outliers, despite the fact that some sequences might be more dissimilar than others. In our work, the "outlier" is a randomly chosen sequence injected to the group that comes from the same dataset as the group, e.g. a chord transcription of "Let it be" injected in a group of "Yellow submarine" transcriptions by The Beatles.

4   Basic Outlier Detection

There are many outlier detection methods in the literature; however their applicability is dependent on the nature of the data, e.g. number of samples, number of outliers and so on. Therefore, picking the appropriate one, in the context of music documents, is not trivial. In this work we decided to focus on the interpretability criterion, which is of crucial importance for the analyst, since it can answer the question of why a sample is considered an outlier [6].

Extreme value analysis is the most basic and interpretable outlier detection method and has two components (see Fig. 1). It works on one-dimensional data

---

[6] www.secondhandsongs.com.

[7] www.echonest.com.

[8] www.isophonics.net/content/reference-annotations-beatles.

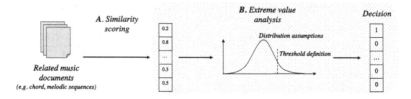

Fig. 1. The basic outlier detection pipeline using extreme value analysis.

(first component) and it assumes that outlier values are too small or too large with regard to the rest of the value distribution. In the music context, the input values correspond to the similarity $\in \mathbb{R}$ of each music document to the rest. Defining what constitutes an extreme value is typically performed by modelling the data distribution and its statistical tails (second component). The problem comes down to finding the best combination of the two components, which we analyse in the following sections. Each combination is an *outlier detection setting*.

5 Similarity Scoring Methods

We will now present the similarity scoring methods that we consider in our work (see Fig. 1, A). All are generic in the sense that they do not incorporate music heuristics. As a consequence they can be used for any form of sequences.

5.1 Pairwise Alignment

Sequence matching is typically performed using pairwise alignment (PW), the process of making two sequences have the same length by introducing gaps "-" while aligning related symbols. Sequence alignment via dynamic programming [31] is widely used for approximate string matching, and found early application in MIR. The quadratic-time Needleman and Wunsch [30] algorithm finds an optimal alignment of two sequences and returns a score that represents the cost, and can be interpreted as a quality measure. It should be noted that the score is highly dependent on the penalties for inserting and extending gaps. In our work, the PW score of a sequence in a group is the average of its pairwise scores.

In contrast to pairwise alignment, the following methods base their scoring on a multiple sequence alignment rather on multiple pairwise comparisons.

5.2 Multiple Sequence Alignment Based Methods

Before computing the similarity of each sequence in an MSA, the MSA itself needs to be computed. The optimal MSA of a group of sequences has exponential-time complexity, therefore it cannot be used in practice. Instead, the focus is on heuristic approaches that give good alignments not guaranteed to be optimal. Our work considers two algorithms: progressive alignment (PA) and MAFFT.

The most popular approach, PA, starts by building a pairwise similarity tree. Working from the leaves of the tree to the root, PA aligns the alignments, until reaching the root of the tree, where a single MSA is built. MAFFT [25] on the other hand, uses the fast Fourier transform to identify short subregions of one sequence or intermediate alignment that are high-scoring matches with same-length subregions from another sequence or alignment. Based on the MSA, our work considers the following similarity scoring methods:

Percentage Identity (PID). A popular similarity scoring between pairs of sequences is the Percentage Identity (PID). It corresponds to the number of identities, meaning the number of same characters divided by the number of characters compared (gap positions excluded). In our case, the PID of a sequence in an MSA is the average of the its pairwise PID scores.

Neighbor-Joining Tree (NJ-T). Neighbor-Joining Tree, is a clustering method for building phylogenetic trees of biological sequences [22]. NJ-T takes as input a distance matrix, typically based on PID, and at each stage the two nearest nodes of the tree are chosen and joined. The process is performed recursively until all of the nodes are paired and the tree is constructed. We ommit the details of the formulation of "nearest" nodes due to lack of space. We use the branch length from each leaf (sequence) to the root as a measure of similarity.

Majority-Vote Consensus (MJ-CONC). A multiple alignment can be summarized to generate a single sequence that we call *consensus*. For each column, the voting process determines if the frequency of the most common symbol is above a threshold. If so, that symbol represents that column in the consensus; if not, the column is represented by an ambiguous symbol. The similarity score of each sequence is the pairwise alignment score of itself with the consensus.

Data Fusion (DF). Data Fusion can be seen as an extension of majority voting in the sense that in addition to finding the most common symbol per column, it also uses the agreement between rows as a weight to favor values of rows with higher agreement [3]. The data fusion consensus happens in two steps: after computing the probabilities for each symbol for each column, a *source accuracy* is computed for each row by taking the mean of its column probabilities. The values of each row are then weighted by multiplying them with their source accuracy. The intuition is that rows with higher agreement with other rows will be more trustworthy. The process of computing symbol probabilities and source accuracy is repeated until the probabilities of the values converge. For each row, the value with the highest probability is taken as the output value. In our work we consider the source accuracy of each sequence as its similarity to the rest.

Profile Hidden Markov Models (HMM-P). Profile Hidden Markov Models (HMMs) are essentially generative probabilistic automata that represent a multiple sequence alignment as a probabilistic, position-dependent scoring system [4]. Profile HMMs contain five types of states whose details are omitted due to lack of space. A query sequence can be represented as a stochastic traversal of the profile's states. As a consequence, comparing an arbitrary sequence to a profile HMM is performed by using the Viterbi or the Forward algorithm [5] for Markov models. The similarity of each sequence is the output of its comparison to a profile HMM built from the remaining sequences.

Alignment Gap Metric (GAP-BASED). Jehl *et al.* [15] presented an MSA-based outlier detection pipeline (called OD-seq) that majorly focused on efficiency since it was aimed to be used on large number of biological sequences. As such it used a gap-based distance metric that considered gap-less pairwise alignments to be of high quality and vice versa. Variations that distinguished between fewer, longer gaps or more, shorter gaps were also presented. In our work, we consider only the linear metric that treats all gaps equally.

5.3   Inlier-Outlier Separation

Although the similarity scoring methods are merely components of the whole outlier detection pipeline, we are interested in answering to which extent each similarity can better separate between outlier and inlier music sequences. Optimally a scoring method should assign high similarity scores to inliers and low scores to outliers with as minimal overlap as possible.

For each similarity method (besides PW) we compute the two score distributions (outlier and inlier sequences) over each dataset and MSA method. PW is computed only for each set since it is independent of the MSA. However, since pairwise alignment is dependent on the gap penalties, we try three different gap open and gap extend settings (.5, .2), (.8, .4) and (1, 0) with an identity substitution matrix (matches get a score of 1, while mismatches a score of -1). Table 2 presents the area under the receiver operating characteristic (ROC) curve, typically called AUC, for each scoring over each dataset and MSA algorithm.

We can make the following observations: first, there is at least one similarity method for each dataset that has higher AUC score than pairwise alignment.

Table 2. Area Under the Curve (AUC) % for each similarity method over each dataset and MSA type.

	Csv-60		TuneFam-26		Shs-50		Beatles	
	MAFFT	PA	MAFFT	PA	MAFFT	PA	MAFFT	PA
PW-.5-.2	95.95	95.95	96.52	96.52	61.66	61.66	86.34	86.34
PW-.8-.4	96.44	96.44	96.66	96.66	62.64	62.64	86.86	86.86
PW-1.0-.0	97.16	97.16	96.73	96.73	61.35	61.35	91.18	91.18
PID	97.47	91.69	91.82	79.1	60.55	54.38	87.96	88.06
NJ-T	98.5	95.07	89.56	82.09	60.55	55.96	84.9	85.34
MJ-CONC	99.07	95.09	91.68	88.81	57.7	54.42	83.74	80.39
DF	**99.22**	95.65	92.12	83.14	63.79	57.52	87.97	88.25
HMM-P	94.88	91.11	**98.29**	96.84	**75.62**	66.7	**94.76**	93.25
GAP-BASED	91.65	89.46	92.99	90.55	65.55	58.19	83.5	88.17

Therefore, we have shown that one can use MSA as a more reliable basis for outlier detection. Secondly, profile HMMs show generally the highest separation between outliers and inliers over all datasets. Thirdly, it becomes obvious that the MAFFT-based MSA results to generally higher separation than progressive alignment. Finally, Csv-60 and Shs-50 seem to be the "easiest" and most difficult datasets respectively to perform outlier detection on.

6 Extreme Value Analysis Algorithms

In the previous section we presented a number of similarity scoring methods. In this section we briefly present five extreme value analysis algorithms that work on top of the similarity scores (see Fig. 1, B). The list is definitely incomplete, however it captures a wide range of algorithms.

Thresholding, classifying as outliers those documents with similarity smaller than a predefined threshold θ, is the simplest form of extreme-value outlier detection. The **Z-score** is one of the simplest ways to avoid using a fixed absolute threshold. It represents the amount of standard deviations σ a value is from the mean. However, one should decide on a threshold θ_z, above which a value would be considered anomalous. The **Grubb's test** [11] is used to detect single outliers on data following approximately a normal distribution. The Grubb's test statistic is the largest absolute deviation from the sample mean in units of the sample standard deviation. Grubb's test requires us to decide on the significance level a. Since the presence of outliers is likely to affect the mean and the standard deviation, Z-score and Grubb's test can become unreliable. The median, however, is typically more robust to outliers. The **Median Absolute Deviation (MAD)** score is the median of the absolute deviations from the data's median. A data point x_i is considered an outlier if $|x_i - median(X)|/MAD$ is larger than θ_{MAD}. Finally, the **percentile** is used in statistics to indicate the value below which a given percentage of observations fall. For large normally-distributed populations, percentiles represent the area under the normal curve, increasing from left to right. A percentile based outlier detection method requires us to set a percentile threshold θ_p above which a value is considered anomalous.

6.1 Evaluation

In order to evaluate each extreme value analysis algorithm, one should analyse their behaviour with respect to their corresponding threshold and each similarity scores they were applied on. However, we are rather interested in validating our initial hypothesis; similar to any classification problem, it comes to reason that different *outlier detection settings* (similarity scoring and outlier algorithm) would behave differently depending on the nature of the dataset.

For each of our music datasets, we brute-force find a single *outlier detection setting* that results to the best overall outlier prediction (measured with the $F1$ score). We then apply the same setting on the remaining datasets. The $F1$ scores

Table 3. The $F1$ score for each optimised setting applied to all datasets. The standard deviation (std) for each dataset is also presented. The optimised settings per dataset are: $CSV-60_{setting}$: NJ-T & z-score & $\theta_z = 1.56$, $TUNEFAM-26_{setting}$: HMM-P & z-score & $\theta_z = 1.8$, $SHS-50_{setting}$: HMM-P & thresholding & $\theta = .22$, $BEATLES_{setting}$: HMM-P & thresholding & $\theta = .13$.

	CSV-60	TUNEFAM-26	SHS-50	BEATLES
$CSV-60_{setting}$	**0.945**	0.676	0.588	0.725
$TUNEFAM-26_{setting}$	0.639	**0.889**	0.630	0.610
$SHS-50_{setting}$	0.665	0.820	**0.668**	0.683
$BEATLES_{setting}$	0.501	0.684	0.626	**0.769**
std	0.161	0.090	0.028	0.058

for each setting applied to each dataset are presented in Table 3. It becomes obvious that different optimised settings applied to different datasets result in major fluctuations in performance.

7   Independent Ensembles

In contrast to the previously presented methods, we are interested in an unsupervised outlier detection method that is parameter-free and generalizable so that it can be applied almost "out-of-the-box" in various music collections.

Independent ensembles [6] are based on different instantiations (parameter settings) of one or more outlier detection algorithms. The parameters even the algorithms themselves can be randomly selected and the output of their execution is combined to form the final decision. The principle behind independent ensembles is to achieve robustness by avoiding dependencies on specific artifacts related to a particular algorithm or data set. In addition, it is assumed more difficult to design a single sophisticated algorithm than to optimize the combination of algorithms with relatively lower complexity [9].

Independent ensembles, depending on the task and context, appear with different names in the literature ranging from "committees" and Multiple Classifier Systems (MCS) to clustering or classification ensembles. However, the fundamental problem is generally the same: given an unlabeled data set $D = \{x_1, x_2, ..., x_n\}$ and a set of classification solutions $\{C_1, C_2, ..., C_k\}$ that map the data to a class $f_j(x) = m$ we are interested in a single "resultant" solution f^* that combines the classification solutions. The "accuracy" of the ensemble is measured by the match between the solution produced and the reference ground-truth. The problem, in our outlier-detection-for-music context, can be considered a special, binary case with $m \in \{0, 1\}$ (where 0 corresponds to "inlier" and 1 to "outlier") and with the classification solutions being the output of different *outlier detection settings*.

7.1   Diversity

An ongoing issue with ensembles in general is how to select the set of classification solutions. A set of similar classification solutions is not guaranteed to lead to the best solution, therefore the concept of *diversity* has been introduced. Although diversity has been shown to be fundamental for an ensemble's success [9], a consistent relation between the ensemble's diversity and the solution's accuracy has not been shown. In addition, diversity can be formulated in various ways. Hadjitodorov *et al.* [12] have shown the potential of moderate diversity based on Adjusted Rand Index [21], but avoided generalizing their observations.

7.2   Diversity Experiment

In this section we describe an experiment to investigate the behaviour of different diversity measures on music datasets. We also answer whether Hadjitodorov's hypothesis, denoted as Soft-Correlation Rule by [9], holds for our particular task.

Each dataset is split into a *training* and a *test set* (70 %–30 % split). The training sets combined form the *grouped training set*. For each group of related music documents in the *grouped training set*, we compute the output-solution of every possible *outlier detection setting*. We shuffle them and randomly select 10 ensembles of 25 solutions each, similar to [12]. We finally compute the diversity and quality of each ensemble. There are three important factors to consider.

As Zimek [23] mentions, an important issue with outlier detection ensembles is how to combine the different individual solutions to derive a consensus or ensemble result. The problem is not trivial especially when the outlier detectors output solutions that are ranked lists or score vectors. In our case however, each solution is a binary vector we therefore employ the intuitive majority voting.

The second important issue is the question of how to measure an ensemble's quality. Given a ground truth it is typical to use measures such as Rand index and Adjusted Rand index [23]. Similarly one can compute the average F-measure score over all solutions as we do in our paper. Yet using ground truth information to assess quality is debatable [22], especially considering we are aiming for an unsupervised outlier detection method, however as [23] states "there is probably no better approach available to assess clustering quality w.r.t. external knowledge." The third issue is which diversity measures should be investigated. Our work considers all diversity measures described [12] which are divided into two general groups, pairwise and non-pairwise. The pairwise method D_P, is based on the Adjusted Rand Index and on pairwise comparisons between all the solutions in the ensemble. Four of the non-pairwise measures $D_{np-1}, D_{np-2}, D_{np-3}$ and D_{np-4} are based on the difference of each solution from the final ensemble decision. The measure proposed by [10], denoted by [12] as H, is based on the entropy of the "consensus" matrix, which stores the co-agreement between all ensembles solutions. In our work, we consider two additional diversity measures, denoted Df_{sa} and E, which are based on data fusion and entropy respectively. In Sect. 5.2 we presented data fusion, a byproduct of which is a *source accuracy* measure for every sequence in the set that we aim to "fuse." Considering that high source accuracy corresponds to all sequences being similar and vice versa, one can use their average complement as a measure of diversity. The E measure makes use of the binary nature of our task which allows us to convert each solution vector to a decimal representation, thus represent the ensemble as a sequence of decimal numbers. The entropy of the sequence can act as a measure of diversity.

Scatter plots of the different diversity measures plotted against the ensemble quality are presented in Fig. 2 (four omitted due to lack of space). The Pearson and Spearman's coefficients for the linearity and monotonicity tests respectively are presented in Table 4. It becomes obvious that the entropy-based diversity measure E is the most correlated to the ensemble quality. Df_{sa} follows (negative correlation), while the rest do not show any particular pattern of correlation (e.g. the closer to the median the better). Therefore, the Soft-Correlation rule cannot be applied to our particular case. Based on the results we can hypothesize that the best strategy for music outlier detection using independent ensembles is picking the ensemble with the highest diversity as measured by the measure E.

Table 4. The Pearson's and Spearman's coefficients for linearity and monotonicity tests.

	D_P	D_{np-1}	D_{np-2}	D_{np-3}	D_{np-4}	H	E	Df_{sa}
Pearson's	−0.12	−0.13	0.08	0.10	0.09	0.05	0.48	−0.23
Spearman's	−0.11	−0.11	0.02	0.06	0.09	0.03	0.51	−0.21

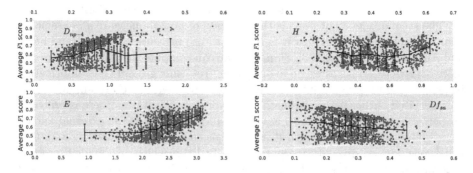

Fig. 2. Diversity (x-axis) versus quality (measured as average $F1$ score) scatter plots for four of the eight different diversity measures.

8 Independent Ensembles Evaluation

We are interested in evaluating our independent ensembles method with regard to its outlier detection generalization ability. We compare it against two supervised approaches that do not treat all datasets independently: (a) a system that applies the same, but optimised, outlier setting to all datasets, and (b) a system that applies a setting to each dataset i, optimised for a different dataset j.

The first supervised approach learns the *outlier detection setting* that maximizes the prediction ability (measured using the $F1$ score) on the *grouped training set*. This model, denoted *Full*, is therefore trained on all music datasets but aims at finding the setting that balances among them. The learned setting is applied on each music dataset in the *test set*. The second supervised method, called *Individual*, firstly learns the optimal *outlier detection settings* for each music dataset in the *training set*. Secondly, for each example in a music dataset belonging to *test set* it randomly applies a learned setting from a different set.

For each example in the *test set*, our ensemble approach that uses an entropy-based diversity (called *Ensemble-E*) starts by computing the output-solution of every possible *outlier detection setting*. We shuffle them and randomly select 50 ensembles of 25 solutions each. We finally select the ensemble with the highest diversity E and compute the final prediction.

The $F1$ scores of a 40-fold, cross-validation are presented in Table 5. We compare the methods' distributions using the Wilcoxon rank-sum test (the scores do not follow a normal distribution). The results show that the *Ensemble-E* method significantly outperforms the rest for all datasets not only as a whole (see "Grouped" on Table 5) but individually also. For Csv-60 the *Full* method shows better performance than *Ensemble-E* but not at a significant level.

Table 5. $F1$ scores for the three outlier detection approaches. $p_{i,e}$ and $p_{f,e}$ are the p-value for the statistical significance tests between *Ensemble-E* vs *Individual* and *Ensemble-E* vs *Full* respectively.

	Individual	Full	Ensemble-E	$p_{i,e}$	$p_{f,e}$
Csv-60	0.61 *(0.09)*	**0.92** *(0.04)*	0.91 *(0.02)*	$< 10^{-7}$	0.0526
TuneFam-26	0.72 *(0.07)*	0.79 *(0.05)*	**0.81** *(0.04)*	$< 10^{-7}$	0.0141
Shs-50	0.56 *(0.04)*	0.55 *(0.03)*	**0.58** *(0.04)*	0.0187	$< 10^{-5}$
Beatles	0.68 *(0.04)*	0.74 *(0.03)*	**0.77** *(0.02)*	$< 10^{-7}$	$< 10^{-7}$
Grouped	0.67 *(0.02)*	0.75 *(0.02)*	**0.77** *(0.02)*	$< 10^{-7}$	$< 10^{-7}$

Our experiments so far were based on the assumption that only one outlier can exist in a group of related music documents. In reality the outliers can be more (but definitely less than half the number of documents). Table 6 presents the results for the same experiment applied on the same train and test sets but now with two outliers per group. Our ensemble approach shows again superior performance for all sets individually (except for Csv-60) and as a whole.

9   Discussion and Conclusions

Working towards an unsupervised outlier detector for music sequences we presented an ensemble approach that outperformed supervised approaches. However,

Table 6. $F1$ scores for the three outlier detection approaches on the datasets with two outliers per group.

	Individual	Full	Ensemble-E	$p_{i,e}$	$p_{f,e}$
Csv-60	0.73 (0.05)	0.80 (0.03)	**0.81** (0.03)	$< 10^{-7}$	0.5908
TuneFam-26	0.68 (0.05)	0.81 (0.04)	**0.84** (0.04)	$< 10^{-7}$	$< 10^{-4}$
Shs-50	0.55 (0.03)	0.53 (0.04)	**0.58** (0.03)	$< 10^{-5}$	$< 10^{-7}$
Beatles	0.65 (0.04)	0.75 (0.04)	**0.77** (0.02)	$< 10^{-7}$	$< 10^{-2}$
Grouped	0.66 (0.02)	0.74 (0.03)	**0.76** (0.01)	$< 10^{-7}$	$< 10^{-2}$

we should be careful before generalizing our observations. The diversity measure employed was selected based on ground truth knowledge. And although there is currently no other way to assess the relationship between a diversity measure and quality, we should avoid calling our approach strictly "unsupervised." In addition, the effect of the ensemble size (beside the number of ensembles from which we pick one) was not investigated and should be addressed in future work.

Despite these reservations our approach shows great potential due to the following: it is based on interpretable components, namely MSA-based similarity and extreme value analysis. MSA is a structure that potentially holds information that is left unexplored using pairwise comparisons. Extreme value analysis is intuitive and extremely efficient. Combining outlier detectors using ensembles, renders the approach domain-agnostic. This advantage is not be taken lightly: modelling the domain is fundamental for any outlier detection algorithm [6], therefore avoiding it for an unknown music dataset can be extremely beneficial.

References

1. Bertin-Mahieux, T., Ellis, D.P., Whitman, B., Lamere, P.: The million song dataset. In: Proceedings of the 12th International Society for Music Information Retrieval Conference, pp. 591–596 (2011)
2. Bountouridis, D., Van Balen, J.: The cover song variation dataset. In: The International Workshop on Folk Music Analysis (2014)
3. Dong, X.L., Berti-Equille, L., Srivastava, D.: Integrating conflicting data: the role of source dependence. Proc. VLDB Endow. **2**(1), 550–561 (2009)
4. Eddy, S.R.: Profile hidden Markov models. Bioinformatics **14**(9), 755–763 (1998)
5. Eddy, S.R.: Accelerated profile HMM searches. PLoS Comput. Biol. **7**(10), e1002195 (2011)
6. Aggarwal, C.C.: Outlier analysis. In: Aggarwal, C.C. (ed.) Data Mining, pp. 237–263. Springer, New York (2015)
7. Flexer, A., Pampalk, E., Widmer, G.: Novelty detection based on spectral similarity of songs. In: ISMIR, pp. 260–263 (2005)
8. Flexer, A., Schnitzer, D.: Using mutual proximity for novelty detection in audio music similarity. In: Proceedings of 6th International Workshop on Machine Learning and Music (MML), pp. 31–34. Citeseer (2013)

9. Freitas, C.O.A., Carvalho, J.M., Oliveira, J.J., Aires, S.B.K., Sabourin, R.: Confusion matrix disagreement for multiple classifiers. In: Rueda, L., Mery, D., Kittler, J. (eds.) CIARP 2007. LNCS, vol. 4756, pp. 387–396. Springer, Heidelberg (2007). doi:10.1007/978-3-540-76725-1_41

10. Greene, D., Tsymbal, A., Bolshakova, N., Cunningham, P.: Ensemble clustering in medical diagnostics. In: 17th IEEE Symposium on Computer-Based Medical Systems, CBMS 2004, Proceedings, pp. 576–581. IEEE (2004)

11. Grubbs, F.E.: Sample criteria for testing outlying observations. Ann. Math. Stat. **21**, 27–58 (1950)

12. Hadjitodorov, S.T., Kuncheva, L.I., Todorova, L.P.: Moderate diversity for better cluster ensembles. Inf. Fusion **7**(3), 264–275 (2006)

13. Hansen, L.K., L.-Schioler, T., Petersen, K.B., Arenas-Garcia, J., Larsen, J., Jensen, S.H.: Learning and clean-up in a large scale music database. In: 2007 15th European Signal Processing Conference, pp. 946–950. IEEE (2007)

14. Hawkins, D.M.: Identification of Outliers, vol. 11. Springer, Netherlands (1980)

15. Jehl, P., Sievers, F., Higgins, D.G.: OD-seq: outlier detection in multiple sequence alignments. BMC Bioinf. **16**(1), 269 (2015)

16. Livshin, A., Rodet, X.: Purging musical instrument sample databases using automatic musical instrument recognition methods. IEEE Trans. Audio Speech Lang. Process. **17**(5), 1046–1051 (2009)

17. Lukashevich, H., Dittmar, C.: Improving GMM classifiers by preliminary one-class svm outlier detection: application to automatic music mood estimation. In: Locarek-Junge, H., Weihs, C. (eds.) Classification as a Tool for Research, pp. 775–782. Springer, Heidelberg (2010)

18. Macrae, R., Dixon, S.: Guitar tab mining, analysis and ranking. In: ISMIR, pp. 453–458 (2011)

19. Markou, M., Singh, S.: Novelty detection: a reviewpart 1: statistical approaches. Signal Process. **83**(12), 2481–2497 (2003)

20. Panteli, M., Benetos, E., Dixon, S.: Automatic detection of outliers in world music collections. In: Fourth International Conference on Analytical Approaches to World Music (AAWM 2016) (2016)

21. Rand, W.M.: Objective criteria for the evaluation of clustering methods. J. Am. Stat. Assoc. **66**(336), 846–850 (1971)

22. Saitou, N., Nei, M.: The neighbor-joining method: a new method for reconstructing phylogenetic trees. Mol. Biol. Evol. **4**(4), 406–425 (1987)

23. Zimek, A., Campello, J.G.B., Sander, J.: Ensembles for unsupervised outlier detection: challenges and research questions a position paper. ACM SIGKDD Explor. Newsl. **15**(1), 11–22 (2014)

24. Gómez, E., Klapuri, A., Meudic, B.: Melody description and extraction in the context of music content processing. J. New Music Res. **32**(1), 23–40 (2003)

25. Katoh, K., Misawa, K., Kuma, K.-I., Miyata, T.: MAFFT: a novel method for rapid multiple sequence alignment based on fast fourier transform. Nucleic Acids Res. **30**(14), 3059–3066 (2002)

26. Krumhansl, C.L., Kessler, E.J.: Tracing the dynamic changes in perceived tonal organization in a spatial representation of musical keys. Psychol. Rev. **89**(4), 334 (1982)

27. Li, S.Z.: Content-based audio classification and retrieval using the nearest feature line method. Speech Audio Process. **8**(5), 619–625 (2000)

28. Malt, B.C.: An on-line investigation of prototype and exemplar strategies in classification. J. Exp. Psychol. Learn. Mem. Cogn. **15**(4), 539 (1989)

29. Martin, B., Brown, D.G., Hanna, P., Ferraro, P.: Blast for audio sequences alignment: a fast scalable cover identification. In: 13th International Society for Music Information Retrieval Conference, p. 529 (2012)
30. Needleman, S.B., Wunsch, C.D.: A general method applicable to the search for similarities in the amino acid sequence of two proteins. J. Mol. Biol. **48**(3), 443–453 (1970)
31. Sankoff, D., Kruskal, J.B.: Time warps, string edits, and macromolecules: the theory and practice of sequence comparison. Addison-Wesley Publishing Company, Reading (1983)
32. van Kranenburg, P., de Bruin, M., Grijp, L., Wiering, F.: The shs-50 tune collections. In: Shs-50 Online Reports (2014)

Scalable Similarity Search in Seismology: A New Approach to Large-Scale Earthquake Detection

Karianne Bergen[1](✉), Clara Yoon[2], and Gregory C. Beroza[2]

[1] Institute for Computational and Mathematical Engineering,
Stanford University, Stanford, CA 94305, USA
kbergen@stanford.edu

[2] Department of Geophysics, Stanford University, Stanford, CA 94305, USA

Abstract. Extracting earthquake signals from continuous waveform data recorded by networks of seismic sensors is a critical and challenging task in seismology. Earthquakes occur infrequently in long-duration data and may produce weak signals, which are challenging to detect while limiting the number of false discoveries. Earthquake detection based on waveform similarity has demonstrated success in detecting weak signals from small events, but existing techniques either require prior knowledge of the event waveform or have poor scaling properties that limit use to small data sets. In this paper, we describe ongoing research into the use of similarity search for large-scale earthquake detection. We describe Fingerprint and Similarity Thresholding (FAST), a new earthquake detection method that leverages locality-sensitive hashing to enable waveform-similarity-based earthquake detection in long-duration continuous seismic data. We demonstrate the detection capability of FAST and compare different fingerprinting schemes by performing numerical experiments on test data, with an emphasis on false alarm reduction.

Keywords: Similarity search · Locality-sensitive hashing · Time series · Data mining · Earthquake detection · Template matching · Signal processing

1 Introduction

Seismology is an observational science that relies on data collected from seismic sensors to study and interpret processes within the earth. Earthquake detection, the use of signal processing to identify seismic signals in continuous ground motion measurements, is critical for enabling discoveries in the field. Modern seismic networks include hundreds to thousands of sensors, each recording data continuously. As the volume of available data grows, the seismology community is increasingly recognizing the need to adopt state-of-the-art algorithms and data-intensive computing techniques to process large seismic data sets.

There are a number of challenges and requirements for the earthquake detection problem. The events of interest, earthquakes, occur infrequently and their signals are short in duration (seconds to tens of seconds). Therefore, earthquake

L. Amsaleg et al. (Eds.): SISAP 2016, LNCS 9939, pp. 301–308, 2016.
DOI: 10.1007/978-3-319-46759-7_23

detection requires processing months to years of data, most of which contains only background signals, including local, persistent noise sources. A practical earthquake detection system should be able to detect weak signals from small earthquakes while controlling the false alarm rate; a large number of false detections could easily overwhelm true events, so maintaining high precision is critical when processing large data sets. Small, low signal-to-noise events are hard to detect and can often only be confidently distinguished from noise by identifying coherent signals across an array of sensors. Sensor dropout and changes in sensor array configuration are not uncommon, so we focus on network-based detection approaches that detect independently on each channel as an initial step. This paper will focus on the single-channel detection problem.

The STA/LTA algorithm [1], widely used for general earthquake detection, identifies rapid increases in the signal energy to detect events with impulsive wave arrivals. This approach is attractive because it can be easily applied in near real-time to streaming data, but the simplicity of the detection statistic does not take advantage of the shape of the recorded waveforms.

Earthquake waveforms contain valuable information for detection; it has been widely observed that earthquakes originating at neighboring locations generate similar waveforms at a fixed sensor (Fig. 1). In recent years, seismologists have exploited waveform similarity, measured by the normalized cross-correlation, to detect small earthquakes with weak signals similar to those of known template events [6]. However, the performance of template matching is limited by the quality and availability of template waveforms from earthquake catalogs, which are known to be incomplete, especially for low magnitude events. We seek a general similarity-based earthquake detector that can identify similar earthquake waveforms without templates. Previous efforts toward that goal have proposed a brute-force blind search for similar waveforms [5], but the quadratic scaling of this approach makes it infeasible for large data sets.

In this paper we present on-going work to incorporate similarity search into a modern, scalable earthquake detection pipeline. We have introduced a new earthquake detection approach called Fingerprint and Similarity Thresholding (FAST) [8] to detect earthquakes by identifying similar waveforms in continuous

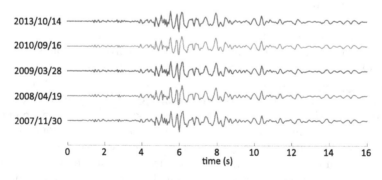

Fig. 1. Similar earthquake waveforms recorded during five distinct events over a period of years at a fixed sensor, station CCOB in Northern California

seismic data. FAST is modeled after scalable content-based audio identification systems [3]. Given continuous waveform data recorded by a single sensor, we extract a binary waveform fingerprint for each short-duration time interval. Then we perform an approximate similarity search using locality-sensitive hashing (LSH) to identify similar waveforms, which are labeled as candidate earthquakes. Below we describe our similarity-search-based approach for large-scale earthquake detection and discuss strategies to lower the false alarm rate and enable the detection of low signal-to-noise events.

2   Our Approach: FAST Earthquake Detector

The Fingerprint and Similarity Thresholding earthquake detection method identifies earthquakes using an efficient blind search for similar waveforms. The two key steps in the FAST detector are feature extraction and approximate similarity search. Feature extraction maps each short-duration waveform segment into a sparse binary fingerprint. The approximate similarity search, which employs locality-sensitive hashing [2] for computational efficiency, identifies similar pairs of fingerprints. Waveform segments corresponding to similar fingerprint pairs are classified as candidate earthquake signals.

2.1   Data

FAST operates on single-channel, continuous, high frequency (up to 100 Hz) data recorded by seismometers that measure ground motion at fixed locations. The data contain seismic signals embedded in background noise. We apply a 1–10 Hz bandpass filter and use a 10 s event window, corresponding to the predominant frequencies and duration of seismic waves for small local earthquakes.

2.2   Feature Extraction

Earthquake waveforms are searched using sparse binary waveform fingerprints. The feature extraction approach used in FAST is adapted from the Waveprint [3] method for audio fingerprinting. Audio fingerprinting provides a good starting point for the development of earthquake waveform fingerprints – there is structural similarity between the data and both applications require fingerprints that are robust to small variations and additive noise. The feature extraction process converts short-duration waveforms into sparse binary fingerprints (Fig. 2) and is described in the following steps.

1. **Spectrogram.** We convert the time series data to the spectrogram, a time-frequency representation computed with the short-time Fourier transform.
2. **Spectral Images.** We divide the spectrogram into short (10 second) overlapping segments, and resize spectral images to fixed dimensions: 32 frequency bins and 64 time bins. The spectral domain provides some shift invariance, unlike the time domain where waveforms must be precisely aligned; this allows a larger lag between adjacent intervals (1.0 vs. 0.05 s) and fewer fingerprints total, but the trade-off is reduced detection sensitivity.

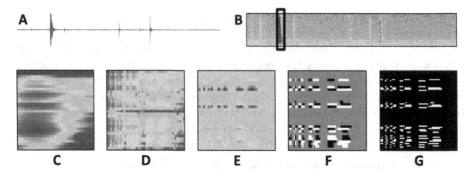

Fig. 2. Feature Extraction process in FAST: (A) continuous data, (B) spectrogram, (C) spectral image, (D) discrete Haar wavelet transform, (E) adjusted wavelet coefficients, (F) coefficient selection, (G) conversion to binary fingerprint

3. **Haar Wavelet Transform.** For each spectral image, we compute the two-dimensional discrete Haar wavelet transform.
4. **Coefficient Selection and Conversion to Binary.** We select the K *most anomalous* Haar coefficients (as described in Sect. 2.4) for each spectral image. K is typically selected in the range of 200–800 (out of 2048). For the selected coefficients, we retain only the sign value and set all other coefficients to zero. We convert the sign values to binary using two bits per coefficient, resulting in sparse binary fingerprints of dimension 4096 with K non-zeros.

2.3 Similarity Search

The computational efficiency of FAST comes from the use of locality-sensitive hashing [2] to perform a fast approximate similarity search. The Jaccard similarity coefficient quantifies the similarity between fingerprints. In the similarity search step, hash signatures are generated using MinHash [4] to preserve the Jaccard similarity, and LSH is used to identify fingerprints with similar signatures. The use of MinHash and LSH provides a significant improvement over the quadratic scaling of a brute-force all-to-all search. For instance, when applied to one week of continuous data, FAST has demonstrated a factor 140 speed-up over the brute-force search and detected 89 events compared with 24 events in the earthquake catalog (see [8] for details).

2.4 Haar Coefficient Selection

The effectiveness of similarity search is highly dependent upon the data representation. Fingerprints must be discriminative, that is similar waveforms map to similar fingerprints under the Jaccard metric. The imbalanced data set poses an additional challenge; the signals of interest, similar earthquake waveforms, appear infrequently in data dominated by background noise. In our template-free search, the potential for false detections is high because we search the full seismic data record for similar pairs of waveforms to identify weak earthquake signals. Therefore we require fingerprints corresponding to background signals to be mutually dissimilar, even in the presence of persistent noise sources, to distinguish weak seismic signals while also limiting the number of false alarms.

The original feature extraction approach, following Waveprint, creates a compact representation using Haar wavelets by retaining the coefficients that are largest in magnitude. While this approach has been successfully applied in audio fingerprinting, when it is applied to seismic data the resulting fingerprints provide an inefficient representation; the largest magnitude coefficients often belong to a subset of frequently selected coefficients, while the majority of the coefficients are rarely selected. For instance, on a test data set with $K = 400$ selected coefficients, 16 % of the coefficients are "frequently selected" (i.e. active in at least 25 % of fingerprints) while 50 % are "rarely selected" (active in fewer than 1 % of fingerprints). This inefficiency impacts the performance of earthquake similarity search by increasing the average similarity between "background" fingerprints, thus making it more difficult to distinguish weak earthquake signals. Therefore we adjust our approach to select coefficients that are more discriminative with respect to background signals.

We select the Haar coefficients that the most discriminative or anomalous, rather than those that are largest in magnitude. To achieve this, we compute adjusted Haar coefficients by standardizing each coefficient based on its distribution across the full, background-dominated data set. We model the unknown coefficient distributions using simple statistics: with mean and standard deviation (Z-score), or with the median and median absolute deviation (MAD). These metrics allow us to choose coefficients that are not largest in magnitude, but farthest on the tails of the distribution. Empirically, this approach suppresses detections of persistent noise sources while maintaining high accuracy on earthquake signals (Fig. 3). We compare these fingerprinting schemes in Sect. 3.1.

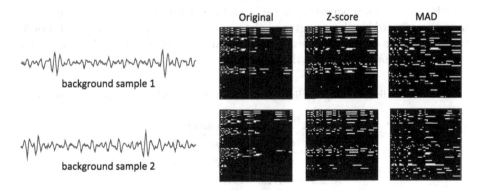

Fig. 3. Comparison of fingerprinting schemes applied to background noise. The Jaccard similarities between the fingerprints are: 0.266 (original), 0.117 (Z-score), and 0.040 (MAD).

3   Experiments

We compare the performance of the fingerprinting schemes described in Sect. 2.4 and demonstrate their accuracy for earthquake waveforms, then demonstrate the performance of FAST on a planted waveform test set in which earthquake waveforms are embedded in recorded background signals at known times and signal-to-noise ratio. All data used in the tests below were recorded at Northern California Seismic Network station CCOB, and sample earthquake waveforms were selected using the Northern California Earthquake Catalog.

3.1   Performance of Feature Extraction

We compare the three feature extraction schemes described in the previous section: (1) original, (2) Z-score-, and (3) MAD-adjusted fingerprints.

We test two criteria to measure the quality of fingerprints for our earthquake detection problem: fingerprint accuracy and baseline similarity. Accuracy is a measure of the quality of the fingerprints of earthquake waveforms for similarity-based detection under additive noise. Baseline similarity quantifies the similarity between background fingerprints in the presence of persistent noise to estimate false detection rates.

To assess accuracy, we compare the fingerprints of clean earthquake waveforms to low signal-to-noise versions of the same waveform embedded in noise:

$$\text{accuracy}(i, j) = \text{jaccard}\left(\mathcal{F}_\mathcal{P}(x^{(i)}), \mathcal{F}_\mathcal{P}(\alpha x^{(i)} + n^{(j)})\right), \tag{1}$$

where $\mathcal{F}_\mathcal{P}$ is the feature extraction operation, $x^{(i)}$ is the i-th earthquake waveform, $n^{(j)}$ is the j-th background waveform, and α is a scaling factor to control the signal-to-noise ratio (SNR). We use waveforms from 300 known earthquakes and embed each one in 10 noise segments at a low SNR ranging from 1.0 to 5.0. To test the robustness of the fingerprints, the signals were bandpass filtered to 1–10 Hz and include persistent noise in the 1.5–3.5 Hz range. We directly compute the Jaccard similarity between the clean and noisy fingerprints and report the median for each feature extraction scheme in Table 1. The MAD-adjusted fingerprints consistently have the highest accuracy.

Table 1. Median Jaccard similarity of clean and low-SNR earthquake waveforms

SNR	Fingerprint accuracy		
	Original	Z-score	MAD
1.0	0.3093	0.3629	0.4760
2.0	0.5123	0.6736	0.7279
4.0	0.7354	0.8561	0.8735

The baseline similarity distribution is estimated from the Jaccard similarities for 5000 pairs of background fingerprints:

$$\text{baseline}(k, \ell) = \text{jaccard}\left(\mathcal{F_P}(n^{(k)}),\ \mathcal{F_P}(n^{(\ell)})\right).\tag{2}$$

The similarity between background fingerprints is substantially lower for MAD- and Z-score adjusted fingerprints than for the original top magnitude fingerprints, with median Jaccard similarities of 0.047, 0.071, and 0.185, respectively.

'In order to maintain high overall precision in an imbalanced data set, we require both high accuracy for fingerprints and low baseline similarity to limit false detections. We characterize the trade-off between false detections and missed detections, specifically for the case of identifying low SNR earthquakes similar to clear earthquake waveforms, using in a ROC curve (Fig. 4a). For a given Jaccard similarity threshold, the *true positive rate* is defined as the rate at which the accuracy exceeds this threshold, and the *false positive rate* is the rate at which the baseline similarity exceeds the same threshold. We also consider the more challenging and relevant case in which we seek to identify pairs of similar low SNR earthquake waveforms, i.e. both instances of the waveform include additive noise in a modified accuracy formula (Fig. 4b).

3.2 Detection Performance

To have a clear ground truth for measuring detection performance, we inject real earthquake waveforms into a dataset consisting of 16 hours of recorded background signal. Twelve pairs of known event waveforms are embedded in the background at low SNR. We report the results for MAD-adjusted fingerprints with K = 400 non-zeros, and 100 hash tables with 4 hash functions per table in the LSH search. The detection statistic is the fraction of hash tables in which a fingerprint appears in the same hash bucket as its nearest neighbor. FAST successfully identifies all 24 low SNR events with only 4 false detections (85.71 % precision). FAST has shown promising initial results on real earthquake sequence

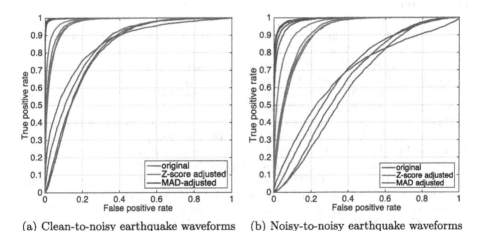

(a) Clean-to-noisy earthquake waveforms   (b) Noisy-to-noisy earthquake waveforms

Fig. 4. Trade-off between detection rate for weak signals (SNR 1.0) and false detections. Multiple lines represent results for several different values of K.

data, detecting previously unknown events with a manageable number of false detections in months of continuous data.

4 Discussion

In this paper, we present an application of approximate similarity search with LSH to the problem of earthquake detection in continuous seismic data. This work represents a new direction for waveform-similarity-based earthquake detection that does not require prior knowledge of event waveforms and has sufficient computational efficiency to allow for application to long-duration data that would not be feasible using a brute-force search. Our initial experiments with FAST demonstrate that this approach can successfully detect previously unknown small earthquakes using blind similarity search. Furthermore, we have demonstrated that modifications to audio fingerprinting methods based on the empirical data distribution can improve accuracy on imbalanced data sets, which contain relatively few pairs of moderate-to-high similarity. Scalable similarity search has the potential to impact both the study of earthquakes and earth and environmental monitoring more broadly. Imbalanced data sets appear in many of these applications, such as acoustic recordings used for mining bioacoustic soundscapes in ecological studies [7], and we believe the techniques developed for FAST can be applied in these domains.

Acknowledgments. This research was supported by NSF grant EAR-1551462 and by the Southern California Earthquake Center (contribution no. 6325). Waveform data, metadata, or data products for this study were accessed through the Northern California Earthquake Data Center, doi:10.7932/NCEDC. We thank Ossian O'Reilly for his assistance with the hashing techniques used in this work.

References

1. Allen, R.: Automatic phase pickers: their present use and future prospects. Bull. Seismol. Soc. Am. **72**(6B), S225–S242 (1982)
2. Andoni, A., Indyk, P.: Near-optimal hashing algorithms for approximate nearest neighbor in high dimensions. Commun. ACM **51**(1), 117–122 (2008)
3. Baluja, S., Covell, M.: Waveprint: efficient wavelet-based audio fingerprinting. Pattern Recogn. **41**(11), 3467–3480 (2008)
4. Broder, A.Z., Charikar, M., Frieze, A.M., Mitzenmacher, M.: Min-wise independent permutations. J. Comput. Syst. Sci. **60**(3), 630–659 (2000)
5. Brown, J.R., Beroza, G.C., Shelly, D.R.: An autocorrelation method to detect low frequency earthquakes within tremor. Geophys. Res. Lett. **35**(16), L16305 (2008)
6. Gibbons, S.J., Ringdal, F.: The detection of low magnitude seismic events using array-based waveform correlation. Geophys. J. Int. **165**(1), 149–166 (2006)
7. Servick, K.: Eavesdropping on ecosystems. Science **343**(6173), 834–837 (2014)
8. Yoon, C.E., O'Reilly, O., Bergen, K.J., Beroza, G.C.: Earthquake detection through computationally efficient similarity search. Sci. Adv. **1**(11) (2015)

Scalable Similarity Search

Feature Extraction and Malware Detection on Large HTTPS Data Using MapReduce

Přemysl Čech[1]([✉]), Jan Kohout[2], Jakub Lokoč[1], Tomáš Komárek[2], Jakub Maroušek[1], and Tomáš Pevný[2]

[1] SIRET Research Group, Faculty of Mathematics and Physics, Department of Software Engineering, Charles University in Prague, Prague, Czech Republic
{cech,lokoc}@ksi.mff.cuni.cz, marousej@artax.karlin.mff.cuni.cz
[2] FEE, Cognitive Research Center in Prague, Czech Technical University in Prague, Cisco Systems, Inc., Prague, Czech Republic
{jkohout,tpevny}@cisco.com, komartom@gmail.com

Abstract. Secure HTTP network traffic represents a challenging immense data source for machine learning tasks. The tasks usually try to learn and identify infected network nodes, given only limited traffic features available for secure HTTP data. In this paper, we investigate the performance of grid histograms that can be used to aggregate traffic features of network nodes considering just 5-min batches for snapshots. We compare the representation using linear and k-NN classifiers. We also demonstrate that all presented feature extraction and classification tasks can be implemented in a scalable way using the MapReduce approach.

Keywords: Hadoop · MapReduce · HTTPS data · Intrusion detection · Approximate similarity join

1 Introduction

The detection of secure HTTP (HTTPS) connections related to malicious activity is a pressing problem for increasing volumes of HTTPS traffic on the Internet. Unlike traditional malware detection systems that can rely on known byte sequences in packets [14], information about HTTPS connections—because of the encryption—are limited just to very high-level features such as the number of uploaded/downloaded bytes and a duration of the connection. Even the visited URLs might not be available to the detection system if the privacy of the users has to be fully preserved. For these reasons, the encrypted HTTPS communication has been employed also by malware to prevent its detection. Nevertheless, recent works [7,8] demonstrate that statistical descriptors[1] $x \in \mathbb{R}^d$ of servers based on the available information for HTTPS connections can be effectively used to detect malware or group servers that are running similar applications. Whereas most of the approaches that employ statistical modeling

---

[1] The statistical descriptor is a d-dimensional vector x capturing statistical properties of the communication. For more details see Sect. 2.

© Springer International Publishing AG 2016
L. Amsaleg et al. (Eds.): SISAP 2016, LNCS 9939, pp. 311–324, 2016.
DOI: 10.1007/978-3-319-46759-7_24

of network traffic work on the packet level [4,6,15], our work is based on descriptors extracted from high-level web proxy logs that does not require packet level captures. Although the descriptors investigated both for linear and k-NN-based classification [10] showed promising results, the efficiency and effectiveness of statistical decriptors require further improvements for practical use cases.

Since the volume of network communication has been constantly rising and also new malware variants are emerging every day [1], the malware detection systems have to face scalability issues because they have to process more and more data effectively. Therefore, traffic analysis systems require distributed approaches [9] often relying on the *MapReduce* programming model [5] - a paradigm often used in large-scale data processing.

Each MapReduce program consists of a *Map* procedure performing filtering and sorting, and a *Reduce* method evaluating a summary operation. The MapReduce model is suitable for the creation of statistical descriptors as it naturally supports selection of high-level communication features in the Map phase and their aggregation into one descriptor in the Reduce phase (for more details see Sect. 3). Furthermore, the MapReduce approach enables efficient implementation of a linear classification on the top of the descriptors. Hence, after the feature extraction job is finished, the classification job can be evaluated. However, the descriptors representing malware and benign communications might not be linearly separable [10]. In such cases, a k-NN classifier that employs an annotated reference database of statistical descriptors can be utilized. Since the numbers of classified descriptors and reference descriptors can become very large, a scalable distributed similarity join operation for efficient k-NN classification is required [11].

In this paper, we address both effectiveness and efficiency of malware detection systems relying on communication descriptors extracted from HTTPS connections. We investigate related work which used grid-based histograms [7,8] as the descriptors. Since modern malware detection systems should collect and analyze a vast amount of network traffic data, we investigate also the robustness and stability of k-NN classifiers with growing reference datasets of descriptors. Last but not least, considering huge datasets we focus also on the scalability of malware detection systems based on descriptors. More specifically, we investigate the MapReduce approach for descriptors creation and classification. In the experiments, we demonstrate that the MapReduce approach is suitable for descriptors creation from HTTPS connections and also that the MapReduce approach can handle expensive similarity joins for efficient k-NN classification.

The paper is organized as follows. Section 2 presents the adopted approach to create descriptors of HTTPS communication. Section 3 focuses on MapReduce implementation of the feature extraction and the classification. Since for the k-NN classification of descriptors the exact similarity join algorithm is not sufficiently fast, we investigate approximate similarity join approach in Sect. 4. Section 5 presents experimental evaluation and finally Sect. 6 concludes the paper.

2 HTTPS Feature Extraction and Classification

The goal of this work is to be able to identify users whose machines are infected with malware communicating over HTTPS. This involves representation of the snapshots of the users' communication such that these snapshots can be stored as (preferably compact) objects and effectively compared to each other and assessed by detectors of malware. To achieve this, we split the continuous stream of HTTPS traffic from each client's machine into 5-min batches to create so called *communication snapshots*. One communication snapshot is defined as a set of all requests for establishing an SSL tunnel (which is used for the HTTPS communication) issued by the same user during one 5-min interval. To effectively represent this set of requests, we need to define a transformation which transforms the set into one feature vector of fixed dimension. This fixed-size real vector is called a *descriptor* of the communication snapshot. Transforming the sets into descriptors makes them easy to compare and classify. However, the transformation has to be chosen carefully such that it retains maximum possible information. The approach that we have adopted is following: We see each communication snapshot as a set of *messages* interchanged in the communication. In our case, the messages are the individual requests for establishing an encrypted SSL tunnel. These requests are typically logged as separate proxy log lines by the web proxies. Furthermore, we assume that each message can be represented by a real vector $m \in \mathbb{R}^n$. This assumption is realistic as the web proxies usually log multiple data fields that can be used as numerical features - e.g., amounts of transferred bytes within the tunnel. In this paper, we use four numerical features to represent the messages (i.e., $n = 4$), namely:

1. **bytes sent:** m_{up} through the tunnel from the user's machine to the target server,
2. **bytes received:** m_{down} by the client's machine from the server,
3. **duration:** m_{dur} of the tunnel, i.e., the length of the time interval for which the tunnel was active,
4. **inter-arrival time:** m_{ia} (in seconds) elapsed between two consecutive requests for establishing a tunnel from the same user and server.

A message m (one SSL tunnel) is then represented as a 4-tuple

$$m = (\log(1 + m_{\mathrm{up}}), \log(1 + m_{\mathrm{down}}), \log(1 + m_{\mathrm{dur}}), \log(1 + m_{\mathrm{ia}})).$$

The logarithmic scale is used to suppress noise and decrease ranges of the features. Having the numerical representation of messages defined, each communication snapshot is then treated as a finite sample from an n-dimensional random variable with unknown probability distribution p. The messages are treated as individual realizations of this random variable. The question of building the descriptor of the communication snapshot then becomes a question of building the descriptor of the distribution p based on a finite sample from it. For such task, there has been a large amount of approaches proposed in the literature that aim at representation of probability distributions. In this work, we chose an approach which is based on smoothed histograms capturing frequencies of the observed features. This approach is described in the following subsection.

2.1   Communication Descriptors Based on Histograms

Histograms have been widely used to capture frequencies of observations and represent them as a real vector of fixed dimension in many different domains. The way how we utilize histograms for creating descriptors of the communication snapshots was described in [7] and also used for similar purpose in [10]. The joint histogram is formed from 4-dimensional messages belonging to the same communication snapshot. The contribution of each message is distributed among multiple bins. The bins are fitted and centered into a lattice $L = \{0, ..., 11\}^4$, thus final descriptor is represented by a feature vector of fixed dimension $11^4 = 14641$. Thanks to all this, the histogram-based representation is purely unsupervised and can be constructed without fitting any parameters.

2.2   Classification of Communication Descriptors

The classifiers assign a confidence value to all analyzed communication descriptors $x \in \mathbb{Q} \subset \mathbb{R}^d$. Given the confidence value, all the analyzed histograms can be ordered, and the objects that are highly ranked in that ordering are assumed to correspond to descriptors of communication belonging to hosts infected by malware. Since the problem of malware detection is known to be a highly imbalanced with a strong prevalence of legitimate traffic, the quality of the ordering is measured by the false alarm rate at 50 % recall on malware [13] (further called FP-50). The FP-50 error measure was designed for domains where significant emphasis is put on very low false-positive, such as steganalysis and anomaly detection in security. This makes it a good error measure for our problem, too (because of the strong prevalence of legitimate traffic, the false-positive rate is very important if the method is used in practice). For these reasons, we prefer the FP-50 measure to other error measures, such as average accuracy. Formally, the value of the FP-50 error e is defined as follows:

$$e = \frac{1}{|\mathcal{I}^-|} \sum_{i \in \mathcal{I}^-} \mathrm{I}\left[f(x_i) > \mathrm{median}\{f(x_j)|j \in \mathcal{I}^+\}\right],$$

where $\mathcal{I}^-$ are indexes of the negative (benign) training samples (descriptors), $\mathcal{I}^+$ are indexes of the expositive training samples (descriptors of communication snapshots that contain malicious traffic), $f(x)$ is the output of the classifier (the confidence value) for the descriptor $x \in \mathbb{Q}$ and I is an indicator function. The motivation for FP-50 is that missing 50 % of malware is acceptable for the benefit of having extremely low false positive rate. The exponential Chebyshev Minimizer (ECM) proposed in [13] is a suitable linear classifier optimizing the FP-50 measure. We use the ECM in the experiments.

For problems that are not linearly separable, this paper considers also a simple nearest-neighbor based classifier [10] employing the Euclidean norm L_2 as a distance between two samples. The classification rule for a test sample $x \in \mathbb{Q}$ is based on its k nearest neighbors from a given training reference set $\mathbb{S} \subset \mathbb{R}^d$. The k-NN query is defined for $k \in \mathbb{N}^+$, $x \in \mathbb{R}^d$ and $\mathbb{S}$ as:

$$kNN(x) = \{\mathbb{X} \subset \mathbb{S}; |\mathbb{X}| = k \wedge \forall y \in \mathbb{X}, z \in \mathbb{S} - \mathbb{X} : L_2(x, y) \leq L_2(x, z)\}.$$

In order to enable ordering of test samples $x \in \mathbb{Q}$, the k-NN classifier assigns two values to each test sample $x \in \mathbb{Q}$. The first value v_1 represents the number of malicious objects in $kNN(x)$, while the second value v_2 aggregates the sum of distances to the malicious objects in $kNN(x)$. The samples then supports multi-value sorting by v_1 in descending order followed by v_2 in the ascending order. This ordering is the input for the FP-50 measure.

3   Scalable Processing of HTTPS Data

In this section we describe architecture of our MapReduce framework which is composed of a feature extraction job and a classification job. Note that we assume that data are already in a tabular form and that parameters of feature extraction and classifiers are trained in a preprocessing phase.

3.1   Feature Extraction

The first MapReduce job performs transformation of raw data collected by Cisco cloud to descriptors using the histogram approach presented in Sect. 2.1. The job assumes that raw input data which are initially stored in text log files are converted to a table format with eight columns: ID of five minute interval, client ID, domain (server) ID, bytes sent, bytes received, duration of a request, time between requests for the fixed client-domain pair and a label describing whether a request is considered malicious or not. To be more specific, feature columns are already precomputed from pure text logs, values are stored in a logarithmic scale and inter-arrival time between requests was evaluated (see [7]). This preprocessing is performed separately from the MapReduce job and the precomputed table is the input for the map phase.

In the map phase $<key; value>$ pairs for further processing are generated. In this paper, the *key* is composed of the client ID and five minute interval ID column. The usage of five-minute intervals has been widely adopted in a network security, hence, we also adopt this size for our experiments. However, the designed method would be able to work with an arbitrary size of the time interval. Note that the presented malware detection focuses only for clients and ignores domain part of communications (not completely available for HTTPS data). Thus all records which has the same value in these two key columns will be grouped together and will form one histogram. All features and the infected label column belong to a *value* part.

The reduce phase takes sets with the same key. For each key, transformation of features (quadruplets) into a histogram is performed using the algorithm described in Sect. 2. Also every histogram is labeled infected or not if any request (from which histogram was created) was labeled infected. Because a lot of histogram bin values are usually empty (equal to zero) we use a space-saving format for storing the result histograms. More specifically, each sparse histogram is represented as a set of pairs *binID:value*, where pairs with *value* = 0 are omitted.

The output *key* of the reduce phase remains the same and the *value* is a computed descriptor. Format is also depicted in the Fig. 1 where output descriptors are formed. The first number represents infected label and then all non zero pairs *binID:value* separated by semicolons follow.

The whole feature extraction job scheme is also summarized in the Fig. 1.

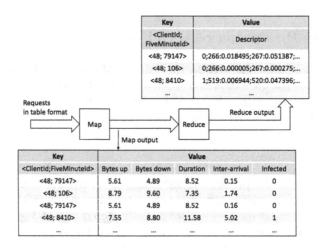

Fig. 1. A scheme of the feature extraction job.

3.2   Classification

The second MapReduce job is dedicated to a classification task of communication histograms from the database $\mathbb{Q} \subset \mathbb{R}^d$ obtained from the previous job. Since the final sorting for the FP-50 measure is not a bottleneck task, the main goal of the second MapReduce job is to efficiently classify a large number of query objects stored in HDFS.

Given weights trained [13] in a preprocessing step, the MapReduce implementation of the linear classifier is straightforward. Besides histograms, the MapReduce job also takes trained weights for the classifier and uses only the map phase to compute a classification score for all histograms. The final classification score is the result of a scalar product of the histogram and weights plus addition of a hyperplane offset w_0. In the linear classifier scenario a reduce phase is not needed.

For the k-NN classification of HTTPS data, Lokoc et al. [10] have presented a centralized memory based approach that uses Voronoi partitioning and metric filtering rules for efficient approximate k-NN search [3,12,18]. However, for large sets Q of query objects and S of reference objects available in advance, an efficient implementation of similarity joins is necessary [2,16,17]. Recently, a MapReduce implementation of exact similarity joins has been introduced [11], where Lu et al. propose a method employing Voronoi partitioning, metric filtering rules and the replication of reference database objects from S for efficient exact similarity join processing. However, the good efficiency results of exact similarity joins are reported just for data with different properties than our communication descriptors. Whereas in the paper Lu et al. used low-dimensional vectors, the communication descriptors investigated in this work are sparse high-dimensional vectors, with imbalanced sparsity settings across the whole dataset. Furthermore, the distance space (i.e., descriptors in connection with the Euclidean distance) suffers from high intrinsic dimensionality. Therefore, in the following section we present a new approximate similarity join algorithm for MapReduce programming model inspired by both works of Lu et al. [11] and Lokoc et al. [10].

4   Approximate Similarity Join Using MapReduce

The work of Lu et al. [11] focuses on distributed exact similarity joins of two large sets Q and S. The proposed method consists of two steps - preprocessing and a similarity join k-NN job. In the preprocessing phase, data from Q and S are split into partitions in order to exploit efficient algorithm parallelization and evaluation using Hadoop MapReduce distributed environment. The metric space Voronoi partitioning is used to split data descriptors into Voronoi cells c_i corresponding to pivots $p_i \in S$. Since the number of reducers is limited, the cells c_i are grouped into bigger parts called groups G_j, which are later processed by reducers in the k-NN job. With small changes, our new approximate similarity join algorithm takes the preprocessing phase and the organization of the dataset into groups[2] as presented by Lu et al. [11]. However, the similarity join k-NN job requires approximate approaches for both data replication method and also for k-NN query evaluation at a particular reducer. The approximate k-NN query processing at a reducer uses approximate search strategy presented by Lokoc et al. [10].

4.1   Preprocessing Phase

For the Voronoi-based partitioning, pivots p_i from the database S have to be selected. Since we consider a high number of pivots, a random pivots selection is employed as a sufficient and not computationally intensive method. In our work we use 2000 pivots. Given a set of pivots, the partitioning of descriptors to cells c_i according to the selected pivots p_i has to be performed. This task is done only by the map job mentioned in [11], where input objects (descriptors) are processed by mappers and for each descriptor the nearest pivot is found. To each descriptor we add and save the nearest pivot ID information plus distance to this pivot [11]. Finally, also statistics for each cell c_i are stored including number of objects, size of objects and distance from the pivot p_i to the furthest object from S assigned to the cell c_i, which comes in handy for filtering techniques in the k-NN job. The partitioning is computed for objects in both sets Q and S, so all objects store ID of their nearest pivot.

---

[2] We would like to thank Lu et al. [11] for sharing their codes with us.

After the partitioning is performed, the Voronoi cells are organized into groups $G_j = \{c_{i_1}, \ldots, c_{i_m}\}$ using the geometry grouping algorithm [11]. More specifically, the Voronoi cells whose corresponding pivots p_i are near to each other are put into the same group. To better balance the computation workload of similarity joins, groups are eventually even up to contain the same number of objects from S [11]. However, for high-dimensional sparse descriptors with different sparsity settings, the count criterion can lead to an imbalanced workload. Therefore, we investigate also a heuristic that balances the groups to reach a similar sum of sizes of objects in the groups. The pseudo code for the grouping technique is depicted in the Algorithm 1.

Algorithm 1. GeoGroupingBySize

1: N = number of groups
2: $p_k = \max\limits_{p_j \in pivots} \sum_{p_i \in pivots} d(p_i, p_j)$ //most distant pivot from others
3: $P = pivots - \{p_k\}$ //all pivots except the initial one
4: $U = p_k$ //used pivots
5: $G_1 = \{p_k\}$ //first group contains initial pivot
6: $G_2, \ldots, G_N = \emptyset$ //rest result groups are empty
7: **for** $(i = 2; i < N; i{+}{+})$ **do**
8:     $p_k = \max\limits_{p_j \in P} \sum_{p_i \in U} d(p_i, p_j)$ //most distant pivot from P to all pivots in U
9:     $P = P - \{p_k\}; U = U \cup \{p_k\}; G_i = \{p_k\}$
10: **end**
11: **while** $P \neq \emptyset$ **do**
12:     G_i = group with the smallest size of objects
13:     $p_k = \min\limits_{p_j \in P} \sum_{p_i \in G_i} d(p_i, p_j)$ //nearest pivot to all pivots in G_i
14:     $P = P - \{p_k\}; G_i = G_i \cup \{p_k\}$
15: **end**
16: **return** $G_1, G_2, \ldots, G_N$

4.2   Similarity Join Evaluation Using MapReduce

After the preprocessing phase (finding for all objects from Q and S their corresponding cell c_i, and grouping cells to groups G_j), the similarity join can be performed on j reducers. The similarity join on j-th reducer consists of k-NN searches for objects assigned to group G_j in a Map phase [11]. More specifically, the similarity join at j-th reducer finds for all objects from Q assigned to Voronoi cells in group G_j the k closest database objects from S that are either assigned to Voronoi cells in group G_j or replicated from close Voronoi cells belonging to different groups. Note that the replication was introduced to guarantee exact search results for distributed processing based on partitioning into groups [11].

Detailed MapReduce job for the k-NN search is depicted in Algorithm 3, including also two new approximation parameters *replicationThreshold* and *filterThreshold*. Before the Map phase, the nearest groups G_j for all pivots p_i are computed (see the Algorithm 2). Then in the Map phase, objects from Q and S

are assigned to particular reducers by specifying GroupId as the key part and object's descriptor as the value part. Initially, all objects from Q and S have GroupId of their corresponding Voronoi cell, while some object from S are further replicated with different GroupId. The first proposed approximate search extension is to reduce the amount of replication of objects from S to different groups. Even if we use the upper and lower bound pruning techniques described in [11], the high intrinsic dimensionality of the distance space could lead to massive replication of all database objects to all groups/reducers. In such cases the reducers quickly run out of a random access memory (RAM). Therefore, our method replicates database objects $s \in c_i \subset S$ only to groups G_j that contain pivots that are within the *replicationThreshold* closest pivots to p_i. If all the *replicationThreshold* closest pivots to p_i are from the same group G_j as p_i, then the objects $s \in c_i \subset S$ are not replicated at all. Detailed pseudo code is mentioned in the Algorithm 2.

In the reduce phase of the main k-NN algorithm, incoming objects are parsed and query and database sets for every reducer are established. The database objects S assigned to reducer j are further organized as a list of Voronoi cells c_i^S. Then similar evaluation to the work of Lokoc et al. [10] is performed. For every query object q at reducer j, the method precomputes distances to all pivots determining cells c_i^S and sorts the list of Voronoi cells with respect to $d(q, p_i)$. Then, for each cell query ball overlap check[3] is performed (the Algorithm 3 row number 21) followed by the lower bound filtering for specific database objects in the cell (row number 23). If the database object was not filtered, a distance between query and database object is computed and if the distance is lower than the actual query radius, the result k-NN candidate set and the actual query radius are updated. Also the constant *filterThreshold* is used to evaluate neighbors only in the nearest Voronoi cells c_i^S, with respect to $d(q, p_i)$. At the end, the algorithm produces an approximate k-NN result for every query object and also includes a classification score to the output. As described in Sect. 2.2, the k-NN classification score is composed of two values v_1 and v_2.

5   Experiments

In this section, we experimentally evaluate the proposed MapReduce framework. The emphasis is put on verification of scalability of the solution, while we also present preliminary results showing that the k-NN classifier is able to achieve promising results when compared to the linear classifier. All the experiments ran on a virtualized Hadoop cluster with 20 worker nodes, each with 8 GB RAM and 2 core CPU (Intel(R) Xeon(R) running at 2.20 GHz).

5.1   Dataset Used

The dataset used for the experiments is the same as the one used in [10]. However, the original work extracted descriptors of web servers from the data. In our

---

[3] The cell c_i^S query ball is defined by pivot p_i and radius that equals to $\max d(p_i, o_j)$ for all $o_j \in c_i^S$ determined in the preprocessing phase.

Algorithm 2. ComputeNearestGroups

1: $nearestGroups$ = array of size equals to pivot count
2: **foreach** p_i in $pivots$ **do**
3:      $dist_{p_i} = \emptyset$ //empty set of distances to all other pivots
4:      **foreach** p_j in $pivots$ **do**   //for each pivot combination
5:          $groupID = \text{GetGroupID}(p_j)$ //group id where p_j belongs
6:          $dist = d(p_i, p_j)$ //distance between pivots
7:          add pair $<groupId; dist>$ to $dist_{p_i}$
8:      **end**
9:      sort $dist_{p_i}$ in ascending order by $dist$ in pairs
10:     $nearestGroups = \emptyset$
11:     $pivCount = 0$ //number of considered nearest pivots - for approximation
12:     **foreach** $<groupId; dist>$ in $dist_{p_i}$ **do**
13:         **if not** $nearestGroups$ contains $groupId$ **then**
14:             add $groupId$ to $nearestGroups$
15:         **endif**
16:         $pivCount++$
17:         **if** $pivCount > replicationThreshold$ **then break**
18:     **end**
19:     $pivotID = \text{GetPivotID}(p_i)$
20:     $nearestGroups[pivotID] = nearestGroups$
21: **end**
22: **return** $nearestGroups$

---

case, we used the data to extract descriptors of the communication snapshots, as described in Sect. 2. This presents a complementary approach to modelling behavior of servers, which relies on the context of the complete user's communication to identify malicious activity in it. The dataset is composed of HTTPS proxy logs covering 24 h of traffic in 500 corporate networks. In total, the dataset contains 145 822 799 proxy log lines representing individual requests for establishing SSL tunnels for HTTPS communication. The dataset also contains labels indicating for each SSL tunnel whether it was requested by a malicious binary or not. These labels were used in our experiments for assessment of the classification. A communication snapshot was considered malicious, if it contained at least one request for SSL tunnel issued by a malicious binary. Otherwise, the communication snapshot was considered benign. For more detailed information about this data set, see [10].

From the dataset we totally extracted 8642368 unique descriptors of communication snapshots, out of which were 7591651 (5.3 GB) were used to construct the set S (the training set for the linear classifier and the set of reference objects for the k-NN classifier) and the remaining 1050717 (0.6 GB) descriptors were used to construct the set Q (query objects).

5.2 Feature Extraction Experiment

The aim of the first experiment was to measure the gain of using the MapReduce framework for building the descriptors of communication snapshots.

Algorithm 3. kNNJoinApprox

```
 1: map-setup
 2:     nearestGroups = ComputeNearestGroups()

 3: map (k1, v1)
 4:     if k1.dataset = Q then    //query object
 5:         groupID = GetGroupID(k1.cell)
 6:         output(groupID, (k1, v1))
 7:     else                      //database object
 8:         pivotID = GetPivotID(k1.pivot)
 9:         foreach groupID in nearestGroups[pivotID] do
10:             output(groupID, (k1, v1))
11:         end
12:     endif

13: reduce (k2, v2)
14:     parse objects from Q into D_Q and from S into list L of Voronoi cells c_i^S
15:     foreach q in D_Q do
16:         compute distance to pivots d(q, p_i) and sort Voronoi cells c_i^S in L
17:         kNN = ∅            //kNN result
18:         r = MAX_VALUE      //query radius
19:         foreach c_i^S in L do  // for each Voronoi cell check its objects
20:             if i > filterThreshold then break  //approximation
21:             if d(q, p_i) > c_i^S.r_i + r then continue    //query-cell overlap check
22:             foreach o_S in c_i^S do
23:                 if |d(q, p_i) − d(o_S, p_i)| > r then continue   // lower bound filter
24:                 distance = d(q, o_S)
25:                 if distance ≥ r then continue
26:                 update kNN by o_S
27:                 r = d(q, kNN[k])   //radius = distance from q to k-th object o_S
28:             end
29:         end
30:         output(q, kNN)
31:     end
```

We measured the computational time needed for building descriptors from total amount of 1.28 GB of input data for different numbers of utilized mappers and reducers. The process of building the descriptors was described in Sect. 2. For the map phase, the input data were split into blocks and each block was processed by one mapper. Therefore, the preset block size determined the number of used mappers. The number of reducers was set directly. The amount of time needed for building the descriptors depending on the number of mappers and reducers is shown in Fig. 2. The graph demonstrates that moving the creation of the descriptors to the MapReduce environment is a promising way to go as even the relatively low number of reducers can significantly contribute to the decrease of time needed for the computation.

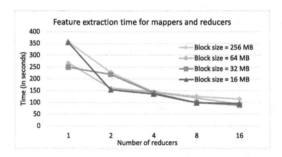

Fig. 2. Time needed for building descriptors from 1.28 GB of input data depending on the number of mappers and reducers used. The number of mappers was determined by the block size.

5.3 Classification Experiment

The second experiment was focused on evaluation of the performance and accuracy of the k-NN classifier working on top of the descriptors. The implementation of the k-NN classifier was described in Sect. 3. This experiment compares the computational time and the classification error (measured by the FP-50 measure) for different settings of the classifier's parameters, which are:

1. ApproxThreshold—the number of considered nearest Voronoi cells for both parameters *replicationThreshold* and *filterThreshold*.
2. Grouping—the method for balancing Groups.
3. K—the number of nearest neighbours used for assigning the confidence value to a given query object.

The number of pivots used by the classifier was always fixed at the value 2000. Furthermore, the FP-50 error of the k-NN classifier was compared to the FP-50 error of the linear classifier trained on the same data that were used as reference objects for the k-NN classifier (see Sect. 2 for details about the implementation of the linear classifier). Both classifiers were tested on the set of query objects Q (containing 1050717 descriptors), while the set S was used as the reference database (i.e., training data) for the classifiers. The results of this experiment are summarized in Fig. 3 and Table 1. We can see that the k-NN classifier is able to achieve superior accuracy (by means of the FP-50 error) over the linear classifier, which was trained specifically to minimize the FP-50 error. The column *Objects of S* presents the number of replicated objects, where for *ApproxThreshold* = 1 no data objects from S are replicated (the value corresponds to the number of objects in S) and only one Voronoi cell is visited. This results in high FP-50 error. However, for growing value of *ApproxThreshold* the FP-50 error improves drastically. Also the number of replications grows significantly and the number of visited Voronoi cells increases. Note that the number of visited cells affects the number of evaluated distance computations and computation time. We may also observe, that for *ApproxThreshold* = 10, the Count grouping strategy ran out of memory on our HW (indicated by dashes).

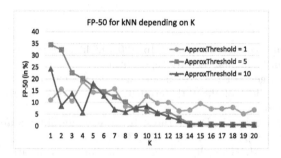

Fig. 3. FP-50 error for different K and ApproxThreshold parameters.

Table 1. Results of the classification experiments on test data.

k-NN search						
Grouping	ApproxThreshold	Time (s)	Objects of S	Dist. computations	K	FP-50 (%)
Size	1	1227	7591651	6999248972	20	6.855
	2	1841	11094657	12055185114	20	0.746
	3	2586	14145252	17351667470	20	0.731
	5	4017	18799234	26952251002	20	0.668
	7	7425	23106879	37155010644	20	0.548
	10	14828	28992207	50691392745	20	0.551
Count	1	2333	7591651	6984716770	20	8.043
	2	4417	11077234	12433727942	20	0.756
	3	5554	14075694	17654051467	20	0.700
	5	8165	18447445	26972030752	20	0.672
	7	10972	22484608	36421226362	20	0.620
	10	–	–	–	–	–
Linear classifier						6.434

6   Conclusion

In this paper, we have introduced descriptors representing communication snapshots of five-minute batches of HTTPS communication from individual users in a network. We have proposed a MapReduce framework for extracting and classification of the descriptors. For the classification, we have adapted the related work MapReduce approach for similarity joins to support approximate k-NN searches. In the experiments, we demonstrate that the framework represents a scalable solution for malware detection systems. In the future, we would like to focus on more effective data replication and k-NN search strategies using repetitive Voronoi partitioning.

Acknowledgments. This project was supported by the GAČR 15-08916S and GAUK 201515 grants.

References

1. Cisco Annual Security Report 2016 (2016). http://www.cisco.com/c/en/us/products/security/annual_security_report.html
2. Bohm, C., Kriegel, H.P.: A cost model and index architecture for the similarity join. In: Proceedings of the 17th International Conference on Data Engineering, pp. 411–420 (2001)
3. Chávez, E., Navarro, G., Baeza-Yates, R., Marroquín, J.L.: Searching in metric spaces. ACM Comput. Surv. **33**(3), 273–321 (2001)
4. Crotti, M., Dusi, M., Gringoli, F., Salgarelli, L.: Traffic classification through simple statistical fingerprinting. SIGCOMM Comput. Commun. Rev. **37**, 5–16 (2007)
5. Dean, J., Ghemawat, S.: MapReduce: simplified data processing on large clusters. Commun. ACM **51**(1), 107–113 (2008)
6. Dusi, M., Crotti, M., Gringoli, F., Salgarelli, L.: Tunnel hunter: detecting application-layer tunnels with statistical fingerprinting. Comput. Netw. **53**, 81–97 (2009)
7. Kohout, J., Pevny, T.: Automatic discovery of web servers hosting similar applications. In: 2015 IFIP/IEEE International Symposium on Integrated Network Management (IM) (2015)
8. Kohout, J., Pevny, T.: Unsupervised detection of malware in persistent web traffic. In: 2015 IEEE International Conference on Acoustics, Speech and Signal Processing (ICASSP) (2015)
9. Lee, Y., Lee, Y.: Toward scalable internet traffic measurement and analysis with hadoop. SIGCOMM Comput. Commun. Rev. **43**(1), 5–13 (2012)
10. Lokoc, J., Kohout, J., Cech, P., Skopal, T., Pevný, T.: k-NN classification of malware in HTTPS traffic using the metric space approach. In: Chau, M., Wang, G.A. (eds.) PAISI 2016. LNCS, vol. 9650, pp. 131–145. Springer, Heidelberg (2016). doi:10.1007/978-3-319-31863-9_10
11. Lu, W., Shen, Y., Chen, S., Ooi, B.C.: Efficient processing of k nearest neighbor joins using MapReduce. Proc. VLDB Endow. **5**(10), 1016–1027 (2012)
12. Novak, D., Batko, M., Zezula, P.: Metric index: an efficient and scalable solution for precise and approximate similarity search. Inf. Syst. **36**(4), 721–733 (2011)
13. Pevny, T., Ker, A.D.: Towards dependable steganalysis. In: IS&T/SPIE Electronic Imaging (2015)
14. Roesch, M.: Snort - lightweight intrusion detection for networks. In: Proceedings of the 13th USENIX Conference on System Administration, LISA 1999, pp. 229–238. USENIX Association, Berkeley (1999)
15. Wright, C., Monrose, F., Masson, G.M.: On inferring application protocol behaviors in encrypted network traffic. J. Mach. Learn. Res. **7**, 2745–2769 (2006)
16. Xia, C., Lu, H., Ooi, B.C., Hu, J.: Gorder: an efficient method for KNN join processing. In: Proceedings of the Thirtieth International Conference on Very Large Data Bases, VLDB 2004, vol. 30, pp. 756–767. VLDB Endowment (2004)
17. Yu, C., Cui, B., Wang, S., Su, J.: Efficient index-based KNN join processing for high-dimensional data. Inf. Softw. Technol. **49**(4), 332–344 (2007)
18. Zezula, P., Amato, G., Dohnal, V., Batko, M.: Similarity Search: The Metric Space Approach. Springer, New York (2005)

Similarity Search of Sparse Histograms on GPU Architecture

Hasmik Osipyan[1,2]($\boxtimes$), Jakub Lokoč[2], and Stéphane Marchand-Maillet[3]

[1] National Polytechnic University of Armenia, Yerevan, Armenia
hasmik.osipyan.external@worldline.com
[2] SIRET Research Group, Faculty of Mathematics and Physics,
Charles University in Prague, Prague, Czech Republic
lokoc@ksi.ms.mff.cuni.cz
[3] University of Geneva, Geneva, Switzerland
stephane.marchand-maillet@unige.ch

Abstract. Searching for similar objects within large-scale database is a hard problem due to the exponential increase of multimedia data. The time required to find the nearest objects to the specific query in a high-dimensional space has become a serious constraint of the searching algorithms. One of the possible solution for this problem is utilization of massively parallel platforms such as GPU architectures. This solution becomes very sensitive for the applications working with sparse dataset. The performance of the algorithm can be totally changed depending on the different sparsity settings of the input data. In this paper, we study four different approaches on the GPU architecture for finding the similar histograms to the given queries. The performance and efficiency of observed methods were studied on sparse dataset of half a million histograms. We summarize our empirical results and point out the optimal GPU strategy for sparse histograms with different sparsity settings.

Keywords: GPU · Similarity search · High-dimensional space · Sparse dataset

1 Introduction

Similarity search in high-dimensional data [27] is a frequently used operation in many areas like multimedia retrieval/exploration, machine learning, computer vision etc. In order to perform similarity search, objects from a particular dataset have to be transformed into a descriptor space $\mathbb{U}$ with a distance function assigning a similarity score for two descriptors (smaller distance, higher similarity, and vice versa). The descriptors are often modeled as vectors (histograms) in $\mathbb{R}^m$ while the similarity function between vectors $o, q \in \mathbb{R}^m$ is usually modeled by means of the Euclidean distance $L_2(o, q) = \sqrt{\sum_{i=1}^{m}(o_i - q_i)^2}$. One of the most popular similarity operations is the kNN query defined for $k \in \mathbb{N}^+$, a query object $x \in \mathbb{R}^m$ and a dataset $D \subset \mathbb{R}^m$ as: $kNN(x) = \{\mathbb{X} \subset D; |\mathbb{X}| = k \wedge \forall y \in \mathbb{X}, z \in D - \mathbb{X} : L_2(x, y) \leq L_2(x, z)\}$. In some scenarios, the dataset D and a set

© Springer International Publishing AG 2016
L. Amsaleg et al. (Eds.): SISAP 2016, LNCS 9939, pp. 325–338, 2016.
DOI: 10.1007/978-3-319-46759-7_25

of query objects Q are both available in advance and the task is to evaluate all the kNN queries within a limited time period (e.g. online kNN classification).

For finding the nearest objects from D to the given query $q_i \in Q$, the sequential search approach can be used. Here, the distances between the query q_i and all objects in D are computed. Then, the distances are sorted and the nearest objects are taken. Whereas the sequential search has been outperformed by various indexing/hashing techniques on classical CPU architectures [8,27,30], novel many core GPU architectures [24] cause renaissance of sequential searching as it constitutes a trivial data parallel problem efficiently applicable on a commodity hardware. Furthermore, brute force kNN search is more robust against high dimensionality of vectors and high number of required nearest objects [16].

In this paper, we investigate brute force kNN search in a dataset consisting of sparse vectors, focusing on various sparsity settings of the dataset. We assume that the query objects are collected dynamically (e.g. online malware detection in persistent web traffic [12]) and they have to be processed within a short time interval, preventing from additional vector space transformation operations. Although additional optimizations considering compact representations of sparse vectors may result in more efficient performance on a CPU platform (see Algorithm 1 in Sect. 4), the same optimizations may suffer from novel GPU designs and specifics. Therefore, we revisit brute force kNN sequential search in sparse vector datasets and confront compact form representation with GPUs.

The rest of the paper is organized as follows. In Sect. 2, we review the literature related to similarity search algorithms on GPUs. Section 3 revises the GPU architecture fundamentals. The analysis of the similarity search algorithm along with the proposed methods are summarized in Sect. 4. In Sect. 5, we describe our experimental setup and perform the empirical evaluation. Section 6 concludes the paper.

2 Related Work

In the literature, many approaches were proposed to improve both exact and approximate similarity search algorithms. For obtaining best speedup, some of these methods even use multi-core CPUs or heterogeneous systems based on the GPUs. In this section, we review the most recent parallel approaches of similarity search algorithms as well as their individual steps for sparse dataset.

Matsumoto et al. [20] presented new exact kNN search algorithm based on the partial heap sort. Implementing distance calculation on the GPU combined with the fast heap sorting on the CPU, authors achieved better performance compared to the existing methods on GPUs. In general, this performance is the result of the new heap sort method based on the minimal overhead threshold and compression that outperformed even the sorting algorithms on the GPU.

For approximate similarity search on GPUs Krulis et al. [13] showed good performance for permutation-based indexing algorithm. This algorithm based on the sequential indexing was presented in the work of Mohammed et al. [21]. In this case, except distance calculation on GPU [14], the postprocessing steps of obtained distances were implemented on GPUs as well. That includes selection of top-k nearest objects and their sorting with bitonic sort algorithm.

Another approach of approximate similarity search on GPUs was suggested by Teodoro et al. [28]. A new parallel framework, Hypercurves, was able to answer approximate kNN queries with high speed. Based on the filter-stream programming paradigm this method divided the dataset into partitions giving the opportunity to access them independently in a parallel manner. Then kNN run on the GPU. Several papers described different kNN approaches on the GPU architectures [7,17,26]. However, the method in the paper [28] outperformed previous approaches by using heaps for selection procedure of top-k points. This dynamic partitioning along with kNN implementation reduced query responses approximately 80× compared to sequential version.

In [16], authors suggested a parallel approach of brute-force kNN for multiple queries. Here, distance matrix calculation was based on the standard dot product calculation on GPUs. Each portion of the divided distance matrix was computed by a block of threads. Then merge sort was implemented on the GPU to sort each portion parallel and the final k points were obtained after merging.

Several works present GPU approaches for individual steps of similarity search algorithms as a standalone problem. Chang et al. [3] obtained 40× faster results on GPU hardware for pairwise distance calculation. The authors presented two different implementations. In the first approach, they used $1D$ grid and $1D$ block and the threads were organized into blocks of 256. In this case, shared memory was used to process one row of the data matrix and to calculate its distances to the 256 rows corresponding to the threads in the block. In the second approach, authors used 16×16 threads in each blocks where one thread computed one entry in the output. This led to better performance than the first implementation. Li et al. [15] suggested another way of Euclidean distance calculation on GPUs which achieved approximately 15× speedup for a dataset comprising million objects. Authors used map-reduce technique to split up the final distance matrix into smaller ones. Then, the partial distance matrices were calculated on the GPU and the final solution was obtained after merging.

All discussed works described different algorithms of similarity search on parallel architectures. However, none of them solved the performance issues arising in the applications working with sparse datasets [9]. Different methods were suggested to learn efficiently similarity measure of sparse dataset [4,18,29]. Nevertheless, the main constraint of these methods remains the computation time. A few papers discuss the suitability of GPUs for the applications of sparse data model but all of them are concentrated on the sparse matrix multiplication.

In their continuous work, Neelima et al. [22,23] presented different sparse matrix formats that increased application performance with respect to GPU. For example, in one representation, they have defined two data structures for non-empty elements, one for data itself and the second for column and row indexes. For the row wise computation values, the better performance is achieved by using the latency hiding mechanism of the GPU. Another approach of sparse matrix vector multiplication was suggested by Ashari et al. [2]. In this approach, the non-empty elements of the rows were grouped into the constant number of blocks, which helped to reduce the thread divergence. Liu et al. [19] provided another storage format, Compressed Sparse Row 5 (CSR5), for sparse matrix. This format is an extension of CSR (Compressed Sparse Row) including the avoiding of structure dependent parameter tuning and the applicability for regular and irregular matrices. In the paper, for sparse matrix vector multiplication [5] the standard segmented sum algorithm was redesigned by prefix-sum scan.

For sparse data processing, the performance of discussed methods strongly depends on the sparsity of the input data. Hence, the implementation of applications for sparse dataset on heterogeneous systems needs the careful understanding of underlying architecture and the usage of the right data format.

3   GPU Architecture

In this section, we present the basics of GPU architecture with particular emphasis on the aspects, which have great importance in the light of the studied problem. We will focus mainly on the NVIDIA Kepler [24] architecture as it was used in our experiments.

A GPU card is a peripheral device connected to the host system via the PCI-Express (PCIe) bus. It consists of several *streaming multiprocessor* units (SMPs), which share only the main memory bus and the L2 cache. The GPUs employ a parallel paradigm called *data parallelism* where the concurrency is achieved by processing multiple data items simultaneously by the same routine (*kernel*). Each thread executes the kernel code, but has a unique thread *ID*, which is used to identify the portion of the work processed by the thread.

The threads are grouped together into blocks of the same size. Threads from different blocks are not allowed to communicate with each other directly, since it is not even guaranteed that any two blocks will be executed concurrently. Furthermore, threads in a block are divided into subgroups (*warps*). The number of threads in warps is usually fixed for each architecture (current NVIDIA GPUs use 32 threads per *warp*).

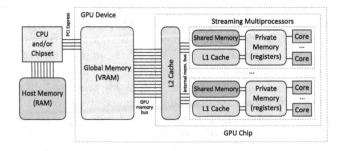

Fig. 1. Host and GPU memory organization scheme

The other main difference from CPU architecture is the memory organization which is depicted in Fig. 1. The *host memory* is the operational memory of the computer, which cannot be accessed by GPU. At first input data needs to be transferred from the host memory (RAM) to the GPU memory (VRAM) via PCI-Express, which is rather slow (8 GB/s) when compared to the internal memory buses. The *global memory* can be accessed from the GPU cores, and it shows both high latency and limited bandwidth. The *shared memory* is shared among threads within one group. It is rather small (tens of kB) but almost as fast as the GPU registers. The shared memory can play the role of a program-managed cache for the global memory, or it can be used to exchange intermediate results by the threads in the block. The *private memory* belongs exclusively to a single thread and corresponds to the GPU core registers. Its size is very limited; therefore, it is suitable just for a few local variables. The L2 cache is shared by all SMPs and transparently caches all access to global memory. The L1 cache is private to each SMP and caches data from global memory selectively.

Another important issue on GPUs is the branching problems. When threads in a warp choose different code branches (*if* or *while* statements), all branches must be executed by all threads. Thread masks instruction execution according to their local conditions to ensure correct results, but heavily branched code does not perform well on GPUs.

Two programming techniques were proposed to work directly with GPU hardware: Compute Unified Device Architecture (CUDA) developed by NVIDIA [25] and Open Computing Language (OpenCL) developed by Khronos Group [11]. Although for some applications OpenCL can be a good alternative to CUDA [6], it is shown that CUDA is the best choice for high performance needs [10]. Hence, in our research, we will base on the CUDA technique.

4   Searching for Nearest Sparse Histograms

In our work, we consider sparse vectors that can be represented in a compact form as an array of pairs *[ID, value]*. Given such compact representation, the Euclidean distance evaluation can be implemented efficiently for CPU processing considering only non-empty bins as presented in Algorithm 1. In this case, the final result is obtained from the sum of distances for the values with the same IDs and the squares of remaining values. However, when considering GPU architectures, the distance evaluation, consisting of many $if-statements$, may represent a new performance bottleneck, despite lower memory requirements. Therefore, we revisit kNN sequential search in sparse vector datasets and confront compact form representation with GPU architectures. Note that the sequential search is quite memory-intensive, so it has to be implemented in a cache-aware manner to achieve optimal performance. For a sparse dataset, the performance of distance calculation on GPU depends also on the internal sparsity settings of given data. Hence, we need to consider the scope of individual parameters of used dataset, so that we can optimize technical details of our implementation.

We examine four possible approaches - a *conditional solution (CDS)* based on the *if-statements*, *naive solution (NS)* based on the data division, *compressing*

Algorithm 1. Distance for compact sparse histograms $q_j \in Q$ and $o_i \in D$

1: $d = 0, k_Q = 0, k_D = 0$
2: **while** $k_Q < q_j.length \wedge k_D < o_i.length$ **do**
3:   **if** $q_j[k_Q].binId == o_i[k_D].binId$ **then**
4:     $d += (q_j[k_Q].value - o_i[k_D].value)^2$, $k_Q += 1, k_D += 1$
5:   **else if** $q_j[k_Q].binId < o_i[k_D].binId$ **then**
6:     $d += (q_j[k_Q].value)^2$, $k_Q += 1$
7:   **else**
8:     $d += (o_i[k_D].value)^2$, $k_D += 1$
9:   **end if**
10: **end while**
11: **while** $k_Q < q_j.length$ **do**
12:   $d += (q_j[k_Q].value)^2$, $k_Q += 1$
13: **end while**
14: **while** $k_D < o_i.length$ **do**
15:   $d += (o_i[k_D].value)^2$, $k_D += 1$
16: **end while**
17: **return** $\sqrt{(d)}$

---

solution (CS) based on the compressing of query and object data and finally, *column-based solution (CBS)* inspired by inverted files. *CDS* method is the baseline GPU implementation of Algorithm 1 where the performance suffers from the branching problems. This method shows worse performance compared even with the standard CPU-only solution. Hence, in the next subsections we will only explain the *NS, CS* and *CBS* methods. As a baseline CPU solutions, we consider standard solution (*STS*) based on the *if-statements* and the inverted files solution (*IFS*) [1]. In IFS, the object database is represented by the list of values with the same IDs, which accelerates the retrieving time of the values for each query.

4.1 Naive Solution

To solve the issues of *CDS*, we have adjusted Algorithm 1 to better use the architecture of GPU. We proposed to keep *if-statements* on the CPU during the reading of the data and send already arranged data on the GPU.

Hence, we have two kernels on GPU. The first one is responsible for distance calculation of two arrays containing only the bins with the same *ID*. The second one calculates the Euclidean norm of the array, which contains the remaining points from query and data object. Then, the results of our two kernels are merged on the CPU side. Simple data division example is shown in Fig. 2.

This method is based on two dimensional thread organization. Shared memory (48 kB) was used to cache the query data points. The cached data are associated with y thread grid dimension while the non-cached data are addressed by x coordinates. Different streams are used for two kernels and the results of each kernel are kept on the CPU while processing the next portion of the data. This helps to avoid the synchronization between two kernels as the results are merged

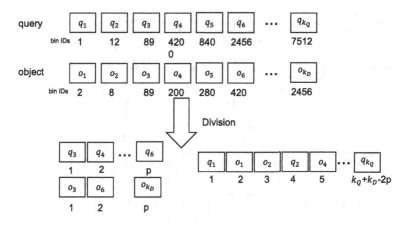

Fig. 2. Data arrangement for *NS*

in the final stage. Each thread/block size was tested for different configurations and the efficient configuration was selected in the final results.

Despite the avoidance of *warp* divergence in this approach, the division of two kernels may reduce the performance. Depending on the number of bins with the same IDs, the time required for transferring the data between CPU and GPU can exceed the time spent on the operation itself, therefore, increasing the total computation time. Theoretically, in the worst case, when there is no bin with the same *ID*, the time complexity required for reading from file/data arrangement is equal to $O(k_Q + k_D)$ and another $O(k_Q + k_D)$ for Euclidean norm calculation, which leads to total $O(2k_Q + 2k_D)$. Here, k_D and k_Q are the number of non-empty bins of object and query respectively. In the average case, when the number of the bins with the same bin *ID* is p, the time complexity becomes $O(2k_Q + k_D - p)(O(k_Q + 2k_D - p)$ in case where $k_D > k_Q$).

4.2 Compressing Solution

To avoid data division, the query or database objects can be changed in the way to fill out the missing bin IDs with 0 values. Hence, for each query/database object the array with size of equal to the (last bin *ID* - first bin *ID*) value can be created. Although this approach avoids *if-statements* in the final distance calculations, it is very sensitive to sparse dataset itself. The higher the value of the (last bin *ID* - first bin *ID*), the more memory is used for keeping all corresponding 0 values. Considering the small amount of available memory on GPU hardware, this approach limits the number of simultaneously processed histograms, which affects on the final performance of the algorithm.

To make the approach less sensitive to the value of bin *IDs*, a new compressing solution (*CS*) was proposed. In the distance calculation, all *if-statements* were avoided by compressing the query/object data based on the bin *ID* information. Our simple data compressing example is shown in Fig. 3.

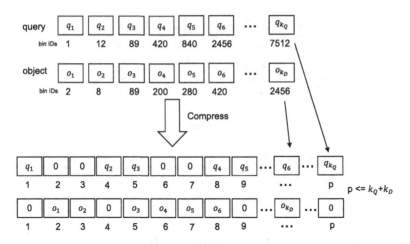

Fig. 3. Compressing for *CS*

The bins, which *ID* exists at least in one histogram, are processed by adding 0 values in the corresponding empty bin. In the worst-case scenario, the memory required for holding query/object data is equal to the number of non-empty unrepeatable participants of both sides. Hence, the performance is also sensitive to the internal structure of the dataset.

This approach could decrease the execution time compared to previously described methods as more data points can be processed simultaneously. After the compression procedure, the distance calculations become completely independent and each distance is computed by the exact same number of arithmetical operations on the GPU. The organization of work among the threads and thread blocks remains the same as for *NS* method. Theoretically, in the worst-case, time complexity for reading from file including data arrangement (compression) is $O(k_Q + k_D)$. For the distance calculation we have $O(k_Q + k_D)$, which leads to $O(2k_Q + 2k_D)$ total time complexity. In the average case, when the number of the bins with the same bin *ID* is p, the time complexity becomes $O(2k_Q + 2k_D - 2p)$.

4.3 Column-Based Solution

Given sparse query vectors, inverted files represent a popular efficient index in document and multimedia retrieval systems. Inverted files are usually coupled with the cosine similarity, hence only a fraction of the data files (corresponding to non-zero query bins) has to be visited to correctly answer a query. Inspired by the inverted files, sparse vectors can be organized in columns such that vector c_i represents i–th dimension from all database vectors (not in compact form). This database organization can be efficiently used with the Euclidean distance if the size of each database vector is precomputed or if all the vectors are normalized.

More specifically, let I_q, I_o represent sets of non-zero bin IDs for vectors of query $|q| = 1$ and data object $|o| = 1$. Then, the squared Euclidean distance can be simply transformed to the following form:

$$\sum_{i\in\{1...d\}} |q_i - o_i|^2 = \sum_{i\in I_q \cup I_o} |q_i - o_i|^2 = \sum_{i\in I_q} |q_i - o_i|^2 + \sum_{i\in I_o - I_q} o_i^2$$

$$= \sum_{i\in I_q} |q_i - o_i|^2 + 1 - \sum_{i\in I_q} o_i^2 = 1 + \sum_{i\in I_q} (|q_i - o_i|^2 - o_i^2)$$

To follow this form, before starting distance calculations the size of each data object need to be computed (o_i^2). Hence, in this approach, we have two kernels. The first one is responsible for computing the size of vectors on GPU. The second one computes the Euclidean distance ($|q_i - o_i|^2$) after compressing the query and object vectors. The main advantage of this approach compared with previous ones is that the Euclidean distance calculation only requires values stored in few bins with indexes from I_q. Thus, in *CBS*, a large portion of data objects can be used for each iteration of GPU calculations. However, depending on the size of the overlap $I_o \cap I_q$, there could be also a calculation overhead. Theoretically, in the worst-case, the time complexity for vector size computation is $O(k_D + k_Q)$, for data compressing - $O(k_Q)$ and for $2 * q_i * o_i$ - $O(p)$ when the number of bins with the same ID is p. Hence, for each query and object the total time complexity is $O(2k_Q + k_D + p)$.

5   Experimental Results

In this section, we present the hardware and dataset used for the experiments along with the results for different configurations.

5.1   Hardware Setup

Our experiments were conducted on a PC with an Intel Core i3-4010U CPU clocked at 1.7 GHz, which have 4 physical cores and 4 GB of RAM. The desktop PC is equipped with NVIDIA GeForce GT 740M (Kepler architecture [24]), which have 2 SMPs comprising 192 cores each (384 cores total) and 4 GB of global memory. The host used Windows 7 as operating system and CUDA 5.1 framework for the GPGPU computations. The experiments were timed using the real-time clock of the operating system. Each experiment was conducted 10× and the arithmetic average of the measured values is presented as the result.

The results of four approaches were measured for random generated sparse datasets (475k objects, 400k queries) with different sparsity settings (query and objects have different number of bins with the same IDs). Dataset is loaded from the file where each line corresponds to one histogram. Each histogram is in the compact form containing only non-empty bins. The bins are separated by the comma and each bin includes the pair of *ID* and value (*ID:Value*).

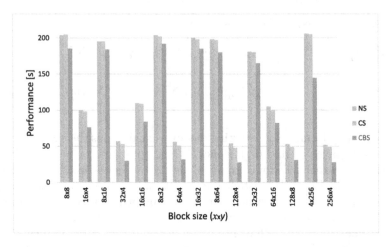

Fig. 4. Performance for various block sizes ($k = k_Q = k_D = 1000, p = k/2, Q = \{10k\}, D = \{475k\}$)

5.2 Results

In Fig. 4, we summarize the results of our three algorithms for different block sizes to find out the optimal GPU configuration. The *CDS* algorithm is not presented in Fig. 4 as it is very slow ($\sim$11$\times$) compared even with the results of the worse GPU configuration (4×256 block size).

Euclidean norm was calculated using CuBLAS library function ($cublasSnrm2$), which automatically picks optimized block size. Let us note that the y component of the block size represents the number of query points cached

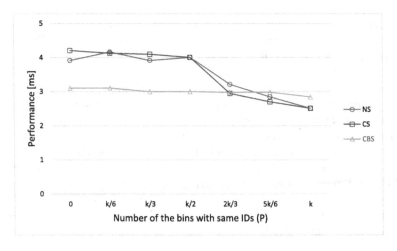

Fig. 5. Performance for different number of bins with the same *ID* ($k = k_Q = k_D = 1000, Q = \{1\}, D = \{475k\}$)

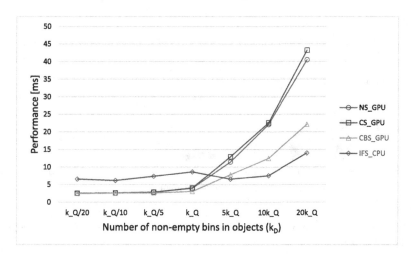

Fig. 6. Performance for different dimensions ($k_Q = 1000, Q = \{1\}, D = \{475k\}$)

in the shared memory. The optimal performance for each algorithm is achieved when smaller amount of query points are cached (4 to 8) while larger number of object points are processed by the thread block. The difference between performances of various block sizes is due to the available amount of shared memory on our GPU (48 kB). Therefore, the algorithms should perform better on a new generations of GPUs, which are expected to have even more shared memory per SMP. For the next experiments, the most efficient block size configuration (256×4) was selected.

Figure 5 presents the measured times for different number of bins with the same *ID* (p), when the number of non-empty bins in the query and data objects are equal. *NS* and *CS* algorithms are very sensitive to the p while the *CBS* algorithm is more stable. We explain the differences of results by the fact that *CBS* algorithm is working only with the non-empty bins of the query. Conversely *NS* and *CS* rearrange query and object data where the number p plays a key role. For large p number, our *NS* and *CS* algorithms give better results as they does not require vector size computation. Hence, *CBS* can outperform other solutions if the vector sizes are precomputed.

In Fig. 6, we show the kernel times obtained for different numbers of non-empty bins in objects (k_D). The experiments were conducted for the number of bins with the same *ID* equal to $k_Q/2$ if $k_Q < k_D$ ($k_D/2$ if $k_D < k_Q$). For small ratios $k_D \ll k_Q$, our three algorithms on GPU give approximately same results outperforming the CPU baseline IFS method approximately 2–3×. For larger ratios $k_D \gg k_Q$, *CBS* algorithm outperforms other GPU algorithms.

This explains by the fact that in our *CBS* method the objects are used only for vector size computation and not all of them participate in distance calculation. We show that IFS method is the best choice only for larger ratios while CBS-GPU is the best for the other cases.

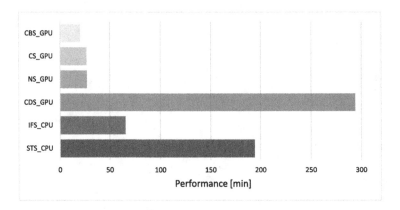

Fig. 7. Performance summary for different algorithms ($k_D = k_Q = 1000, p = k_Q/2, Q = \{400k\}, D = \{475k\}$)

To have final overview, in Fig. 7, we present the total measured times for baseline CPU solutions (*STS*, *IFS*) and four different GPU (*CDS*, *NS*, *CS*, *CBS*) implementations for a huge query/object dataset. The non-empty bins of queries (k_Q) and objects (k_D) are equal to 1000 and the number of same bin IDs $p = k_Q/2 = 500$. For all methods (CPU and GPU), we use the same postprocessing steps (quick sort, top-k selection), which totally take less than $8\,min$. Finally, let us note that *NS*, *CS* and *CBS* algorithms on the GPU give approximately 7−9× faster results than CPU baseline *STS* method. *CBS* method itself provides 2−3× faster results than CPU baseline *IFS* method for huge k_Q and 11−14× faster results than GPU baseline *CDS* implementation.

6   Conclusions

In this work, we have analyzed the performance issues of similarity search for the sparse dataset. In the high-dimensional spaces, the most time consuming operation is the distance calculation. For sparse dataset, this operation becomes more expensive due to the conditional structure. We have studied four hybrid approaches on the GPU and find out the optimal/fastest solution for different sparsity settings.

We showed that *NS*, *CS* and *CBS* approaches on the GPU outperformed *CDS-GPU* and CPU-only *STS* baseline solutions significantly and showed a promising potential for future scaling. Experiments showed that for huge query dimensions our *CBS* method is faster even compared to the *IFS-CPU* method. In addition, the internal structure of the sparse dataset played key role in the final performance. Depending on the number of bins with the same *ID*, the *CBS* solution can be a better choice than *NS* and *CS* solutions and vice versa. We finally note that *CBS* solution is the best choice if the vector sizes are precomputed.

As a future work, we are going to use frequently bin IDs for distance calculations where we will track the occurrence of each bin *ID* and will use it for later queries. After thousands of queries, our learning system would be able to find similar objects in a faster way as only frequently asked bins will be processed on the GPU. In addition, we are going to implement *IFS* GPU method and use it as a baseline solution. We are going also to discuss other potential distance functions and to evaluate their affect on the solutions discussed.

Acknowledgments. This paper was supported by the Czech Science Foundation project 15-08916S and by the project SVV-2016-260331 and in relation to the SNF (Swiss National Foundation) project MAAYA (grant number 144238).

References

1. Amato, G., Savino, P.: Approximate similarity search in metric spaces using inverted files. In: Proceedings of the 3rd International Conference on Scalable Information Systems, pp. 28:1–28:10 (2008)
2. Ashari, A., Sedaghati, N., Eisenlohr, J., Sadayappan, P.: An efficient two-dimensional blocking strategy for sparse matrix-vector multiplication on GPUs. In: ICS 2014, Muenchen, Germany, 10–13 June 2014, pp. 273–282 (2014)
3. Chang, D., Jones, N.A., Li, D., Ouyang, M., Ragade, R.K.: Compute pairwise Euclidean distances of data points with GPUs. In: Proceedings of the IASTED International Symposium on CBB, Florida, USA, 16–18 November 2008, pp. 278–283 (2008)
4. Cui, B., Zhao, J., Cong, G.: ISIS: a new approach for efficient similarity search in sparse databases. In: Kitagawa, H., Ishikawa, Y., Li, Q., Watanabe, C. (eds.) DASFAA 2010. LNCS, vol. 5982, pp. 231–245. Springer, Heidelberg (2010)
5. Dotsenko, Y., Govindaraju, N.K., Sloan, P.J., Boyd, C., Manferdelli, J.: Fast scan algorithms on graphics processors. In: Proceedings of the 22nd Annual ICS, Island of Kos, Greece, 7–12 June 2008, pp. 205–213 (2008)
6. Fang, J., Varbanescu, A.L., Sips, H.J.: A comprehensive performance comparison of CUDA and OpenCL. In: ICPP, Taipei, Taiwan, September 2011, pp. 216–225 (2011)
7. Garcia, V., Debreuve, E., Barlaud, M.: Fast k nearest neighbor search using GPU. In: IEEE Conference on CVPR, Anchorage, USA, 23–28 June 2008, pp. 1–6 (2008)
8. Gionis, A., Indyk, P., Motwani, R.: Similarity search in high dimensions via hashing. In: Proceedings of the 25th International Conference on VLDB 1999, pp. 518–529. Morgan Kaufmann Publishers Inc., San Francisco (1999)
9. Goumas, G.I., Kourtis, K., Anastopoulos, N., Karakasis, V., Koziris, N.: Understanding the performance of sparse matrix-vector multiplication. In: 16th Euromicro International Conference on PDP, pp. 283–292 (2008)
10. Karimi, K., Dickson, N.G., Hamze, F.: A performance comparison of CUDA and OpenCL. CoRR abs/1005.2581 (2010)
11. Khronos OpenCL Working Group: The OpenCL Specification, version 1.0.29, 8 December 2008
12. Kohout, J., Pevny, T.: Unsupervised detection of malware in persistent web traffic. In: IEEE International Conference on ICASSP (2015)

13. Krulíš, M., Osipyan, H., Marchand-Maillet, S.: Optimizing Sorting and top-k selection steps in permutation based indexing on GPUs. In: Morzy, T., Valduriez, P., Bellatreche, L. (eds.) ADBIS 2015. CCIS, vol. 539, pp. 305–317. Springer, Heidelberg (2015)

14. Krulis, M., Osipyan, H., Marchand-Maillet, S.: Permutation based indexing for high dimensional data on GPU architectures. In: 13th International Workshop on CBMI, Prague, Czech Republic, 10–12 June 2015, pp. 1–6 (2015)

15. Li, Q., Kecman, V., Salman, R.: A chunking method for Euclidean distance matrix calculation on large dataset using multi-GPU. In: The Ninth ICMLA, Washington, DC, USA, 12–14 December 2010, pp. 208–213 (2010)

16. Li, S., Amenta, N.: Brute-force k-nearest neighbors search on the GPU. In: Amato, G., Connor, R., Falchi, F., Gennaro, C. (eds.) SISAP 2015. LNCS, vol. 9371, pp. 259–270. Springer, Heidelberg (2015). doi:10.1007/978-3-319-25087-8_25

17. Liang, S., Liu, Y., Wang, C., Jian, L.: A cuda-based parallel implementation of k-nearest neighbor algorithm. In: Cyber-Enable Distributed Computing and Knowledge Discovery, pp. 291–296 (2010)

18. Liu, K., Bellet, A., Sha, F.: Similarity learning for high-dimensional sparse data. In: Proceedings of the Eighteenth International Conference on Artificial Intelligence and Statistics, AISTATS, San Diego, California, USA, 9–12 May 2015 (2015)

19. Liu, W., Vinter, B.: CSR5: an efficient storage format for cross-platform sparse matrix-vector multiplication. In: Proceedings of the 29th ACM on ICS 2015, Newport Beach/Irvine, CA, USA, 8–11 June 2015, pp. 339–350 (2015)

20. Matsumoto, T., Yiu, M.L.: Accelerating exact similarity search on CPU-GPU systems. In: ICDM, Atlantic City, NJ, USA, 14–17 November 2015, pp. 320–329 (2015)

21. Mohamed, H., Osipyan, H., Marchand-Maillet, S.: Multi-core (CPU and GPU) for permutation-based indexing. In: Traina, A.J.M., Traina Jr., C., Cordeiro, R.L.F. (eds.) SISAP 2014. LNCS, vol. 8821, pp. 277–288. Springer, Heidelberg (2014)

22. Neelima, B., Raghavendra, P.S.: CSPR: column only sparse matrix representation for performance improvement on GPU architecture. In: Advances in Parallel Distributed Computing, Tirunelveli, India, 23–25 September 2011, pp. 581–595 (2011)

23. Neelima, B., Reddy, G.R.M., Raghavendra, P.S.: A GPU framework for sparse matrix vector multiplication. In: IEEE 13th International Symposium on Parallel and Distributed Computing, ISPDC, Marseille, France, June 2014, pp. 51–58 (2014)

24. Corporation, N.: Kepler GPU Architecture. http://www.nvidia.com/object/nvidia-kepler.html

25. NVIDIA Corporation: NVIDIA CUDA C programming guide, version 3.2 (2010)

26. Pan, J., Manocha, D.: Fast GPU-based locality sensitive hashing for k-nearest neighbor computation. In: 19th ACM SIGSPATIAL International Symposium on Advances in Geographic Information Systems, Chicago, IL, USA, pp. 211–220 (2011)

27. Samet, H.: Foundations of Multidimensional and Metric Data Structures. The Morgan Kaufmann Series in Computer Graphics and Geometric Modeling. Morgan Kaufmann Publishers Inc., San Francisco (2005)

28. Teodoro, G., Valle, E., Mariano, N., da Silva Torres, R., M Jr., W., Saltz, J.H.: Approximate similarity search for online multimedia services on distributed CPU-GPU platforms. VLDB J. **23**(3), 427–448 (2014)

29. Wang, C., Wang, X.S.: Indexing very high-dimensional sparse and quasi-sparse vectors for similarity searches. VLDB J. **9**(4), 344–361 (2001)

30. Zezula, P., Amato, G., Dohnal, V., Batko, M.: Similarity Search: The Metric Space Approach, 1st edn. Springer, New York (2010)

Erratum to: Pruned Bi-directed K-nearest Neighbor Graph for Proximity Search

Masajiro Iwasaki[✉]

Yahoo Japan Corporation, Tokyo, Japan
`miwasaki@yahoo-corp.jp`

Erratum to:
Chapter 2 in: L. Amsaleg et al. (Eds.)
Similarity Search and Applications
DOI: 10.1007/978-3-319-46759-7_2

In an older version of the paper starting on p. 20 of the SISAP proceedings (LNCS 9939), Fig. 5(b) was represented incorrectly. This has been corrected.

The updated original online version for this Chapter can be found at
DOI: 10.1007/978-3-319-46759-7_2

L. Amsaleg et al. (Eds.): SISAP 2016, LNCS 9939, p. E1, 2016.
DOI: 10.1007/978-3-319-46759-7_26

Author Index

Printed in the United States
By Bookmasters